*The Earth is the Lord's
and all that is in it*

THE PIRACY
OF AMERICA

Profiteering in the Public Domain

edited by

Judith Scherff

Clarity Press, Inc.

ISBN: 0-932863-23-X

In-house Editor: Diana G. Collier
Cover Design: Larry Levy

Cataloguing in Publication Data:

The piracy of America : profiteering in the public domain/
    edited by Judith Scherff ; foreword by Thomas Berry. --
    1st ed.
     p. cm
     Includes bibliographical references and index
     ISBN: 0-932863-28-0

    1. Political ecology. 2. Environmental policy--
United States. 3. Public lands--United States
4. Environmental justice--United States. I. Scherff,
Judith F. II. Berry, Thomas Mary, 1914-

JA75.8.P57 1999   363 7'093
                      AB198-1466

CLARITY PRESS, INC.
Ste. 469, 3277 Roswell Rd. N.E.
Atlanta, GA. 30305
Toll free order: 1-800-626-4330
Fax: 404-231-3899
Internet: http://www.bookmasters.com/clarity
E-mail: clarity@islandnet.com

Table of Contents

DEDICATION

*This collection is dedicated to my family,
biological and extended,
human and nonhuman.*

APPRECIATION

I am grateful not only for the chapters written herein by various authors, but most especially for their determination to spend their lives seeking justice for the issue about which they've written. I appreciate, also, the careers and determination of those writers who submitted material but were not selected for this publication. They, too, have given a good portion of their lives to specific issues in which they believe.

We can all be thankful for:
> • the elected officials who doggedly support "the general welfare" against the pressure of special interests;
> • members of the press who go against popular pressure to present the whole story;
> • those scientists who risk careers reporting scientific data as it really is, unconvoluted, as they have discovered it to be;
> • the clergy, who continue to remind us of Christian teachings in spite of secular and financial pressure within their congregations;
> • and to Clarity Press for printing testimony to the fact that the environment, democracy, civil society and the economy are intertwined and therefore environmental issues are now very much human rights issues.

ACKNOWLEDGMENTS

We wish to thank those whose participation and assistance has been invaluable in the production of this book. We also thank the following authors, publications and publishers for the right to reproduce materials which appear herein.

Addison-Wesley for selected excerpts from Dr. Sandra Steingraber, *Living Downstream: An Ecologist Looks at Cancer and the Environment.*

Carol Publishing Group for excerpt from *Toxic Deception: How the Chemical Industry Manipulates Science, Bends the Law and Endangers Your Health,* by Dan Fagin, Marianne Lavelle and the Center for Public Integrity.

Common Courage Press for chapter from *Toxic Sludge Is Good For You: Lies, Damn Lies and the Public Relations Industry* by John Stauber & Sheldon Rampton.

Conservation Biology for permission to reprint David Orr and David Ehrenfeld, "None So Blind."

Island Press for Jeff DeBonis, "Questioning the Paradigm," in Richard Knight, ed., *A New Century for Resource Management.*

Orion for David Ehrenfeld, "Spending Our Capital."

Peanut Butter Press, for article by Tarso Luís Rmos originally entitled "An Environmental Wedge: The 'Wise Use' Movement and the Insurgent Right Wing," in Eric Ward, ed., *Conspiracies: Real Grievances, Paranoia, and Mass Movements.*

PUBLISHER'S FOREWORD

Why should a human rights publisher, guided since its inception by international human rights law, now publish a book on the environment?

Those who work within the context of international human rights have long understood the indivisibility of the rights involved in human justice issues: how civil and political rights can't be divorced from social, economic and cultural rights — or put another way, how the right to vote is meaningless to those who must sleep in doorways, or how economic security without democratic participation leads to cultural sterility and lassitude.

Now, as the toxic byproducts of human development advance to unprecedented levels, it is clear that a new category of right is arising, equally essential to and indivisible from these human justice rights: the right to a healthy environment. What does it matter if shiny new cars and VCRs are produced in overwhelming number, if the very air we breathe, the water we drink, and the food we eat imperil our lives? If this consumption threatens and despoils not only the earth's natural resource base, but also the physical and spiritual wellbeing of untold millions of the human population? If, even for a snug few in their insulated and shielded cyberdomes, the outside world becomes a hostile, noxious residue where it is not just undesirable but perilous to touch foot?

The writers in this anthology do not argue against development. Nor do they discuss injuries to nature as if these were independent from human needs. What emerges in the reading of these articles, ostensibly about public domain resources in America — our wildlife, our forests, our mineral resources, our air, our climate — is how these are inextricably tied to human physical and spiritual well being, and to human justice issues that relate to the generation, control and apportionment of wealth.

What is emerging, in the struggle for a clean and sustainable environment, is the realization that major domestic and transnational corporations, the central actors in globalization, are also paramount in relation to environmental degradation. It is their actions, their decisions, indeed their *modus operandi* which predominate in most environmental problematics — not the independent actions of masses of unorganized individuals or communities. Yet the message transnational corporations have for the world is that they are the conduit through which all blessings flow, the agency of economic development and prosperity. The rest of us must stand back and let them proceed, their message reads, for it is they who know what is best.

However, the time when people could believe that 'what is good for GM is good for America' is surely gone for good. After all,

> [Their] elected government is controlled by corporations through campaign contributions (which are required because expensive media exposure is the key to election); [they themselves] are made insecure, discouraged and disengaged largely because of corporate policies and practices —downsizing, wage cuts, forced

ix

give-backs, overseas flight, union busting — or simply the fear that any of these tactics will be used.[1]

Clearly, as the articles in this anthology demonstrate, the influence corporations have had upon the environment has been no less unbridled. Not only do they not know what is best for the environment, but this concern is peripheral at best to their fundamental operational paradigms -- more likely to be factored in, where it occurs, as a sop to public opinion, than as a real and equal determinant influencing project development.

We are now at a period of public realization that the hierarchical and hidden decision-making structures of corporations, their fixation on bottom line profits and narrow self-interested approach to development, profoundly conflicts with the broad interests and human and democratic rights of the American people in ways that relate to the public good.

What must be done is so plain it seems absurd that it must not only be stated, but also fought for: we must conserve our natural resources so that they can be available for ongoing generations. We must find new and creative ways to do this, and to develop new resources. We must stop the degradation of our one and only habitat. We cannot wait for corporations to find a way to own solar and wind power before they can be fully developed, or for the petrochemical industry to give its blessing to alternate forms of transportation.

If we fail to question, and indeed to constrain or alter the priorities, policies and practices now promoted by our corporate elites which blindly sap the public wealth in the name of short term profit, the loss will be borne by all humankind. Our heroism lies, not only in the creativity to better conserve our natural heritage, but in the strength to fully recognize and deal with these problems in their socio-economic and political dimensions. We must not permit the future to be mapped by others, then disguised as an inevitability which we can only watch unfold. Rather, we must take part in the evolution of human efforts to try, to test, and to bring into being those institutions, systems and practices which bring about the greater flourishing of both our environment and all humankind — and to discard those which don't.

Do corporations combat environmentalism so fiercely, simply to prevent the protection of trees or the institution of tougher pollution controls? Or is it deeper: because, with a long range foresight keenly attuned to their interests, they sense a threat — not just to their immoderate wealth, but to the prerogative they have seized to unilaterally map the future for all humanity?

This anthology makes visible what should be a commonplace understanding: that sooner or later, the social justice and environmental movements will merge, propelled toward mutual accommodation as they clarify the obstacles they face, and the human-centered goals they share. Indeed, in the 21st century, we may witness the three great dimensions of humankind come together, as the struggle to recover human spirituality joins and underpins the struggle for social and environmental justice.

FOREWORD

Thomas Berry

"In our era, the road to holiness necessarily passes through the world of action."

Dag Hammarskjold

"Piracy" is a powerful word. Yet it is quite appropriate as a description for what has happened to the North American continent, in the past century in particular. This continent was a gorgeous world, Earth's most luxuriant self-expression, when the first colonists came here early in the 17th century. Yet now, as we look back on these earlier centuries from our present situation, we might, indeed, wonder what has happened.

We Europeans had the feeling that we had an historical mission in coming here, the conviction that we brought heaven's blessings wherever we went. We told ourselves that we would elevate the indigenous peoples to a new level of civilization. We would enable them to share in the blessings that we possessed. The continent itself would be redeemed from its aimless existence and brought into the world of meaningful service to the human community.

Now, at the end of the 20th century, almost four centuries after the English Settlement at Jamestown in 1607, we observe that much has happened. We have taken the continent from itself and from ourselves. We have, in a sense, ruined paradise. We have cut the forests, diminished the wildlife, dammed the rivers, exhausted the soils, polluted the air, drained the wetlands, paved vast areas for highways and parking lots until the continent has become degraded in its structural integrity and in its life-giving capacities.

While many causes contributed to the devastation of the North American continent, the truly ruinous period commenced in the last decades of the 19th century when we moved from an ever-renewing agricultural organic economy to a non-renewing industrial economy. The immediate benefits were so overwhelming that we could not resist. Yet the world that we created was no longer composed of mysterious powers that we relate to by evocation but was rather a world of impersonal material forces awaiting our domination.

Our human economy was no longer integral with the ever-renewing processes of nature. It was an economy based on exploitation of the non-renewing natural resources of the Earth.

The difference is critical. For only an organic economy is an enduring economy. An extractive economy is a terminal economy, since all non-renewing resources are limited in their availability and will never again be brought into being on the planet Earth. Even more crucial is the fact that ever-renewing resources can, by abuse, be made non-renewing. The immense numbers of fish in the sea can be exhausted beyond recovery. So with the Passenger Pigeon, a species that once numbered into the billions, a species so abundant that when the flocks would fly overhead the sky would be darkened for hours. Yet they were harvested with nets and shot with rifles and clubbed with poles so extensively that they were driven to extinction by 1917. Already we have cut 95% of the forests of this continent, forests that will never again have the majesty that they had when we came here. Even the soil has been made so toxic with chemicals and eroded so extensively that it is unlikely that it will ever be renewed to its original integrity.

Such is the abundance of nature on this continent and such is the unrestrained manner in which this abundance has been exploited. Our power of exploitation has increased a thousandfold through the scientific and technological discoveries that we have made in the later 19th and throughout the 20th centuries. We travel with previously unimagined speed across the earth or into space. We have knowledge and power such as was never dreamed of in previous centuries. In the process, however, we have emptied the natural world of its inherent meaning. We no longer live in a fascinating world of divine beauty that comes to us through the song of the mockingbird or the sight of a butterfly or the colors of the evening sunset. We still see fragmentary manifestations of these phenomena. Yet we have been so desensitized to their grandeur that we can no longer experience the wonder that we once knew.

As we awaken to the difficulties that we are causing both the planet and ourselves, we begin to realize that the loss of the forests, the pollution of the air, the water and the soil are not simply an economic or aesthetic loss; they constitute a loss of soul. When the night sky is so obscured with light and with pollutants that our children can no longer see the stars in the heaven, this is a diminishment of the soul life of the child. For it is through these experiences that a sense of meaning, a sense of beauty, a sense of the sacred is awakened within us. None of our religious rituals or moralistic teaching can do for us what the natural world does for us. Now, a century later, when the devastation has reached extreme proportions, we are turning back to an intimacy with nature. We begin to appreciate the great wonder of the universe.

While we are all to some extent responsible for what has happened, there is a sense in which those more competent in exploiting the natural world have taken the continent from itself and from us who dwell here, by a kind of piracy. A number of different forces have made this piracy possible. The main condition has been the scientific technologies that began to be systematically developed in the last two decades of the 19th century. This was the period when the contemporary mechanical, electronics and chemical engineering processes came into being.

Possibly even more important than the new technologies that enabled us to exploit the planet in such a ruthless manner has been the development of the modern commercial, industrial and financial corporations. The first of the modern corporations was Standard Oil, incorporated in 1870, so ruthless in its monopolistic functioning that it was later forced to break up by the Interstate Commerce Commission. It is amazing how many of the dominant corporations active even today in the oil, electricity, telephone, steel, chemical, pharmaceutical, paper and automobile industries came into being in the last two decades of the 19th and the early 20th centuries.. Those in command of the corporations have come to wield immense power over both the human community and the natural world. Every aspect of the Earth community has come under the control of the corporation. Scientists work for the corporation. Education from elementary school through university and professional school is designed to serve the interests of the corporation.The benefits of modern invention all came to the service of the corporation. We agreed to this because we have become convinced that our well-being depends on the corporations and the immense profits they bring to their executives and to their stockholders.

In America, the comprehensive power of the corporation has come about by the bonding of entrepreneurial enterprise with the legal profession and the judiciary. This bonding came about quite soon after the American constitution had gone into effect. Thus Morton Horowitz, who holds the Chair of the History of Law at Harvard University, has written in the first volume of his book, *The Transformation of American Law 1780-1860:*

> As political and economic power shifted to merchant and entrepreneurial groups in the post revolutionary period, they began to form an alliance with the legal profession to advance their own interests through a transformation of the legal system... By the middle of the nineteenth century the legal system had been reshaped to the advantage of men of commerce and industry at the expense of farmers, workers, consumers, and other less powerful groups within the society.[1]

All the privileges granted the commercial, industrial and financial corporations, all the subsidies they receive from the government, all the exemptions they receive from government agencies — all these come about through this alliance with the legal, judiciary, legislative and executive establishments. The assumption is that corporate profit is the source of well-being for the nation.

The piracy of America by corporate enterprise is sanctioned by the government, supported by the universities, accepted by the religions. The consequences are not confined merely to humanity or to America; the consequences are devastating for the bio-systems of the Earth. So devastating indeed that biologists such as E.O. Wilson tell us that nothing so damaging has happened to the planet earth in the last 65 million years.

Such is the context in which this collection of essays has been written. The information presented here with such clarity and precision is our basic need at present, since the shaping of a more viable human presence on this continent requires that we first know the issues that we are dealing with. Only in this manner can we bring a healing to the present and get on with creation of a future where humans would be a mutually-enhancing presence on this continent.

THE SCIENCE OF COMMUNION

The Very Reverend James Parks Morton

"Or speak to the earth and it shall teach thee." **Job 12:8**

The Lord of History made a quantum leap in the life of the cosmos when he explicitly placed in human hands and brains both the capacity and the responsibility for the continuing co-creation of the world in partnership with God. In more traditional images, God's abiding focus is the Word-Made-Flesh, what Christians call "the Christ event," wherein the divine-human interchange reached the fullness of incarnation, of co-inherence. A New Day dawned that literally turns old values upside down: the poor in spirit are called "blessed" as are the peacemakers, the meek, and those who are persecuted "for my sake." A New Moment arose when cosmic powers of liberation became available to heal the sick and to raise the dead and to bring together very different peoples and forces and creatures that heretofore were apart.

This coming together of all creation in God's New Day is precisely the mystery celebrated in the Holy Communion: earth rises to heaven, the joys of heaven permeate earth, the future bursts into the present and the past is recovered, forgiven and transfigured. All is made new. In this Holy encounter, God transforms the stuff of everyday reality in such a miraculous way that literally everything — *every thing*: every rock every atom every galaxy every man woman child every flower every wolf every whale become linked, connected, bound together as kin, mutually transparent to the divine radiance. Therefore, the vision of Saint Francis is indeed the vision of Holy Communion, in which the sun becomes our brother and death our sister. Brother Sun! Sister Moon!

Perhaps we should say that ecology is the science of Communion, through which we of the earth community learn our sacred connectedness and that shrines, temples, cathedrals and mosques are those great public sacred spaces that invite all earth's creatures to taste the sweetness of God's Day, present in our midst.

In the last three decades, we have gone through a tremendous re-visioning of what it means to be human beings. We are still in the middle of it, so it is hard to see exactly what it means or where it leaves us. But just consider for a moment the number of momentous changes of heart and mind that have occurred. When in history have there been so many popular movements meant to find out who we are and how we ought to behave towards one another? A corollary of this, perhaps, is the respect that we have begun to accord to indigenous cultures. For the first time, we look towards the recently called "primitive" cultures not to denigrate or idealize them, but to assess how far they are able to live sustainably and honorably on earth.

We have a new capacity to see the human as a species in a larger configuration. The habitual view of *homo sapiens* as the apex of the pyramid of Creation is giving way to the vision of humans as one interdependent (albeit dangerously powerful) presence among many other presences in Creation.

And this view carries over into our relationships to each other. We recognize that humanity encompasses many colors, many traditions, many wishes, and many ways of seeing, all of which call on us to cultivate a new level of acceptance and mutual respect for the value of the human in and for all of its diversity.

There is also a new science! Some of its brightest representatives are heirs to a blossoming that began with Einstein and quantum physicists Max Planck and Werner Heisenberg; gathered force with microbiologists Rene Dubos and co-creators of the Gaia Hypothesis, James Lovelock and Lynn Margulis; and is now in full bloom in the generation that includes physicist Arthur Zajonc; geneticist Wes Jackson; soil scientists Will Brinton and Paul and Julie Mankiewicz; chaos theorist Ralph Abraham; chemist Rupert Sheldrake; cultural historian William Irwin Thompson; and perceptual psychologist Francisco Varela.

The exciting notion that animates all of their words is that we do not live in a universe of billiard balls — a dead mechanical universe — but in a cosmos full of living interconnections. (The word "cosmos," by the way, means "beauty.") In the quantum world, a single particle DOES appear in two places at the same time. In the Gaian planet, organisms are interconnected, active maintainers of the Earth as a living being. And all of these wild facts are not viewed as threatening paradoxes. Instead, they are seen as evidence of the fundamentally dynamic and (w)holistic nature of the universe.

We have realized that ecology is not a fifth wheel in our reflections. It is, or ought to be, the organizing principle of human consciousness, itself.

The fact is, though, that this new consciousness has yet to sink in. It wasn't until the summer of 1988 — that very hot summer when medical wastes washed up on our nearby beaches and a barge full of garbage wandered the globe searching for a place to dump its load — that matters of ecology went from being page-37-kind-of-news to front-page headlines. Only then did ecology and the environment emerge from the thinktanks and appear at the breakfast table all over America.

That was the summer when *Newsweek* ran a cover picture that showed the typical American family under a bell jar representing the Greenhouse Effect (as it if were a new discovery, when in fact our brighter scientists had been warning about it for almost two decades). And at the end of the year, *Time* magazine replaced its usual Man of the Year award with a Planet of the Year award, splashing the photography of a Christo-wrapped earth seen from the Moon across its most-read cover of the year.

Before that time only those perceived as fanatics — members of the Friends of the Earth, the Sierra Club, the NRDC, etc. — regarded the environment as a truly crucial issue. And the churches of America, by and large, were just as backward as the general public. So the media and the church — the two entities most clearly charged with bearing the "Good News"— didn't know it or convey it until after those two important magazine covers in 1988.

At the Cathedral of St. John the Divine, I pushed the importance of ecological thinking for the better part of 20 years. We did all that was in our power to wake up churches and individuals since as long ago as 1974. Figures like John and Nancy Todd, Rene Dubos, Hazel Henderson, Wes Jackson, Tom Berry, Brian Swimme, Carl Sagan, James Lovelock and Lynn Margulis all preached from our pulpit or otherwise participated in Cathedral events. Indeed, there was a time when I was roughly taken to task by my religious confreres for not having "religious" sermons and for neglecting the Gospel.

It is only within the last few years that churches themselves have begun to concur with us on the importance of the issue. Finally, the higher ups in American religious institutions began to hear the message and to meet and consult with top scientists to assess and acknowledge both the depth of the ecological crisis and the need for the matter to be at the center of our spiritual life.

The Joint Appeal in Religion and Science — the pioneer group established at the Cathedral of St. John the Divine that has since evolved into the National Religious Partnership for the Environment — brought and is bringing the issue into at least 53,000 parishes and synagogues around the country.

But what is this dawning ecological consciousness, in itself? One of its components is political and issue-driven, related to the need to address and regulate toxins, soil loss, the pollution of our aquifers, and many more specific issues. A second, equally crucial, component deals with the important concrete life-style changes that must be made. These include the insulation of our houses, the recycling of wastes, the elimination of chlorofluorocarbons, and changes in our diet.

All these tasks require not only knowledge but also discipline. Recycling must become for us as much a matter of daily life as wearing condoms or saying our prayers. It must be a part of our routines: we cannot do it haphazardly. There is nothing mystical about it.

The third — and deepest — component of this new consciousness is our acknowledgment of the interconnectedness of everything. Here, we recognize the fragility of living systems, their mutual strengths and their mutual vulnerability. The word "holistic" has emerged out of the arcana of environmental or new-age publications, and has become a word central to our understanding of who we are and what we must do.

Twenty years ago, the word was not much heard outside of the 60s hippie movement. Indeed, a part of the thoughts and images in our new consciousness had its gestation there. It was the hippies, after all, who turned their thinking toward Native American culture, toward crystals, stars, rainbows, natural foods, herbs and the sun.

But their knowledge was insular and limited. It seldom achieved either the discipline or the mysticism that a true spirituality requires. It took the recognition of the establishment — of community, government, academic, business and religious institutions as well as specific calamities, to bring the issue into the forefront of consciousness and (inter)national priority. And it requires recommitment to the necessity of traditional values and the role of the spirit to achieve an economy for the whole planet.

What this entails is the recognized PRESENCE of the unity of the One and the Many, the staples of faith traditions everywhere. IT is what both St. Paul of the Christian New Testament and the Sufi Persian poet Rumi refer to as the ALL IN ALL. This spirituality is at the core of every religion. When it is intact and active, religion is alive: when the spirit is missing, religion petrifies, turns into stony fundamentalism and dies. Ours is a supra-rational reality, above and beyond, but including, objective explanation. We must, from the start, therefore, recognize the presence of many levels, all active at the same time, all fundamental to existence.

The point is that spirituality brings together opposites under the sign of higher harmony. Nowhere is this seen more clearly than in certain Hindu symbols. In the statues of Kali, for example, the power of the female energy is both creative and destructive, yet viewed without judgment, just as in his Canticle of the Sun, St. Francis praised and welcomed every single aspect of Creation, from Brother Sun to Sister Death. The sacred and the secular, the earth and the stars are not separate, but profoundly structurally related.

In our time, science and technology have brought us the extraordinary view of our island earth from the perspective of the moon. It has also informed us that in fact, our bodies and our entire earth are the residue of exploded stars. So we can understand anew the idea, "Dust thou art." Indeed, we are. But *star dust*.

In our reconfirmed knowledge of the universe, we experience a wonder that is grounded in rhythm and practice. There is ecstasy and mysticism on the one hand, and on the other, there is the nitty gritty compassion that binds us to the slums of Harlem, to the threatened salt marshes of our coasts, and to the suffering neighbor who has passed out in front of Citibank.

We are one in the energy of integration. The spirit must, therefore, be present in all education really worthy of that name. We must study not only with our brains but with our hearts, our bodies, our sexuality, with all that we are and may become — and we must draw on the strengths of all faith traditions.

If, combined with tolerance and diversity, there is a true union of body, mind and spirit, we come to ecstasy, a word that literally means "to stand outside of." This is the wonder, not only of faith and of the arts, but also of real education, those moments when new and unsuspected vistas of both the particular and the whole open up to us, and we recognize the deep, organizing rhythms that discipline all of Creation. In the end, the spirit is one with wholeness, and so the definition of spirituality is very close to the definition of ecology. Both contain the joining together of all systems and the insistence that not one jot be left out, that everything has its role.

In Native American culture, in the Brazilian Rainforest, and among the Bushmen of the Kalahari — as Laurens van der Post described them in his many books — we see a dynamic image of peoples whose religion and ecology are in harmony. Reverence for the Earth is at the heart of their very existence, *and* so without any sentimentality at all, they can serve as models for our own search.

That extraordinary man, writer, anthropologist, my friend, Laurens van der Post — who died December, 1966, just after he celebrated his 90th birthday — understood so well the true ecology of things. In *Mantis Carol*, his writings about the vanishing bushman of his native South Africa, van der Post answers a woman's question about the reasons that a bushman dances. There are two different dances, he tells her, the dance of the Little Hunger and the dance of the Great Hunger.

The first one is of the physical hunger the child experiences the moment he is born and satisfies first at his mother's breast, and which from then on stays with him for the rest of his life on earth. But the second dance is the dance of a hunger that neither the food of the earth nor the way of life possible upon it can satisfy. It is the dance of the Bushman's instinctive intimation that man cannot live by bread alone, although without it he cannot live at all; hence the two.

"Whenever I asked them about this great hunger," he writes, " they would only say, 'not only are we dancing, feeling ourselves to be raising the dust which will one day come blown by the wind to erase our last spoor from the sand when we die, lest others coming and seeing our footsteps there might still think us alive, not only do we feel this hunger, but the stars too, sitting up there with their hearts of plenty, they too feel it and feeling it tremble as if afraid they would wane and their light die, on account of so great a hunger.'"

The bushman's dances, then, link us all.

We stand on a cusp today. Our increasingly interconnected world city is young, desperate and in exile from itself and from the earth herself. We must begin to understand the new ecological vista of society as inclusive of each other and of the world around us.

Yeats wrote years ago of a time when "the best lack all conviction and the worst are full of a passionate intensity." As we read the headlines these days, it often seems that this time has come around again.

The task of every real human being is not to wallow in the worst, but to recognize that we, too, are given an extraordinary energy. We must realize that it is no longer enough for us human beings to dance the dance of the Little Hunger — for all of the things that we know we need— but also the dance of the Big Hunger in which our very bodies resonate with our sisters and brothers above and below us — our planet, our galaxy and the stars from which we come.

INTRODUCTION
Judith Scherff

"In our day, there is a growing awareness that world peace is threatened not only by the arms race, regional conflicts and continued injustices among peoples and nations but also by a lack of due respect for nature, the plundering of natural resources and by a progressive decline in the quality of life. The sense of precariousness and insecurity that such a situation engenders is a seedbed for collective selfishness, disregard for others and dishonesty."

Pope John Paul II, 1989

One of the most prevailing images of the United States is that we are a Christian nation, guided by the best selling book of all time which begins: "In the beginning, God created the heavens and the earth."[1] The major religions of the world share a belief that the earth is a divinely inspired creation, a belief also held by many individuals who claim no organized religious affiliation. The awesomeness of the miracle of that creation is described by Matthew Fox:

> Isn't it surprising that the moon is the exact distance from the earth that it has to be for the tides to happen? And without that being exactly as it is, life would never have emerged from the sea and we would not be here nor any of the wonderful companions with whom we share this globe. Isn't it amazing? The sun is just the right distance from the earth and the ozone layer is just the right thickness to allow sunshine in and a certain amount of radiation out. Isn't it amazing that about 19 billion years ago when this universe began in a ball of fire that in the first second of that fireball the decision was made that the universe would expand at a certain rate? If that rate over 750,000 years ago had been one millionth of a millionth of a second slower or faster than it was, you and I would not be here today. If the overall temperature of that fire ball over 750,000 years ago had been 1 degree warmer or colder than it was, you and I would not be here. About five and a half billion years ago a super nova exploded and gave birth to every element in our bodies.[2]

Likewise, many believe that the process of physical birth is divinely inspired . It is curious to me that birth is celebrated as a miracle at the time it occurs, and then the thought of Godly participation stops. That God gives life when a new creation arrives is recognized by many, but then what happens? That new entity requires air immediately, shortly thereafter and for the rest of its life, water and food. Belief in "the gift of life" as it applies to birth erroneously does not extend to the ecosystems as a continuation of the gift of life because they provide us with life support systems — air, food and water. The ecosystems (biosphere, hydrosphere, atmosphere, lithosphere) were designed to support and sustain the presence of living beings. Life is a unit, the parts of which ingeniously and sometimes miraculously encourage the proper workings of other parts. It is a circular system; an interdependent system; a very well connected system, much of which is still misunderstood, clearly maligned, and enormously undervalued by most of us. However, in 1997, scientists concluded that the value of this life support system to each of us is $33 trillion.[3] And yet, many people consider natural resources no more than a resource from which they can make money.

The Bible goes on to say that "man" shall have dominion over the earth another dubious paradigm. Many people, clergy included, believe that dominion has been perverted. Dominion in the classical sense means "to care for," "to take care of," "to be responsible for." Instead, dominion has become corrupted to mean ownership, possession, control and even destruction of natural resources — in essence, dominion has become domination and domination has been magnified to the point of greed.

In the legal, the human-oriented sense, public domain includes those geographical areas owned by the government, ostensibly the public. However, it may well be argued that the earth — the whole earth — is public domain. While it is true that property rights divide the earth by nations, subdivisions of nations, cities, sections and lots, these political lines are invisible to the workings of those ecosystems upon which we depend. Air blows over land collecting all kinds of particulate matter from human activity regardless of who "owns" it. Water flows in streams past thousands of "property owners." Human "ownership" of natural processes is arrogant if not wholly erroneous. Nevertheless, our political system allows for "ownership" of nature, and most egregiously, allows corporations to degrade, rearrange, diminish, and eradicate members of a natural order for private profit.

We have heard how necessary it is for the extractive corporations to use the environment to create jobs, which in turn created the jobs vs. environment debates. That argument was considered reasonable until reality was introduced. It wasn't the environment that was causing the loss of jobs, but rather that:

> [b]y cutting real wages for two decades, by aggressively working to bust unions, by eliminating full-time jobs and, in their place, giving us part-time temporary jobs without medical or retirement benefits, corporations have created great insecurity among middle class people. Corporations have then used this insecurity as a lever to roll back environmental standards and regulations, and to deflect new environmental initiatives.[4]

There is a vast physical and moral difference between jobs and fortunes. Taking more than is necessary and changing the natural composition of vital elements such as air, water, and food by adding toxic chemicals is what we object to. Acquisition and abuse of earth's life in order for a very few to get rich through a process of buying the government so that these processes become legal (not moral) — that is piracy. Willingness to turn federal and state lands — the public domain — over to corporate and industrial special interests abuses the democratic system that we still erroneously believe is functional.

The collusion of political and corporate manipulation of laws and the abuse of power which denies populations of humans and nonhumans adequate basic resources — while we continue to think of ourselves as a Christian nation and a democracy — is what this collection is about. The threads that weave these chapters into a book illuminate the fact at the environment, democracy, civil society and the economy are the same problem even though we erroneously treat them separately.[5] This book explains why environmental issues have become human rights issues. Integrated with the fact that the rich are getting richer at the expense of all the rest of us is the diminishing health of our environment — the commons of creation we depend upon for life itself.

In 1996, 358 billionaires controlled assets greater than the combined annual incomes of countries representing 45 percent of the world's population (2.5 billion people). Between 1961 and 1991, the ratio of the income of the richest 20% of the world's population to the poorest 20% increased from 30-to-1 to 61-to-1. In 1997, the wealthiest 5% of Americans captured 16.8% of the nation's entire income; by 1989 that same 5% was capturing 18.9%. During the 4-year Clinton presidency, the wealthiest 5% have increased their take of the total to over 21%.[6]

The changes in our economic, environmental, political and social structures have created myths. Although the structure is still in place for a democracy (elections and campaigns and the right to vote), there is slight resemblance to the ideals of this populist nation which was created just over 200 years ago. Numerically, fewer than half of the eligible electorate cast their votes even in a presidential election. A major segment of the voting public believes the act of marking and casting a ballot is futile.

Is it any wonder? During the 1996 election cycle, Political Action Committees (PACs) raised $437.4 million and spent $429.9 million.[7] Children, throughout their school years, have heard that the purpose of government is to do for people what people can't do for themselves. However, that manifestation of "of, by, and for" the people has been replaced by a system of special interests that renders government, as a service institution, dysfunctional. Our founding fathers created a pyramid of political power from the bottom; the people were given the power to direct elected officials to carry out majority will. What we have now is a pyramid of corporate power, driven from the top down, through which elected officials carry out the will of corporations that have funded their campaigns — never mind the public.

As our culture has increasingly moved towards a monetary value system, much of organized Christianity has followed suit by becoming culture oriented rather than Christ oriented. It is in the spirit of following

Christ's teachings — those tenants of the Gospels where many "practicing Christians" and organized religion itself falls short. And certainly, this lack of Christian behavior was most glaringly revealed shortly after the election of the 104th Congress, when the collusion between the so-called Christian Coalition and extremist conservative members of Congress produced the narrow visioned special interest agenda called *The Contract With America.*

The real Contract with America is the Constitution of the United States, which clearly states in its preamble that government was created "to promote the general welfare." The political spectrum has tilted so dramatically that "promoting the general welfare," considering all of the population, doing for people what they cannot do for themselves — those "caring for" ideals — are excoriated as socialist and/or even communist by right wing politicians whose records exemplify self-service. By current political standards, one of the most liberal/left wing commands is in the Gospel of Matthew and instructs people to feed the hungry, clothe the naked, visit the sick and those in prison, welcome the stranger and to "consider the least of these, my brethren."[8] Those people whom Christ would have us lift up are the very ones whom our current Congress has deemed dispensable; they are those for whom Congress withdrew the social safety net. This political behavior underscores the argument that we are practicing neither Christianity nor democracy. It is quite accurate to say that we are a deeply religious nation, but our god is money.

It is only when we recognize that, by its corporate and political actions, this nation is no longer a democracy nor true to the major premises of its spiritual traditions, that we can begin to return to both.

Time Magazine Investigates Corporate Welfare

Commencing with an introductory cover story in its November 9th, 1998 issue, *Time* Magazine has initiated a special investigation into corporate welfare — of major concern in *The Piracy of America*, in relation to corporate plunder of our natural resources. Here is an opening paragraph from *Time's* special report:

"Two years after Congress reduced welfare for individuals and families, this other kind of welfare continues to expand, penetrating every corner of the American economy. It has turned politicians into bribery specialists, and smart business people into con artists. And most surprising of all, it has rarely created any new jobs."

"Corporate Welfare," *Time*
Introduction to a Special Investigation
by Donald L. Barlett & James B. Steele
November 9, 1998, p. 38

Tracking Corporate Greed &

Destruction in the Public Domain

THE KILLING GROUNDS
The Destruction of
Native American Species[*]
Conger Beasley, Jr.

"Just as the canary provides advance warning of toxic air in the mines, so too, endangered species alert us to the extent of the deterioration of our natural environment."

B. Ishmael

Once Upon a Time There Was A Wolf[1]

This is the Gray Wolf, Canis Lupus, Almost all of the wolves left in the world belong to this species.

Weight of the wolf pup at birth	1 lb.
Age pups begin to eat meat	3 wks.
Age pups leave the den	2 mo.
Age pups begin to hunt with the pack	5-6 mo.
Age at which a wolf is fully grown	18 mo.
Weight of an adult wolf	70 - 125 lbs.
Length of the adult wolf	5-61/2feet
Length of time a wolf can survive without eating	2 wks.
Amount of food a wolf can eat at one time	20 lbs.
Size of a wolf's hunting territory	0-200 sq. mi.
Number of distinct howls made by a wolf	6
Usual number of wolves in a pack	8
Number of wolves in a pack that breed	2
Usual number of mates in a wolf's lifetime	1
Breeding season of wolves	3 mo.
Gestation period of young	65 days
Number of pups born to an average litter	6
Weight of a wolf pup at birth	1 lb.
Amount of money that has been spent in the U.S. to exterminate wolves	$100,000,000
Number of wolves in Alaska	6-10,000
Number of wolves in Minnesota	1200
Number of wolves in the rest of the United States	100

Imagine North America swarming with creatures of every description. Imagine the sky dark with birds, the streams and rivers choked with fish, the grasslands teeming with ungulates, the forests thick with fur-bearing animals. Five hundred years ago, this continent was such a place. A trove, a fount, a cornucopia of wildlife. Before contact with Europeans, the human population of North America, excluding Mexico, was probably around 15 million. Needless to say, census-taking then was a non-existent discipline, and a precise count is impossible to determine.

To feed and shelter themselves, these first Americans hunted, fished, and trapped, but despite certain excesses perpetrated against select species, their overall impact on the landscape was minimal. While it's true Plains Indians at times killed buffalo exclusively for their tongues, and the Anasazi overcut the trees growing in the Four Corners region of the desert Southwest, human population pressures in both these instances did not reach sufficient density and weight to threaten the integrity of the habitat in which they lived. Ultimately, the key to the survival of any species lies in the preservation of the environment in which that species dwells. Indiscriminate killing takes a huge toll, of course, but the death knell normally is not sounded until the habitat shrinks to a size where it can no longer support a viable population of reproductive adults.

The list of species driven to extinction by the advance of civilization across the continent in the 18th and 19th centuries forms a melancholy roll call: the Labrador duck, the great auk, the heath hen, the passenger pigeon, the Carolina parakeet, the dusky seaside sparrow; to name but a few.

As near as ornithologists can ascertain, before the European invasion, there were between three and five billion passenger pigeons roving free in tremendous flocks on the North American continent. They inhabited the eastern half of the continent, from about the 100th meridian bisecting the Great Plains states, as far north as the boreal forests of Canada, all the way east to the Atlantic seaboard, south to Florida and the Gulf of Mexico. Their principle nesting area included parts of New England, New York, and Pennsylvania, but was mainly concentrated in the states of the old Northwest territories: Ohio, Indiana, Illinois, Michigan, and Wisconsin. Their favorite roosting sites were the mixed hard- and softwood forests of those regions, which produced tremendous quantities of mast.

Mast is the litter that falls from these trees — acorns, beechnuts, leaves, seeds, dead bark, and twigs. So rich and dense was the duff that collected under these trees, it attracted staggering numbers of passenger pigeons. Their numbers as they migrated from roost to roost darkened the skies of the Ohio Valley. The predominant shape of the flocks was serpentine, a long, wide, curving line composed of millions of birds that materialized out of the sky with loud, whooshing wings. The birds were densely packed; it took hours for the flocks to pass over, sometimes days. According to A.W. Schorger, the normal speed of a migrating passenger pigeon was about 60 m.p.h., with a potential speed of 70 m.p.h. when pressed.

In 1813, John James Audubon, traveling through the Ohio Valley between Louisville and Henderson, Kentucky, witnessed the flight and nightly roost of what he estimated to be over a billion pigeons, capable (in his estimation) of consuming nearly 9 billion bushels of mast a day.

The sheer numbers of birds, the impact of their presence, the efforts people made to bring them down, is mind-boggling to us, today. Flocks sweeping over small towns sent the entire population stampeding into the streets in a jubilant mood. Businesses shut their doors; schools were closed: the grim business of killing pigeons created a virtual holiday atmosphere. Men with poles knocked down as many low-flying birds as they could reach. Muskets, pistols, and rifles were brought out, primed, and triggered into the sky. Old cannons on the town square were loaded with nails and bolts and glass shards; the barrels were then tilted into the air and fired point blank at the dark feathery masses rumbling overhead.

The slaughter was ghastly, but so great were the numbers of birds, no matter how many were knocked down, it seemed inconceivable that all of them might someday be killed. So plentiful were the birds in the 1850s that a measure introduced in the Ohio legislature was turned down by the senate on the grounds that the birds were "wonderfully prolific" and needed no protection. "No ordinary destruction can lessen them or be missed from the myriads that are yearly produced,"[1] the report concluded. But the pigeons were not "wonderfully prolific;" at most, they produced a clutch of only one or two eggs a year— nowhere near enough to keep pace with the accelerating destruction of both their numbers and their habitat.

The density of a flock of passenger pigeons averaged around ten birds per square yard. This made them relatively easy to kill. In the 1870s, a professional hunter from Michigan slew as many as 70 birds with a single shot, especially in the spring when the birds flew in close formation. Schorger says it was not uncommon for this particular hunter to bring down between a thousand and twelve hundred birds before breakfast![2]

By 1889, the overall population of pigeons appears to have dwindled to a few thousand. In spite of numerous laws on the books of several states, no sustained effort by any enforcement official was ever made to protect the pigeons. Public opinion, for the most part, was apathetic to protection, or solidly against it.

Contributing to the pigeon's demise was the destruction of the native forests of the Ohio Valley, especially following the Civil War. A postwar building boom, the demands of industrial furnaces, plank roads, fireplaces, steamboats, and railroads created an insatiable demand for timber. In the early 1800s, sycamores growing along the banks of the Ohio River were judged, by one observer, as "the loftiest and largest trees of the United States."[3] A French botanist collecting samples in the area between 1800 and 1810 reported finding a number of species with circumferences in excess of forty feet. Indiana, at the beginning of the 19th century, was one vast forest of sycamore, oak, maple, beech, dogwood, birch, walnut, and hickory. The preponderance of these same trees made southern Michigan, during the early 1800s, one of the most heavily forested areas in the Union, and gave Illinois plenty of timber and Wisconsin all kinds of wood of the best quality. None, however, could compare with Ohio's woodlands, which presented, according one observer, "the grandest unbroken forest of 41,000 square miles that was ever beheld"[4] on this continent. Immense forests of pine dominated the northern portions of Michigan, Wisconsin, and Minnesota.

Less than century later, only scattered fragments remained of what has been described as "the most extended and valuable forest on the globe." By 1867, five billion cords of wood had been consumed, equivalent to two hundred thousand square miles of woodlands, an area nearly equal to all the land comprising the states of Illinois, Michigan, Ohio, and Wisconsin. "The mass of forests is gradually receding," observed English novelist Frederick Marryat after visiting upstate New York in the 1840s. "Occasionally some solitary tree is left standing, throwing out its wide arms, and appearing as if in lamentation at its separation from its companions, with whom for centuries it had been in close fellowship."[5]

By the end of the century, the lumber industry shifted its focus to the new states and territories of the west, most recently to Alaska and the Olympic Peninsula of the state of Washington, where today we are reaping the fruits of this destructive madness in the spotted owl and old growth controversy.

Meanwhile, back in the timber-devastated region of the Ohio Valley, the numbers of passenger pigeons dwindled alarmingly. As the amount of nestlings declined, competition for squabs — newly fledged pigeons, a lucrative culinary item for big-city restaurants — became keen. Squabs feeding on beechnuts, rich in stored fats from the mast of the forest floor, made for especially fine eating.

In 1882, the retail price for a dozen squabs in Wisconsin varied from 35 cents to a dollar. The fewer the pigeons in any given area, the more persistently they were hunted; persecution was unremitting till the last wild bird had disappeared. By 1900, the fate of the passenger pigeon was sealed. In 1909, the American Ornithologists Union of New York offered a reward for the discovery of a remnant nest or colony. The final report, issued in 1912, confirmed the general opinion that the species was extinct. The last passenger pigeon in captivity in the Cincinnati zoo, a female named Martha Washington, died at approximately two o'clock on the afternoon of September 1, 1914.[6]

The period between the Civil War and 1900 we know as the Gilded Age, a term coined by Mark Twain, the title of one of his lesser-known novels. It was an era of tremendous expansion, with the population more than doubling from 35 million in 1865 to over 70 million by the end of the century (much of it resulting from a tremendous influx of European immigrants). It was also an era of fantastic consumption of natural resources, the product of a willful and systematic slaughter of native species, whose effects we still feel today.

The demise of the buffalo forms another sad chapter of this profligate era. They were brought to near-extinction by a calculated policy initiated by the U.S. military with the full cooperation of the civil authorities to deprive Plains Indians of the material foundation of their culture. Both white and Indian hunters were encouraged to go out and kill. "Unless one owned a ranch," says historian E. Douglas Branch, "the quickest way to make money on the frontier was to turn buffalo hunter."[7] Buffalo blankets, hats, and coats were very much the rage in fashionable circles back east. Both American and English tanneries clamored for buffalo hides, out of which leather belts were fashioned for operating industrial machinery.

A hard-working hunter armed with a lever-action, .55-caliber Sharps rifle with a telescopic sight could bring down as many as 200 buffalo a day. (The weapon had a killing range of fifteen hundred yards.) We have an account of a professional hunter on the plains of northwest Kansas who found himself in the center of a sizeable herd. Steadying his rifle on a tripod, methodically rotating the barrel, carefully planting his feet before squeezing off each shot, he described a complete circle in less than 20 minutes, knocking off 85 buffalo.[8]

Such was the manner in which the mighty bison, once numbering in the millions, were ruthlessly decimated until by 1890 there were less than 500 left in the United States. In the wake of the assassins came the scavengers, men and women driving two-wheeled carts who pulled the bones and horns from the stinking carcasses and bundled them up on railroad sidings in callow, gun-slinging towns in western Kansas. The bones were shipped back east to special factories, where they were broken down to make buttons, stays, hooks, and glue. Fertilizer plants processed the bones into phosphates, sugar refineries into carbon.

There has not been, and hopefully never will be again, an era that for sheer consumption of natural resources can match the excesses of the last half of the 19th century.

Historically, the assault on animals in the United States forms a grim and sorry chapter. The U.S. has the dubious distinction of being one of the few countries in the world whose government has subsidized the slaughter of animals; in the 1870s and 80s, buffalo hunters were sent out to the Great Plains as part of a federal policy to destroy the food supply of nomadic Indians.

Once the buffalo was gone, the consortium between government and hunters turned its attention to animals that preyed on livestock: bears, bobcats, mountain lions, coyotes, and wolves. Especially wolves.

In the Southwest, beginning in the 1890s, a concerted plan to eliminate wolves was put into action. For the next 80 years, thousands of wolves in Texas, New Mexico, and Arizona were systematically trapped, poisoned, and shot. (During the 1920s, there was serious talk of erecting a boundary fence along the border to keep the Mexican lobo from straying onto U.S. soil!) Private hunters took a percentage of the toll; the majority, however, were killed by government employees operating under the aegis of an obscure agency, originally known as the Bureau of Biological Survey, and today as Animal Damage Control (ADC).

ADC has a long and convoluted history. In 1909, Congress provided funds to enable its predecessor, the Biological Survey, to "conduct experiments and demonstrations in destroying noxious animals." Within two years, 1,800 wolves and 23,000 coyotes were eliminated from national forest lands. Such activity was not unusual for a government agency back then; even the National Park Service promoted varmint control. Several of the early icons of the conservationist movement, including John Burroughs and Aldo Leopold, subscribed to the belief that the carnivorous instincts of predators made them less intrinsically worthy than benign animals such as deer. Fewer predators roaming the wilds meant that more desirable creatures would flourish. And flourish they did; to protect 3,000 Rocky Mountain mule deer on Arizona's Kaibab Plateau, the National Park Service in the early 1900s killed thousands of wolves, coyotes, bobcats, and mountain lions. The result was a catastrophe. The deer population rocketed to nearly 100,000. They ate everything green they could find, denuding huge tracts of ground cover. They crowded out other species, and drove their own surplus population to a slow death from starvation.[9]

In 1931, Congress passed the *Predator and Rodent Control Act* (PARC), which drew up the first official "hit" list of so-called "noxious animals;" the list ran the gamut from wolves and mountain lions, through gophers and jackrabbits, to eagles and hawks. The war against predators was now fully licensed and legitimized. Augmented by the use of poisons in the 1940s, the body count soared. A rodenticide developed during World War II called Compound 1080 had a devastating effect. The bait was prepared by killing an old horse or sheep and injecting the still-warm cadaver with a generous dose of 1080. Capillary action circulated the lethal compound throughout the carcass; chunks were then carved out and distributed at select points along the paths that predators were known to use.[10]

By the 1960s, the pervasive use of poisons on western lands had attracted nationwide attention. To counteract a wave of growing concern, Secretary of the Interior Stewart Udall appointed a special committee of wildlife experts to investigate the predator control program.

After a thorough review, the so-called Leopold Committee (headed by A. Starker Leopold, Aldo's son) issued a report that criticized virtually every facet of the federal program. Secretary Udall accepted the Report and moved to implement the recommended changes. Thanks to pressure brought to bear by conservation groups, the use of some toxicants, including Compound 1080, was banned by President Richard Nixon in 1972.

But something got in the way of those other proposed reforms, and that something was the cattle and sheepherding industry of the western states, whose tentacles reach deep, not only into state legislatures, but into congressional committees at the federal level. The 1972 toxicants ban prompted research into other methods of predator control. (The M-44 cyanide ejector device, dubbed the "Coyote Getter," came out of this research.) A thriving black market in banned toxicants sprang up, particularly in states like Wyoming. By the mid-1980s, ADC's annual budget had swelled to $31 million, funded by livestock associations and the American taxpayer. Despite mounting concern about the widespread use of poisons, despite a growing aversion to ADC's methods, despite a marked decrease in the numbers of grazing sheep (from 21 million in the mid-1950s to 11.2 million 30 years later), the agency in the late 1980s seemed to be turning up the heat on predators, especially the coyote. An ominous sign in 1985 was the transfer of ADC from the Department of Interior (where wildlife at least in theory is treated as the equal of domestic livestock), to a branch of the Department of Agriculture (whose sympathy toward cattle growers and sheepherders can be assumed).[11]

Another ominous sign was the 1981 reauthorization by the Reagan Administration of the use of certain poisons. Despite pressure from conservation groups, despite public revulsion at the mass slaying of animals, the war against predators, perpetrated by both ADC and private interests, remained as nasty as ever. During a 1991 sting operation in Wyoming, federal officials bought and seized enough Compound 1080 — strychnine, cyanide, and thallium sulfate; toxicants banned in 1972 — to "probably kill every man, woman, child, and predatory mammal in the western U.S.," according to Galen Buterbaugh, regional director of the Fish and Wildlife Service. "There are some folks out there who do this with a great deal of enthusiasm and pride," Buterbaugh adds. "It goes beyond just protecting their livestock."[12]

Old attitudes die hard. With the virtual elimination of wolves and bears from the western states, attention was turned to coyotes. "Remember Our Slogan: Bring Them In, Regardless of How," a Biological Survey official declared in 1922. Back then, the emphasis on numbers quickly escalated into an obsession. Ben Lilly, the most famous Biological Survey hunter, once lamented that between 1880 and 1904 he managed to slay only 104 bears in Mississippi and Louisiana, whereas in 1832 alone, Davy Crockett had killed 105. The irony was not lost on Lilly's biographer, J. Frank Dobie, who declared, "With renewed assurance of his mission in life, the lover of the wild went on annihilating it."[13]

To current critics of ADC practices, the core of the Ben Lilly mentality has not altered in 70 years. Their primary complaint is the deadly side-effects produced by the use of poisons, namely ancillary or non-target kills. In February 1990, near Folsom, New Mexico, after gobbling a chunk of Faradan-injected meat, a coyote staggered into the bush and died. A golden eagle, feeding off the carcass, also died. Several magpies, feeding off the eagle, then died. Another golden eagle died after eating one of the magpies. When U.S. Fish and Wildlife personnel inspected the scene, they found four more dead golden eagles and a bald eagle lying within 20 yards of the coyote.

Something we need to remember about the coyote: as a species, it is virtually inextinguishable. Between 1915 and 1947, the federal government killed 1,884,897 coyotes.[14] (There is no record of how many were taken by private trappers and hunters.) Despite this fact, and despite the fact that in 1988 alone ADC eliminated 76,033 coyotes, the animal's numbers are booming. Mother Nature has a way of keeping coyotes in circulation. The more they are harassed, the faster they reproduce. Yearling coyotes breed earlier. Litter sizes increase from an average of 3 pups to an average of 8 to 10 pups. Some biologists even feel that predator-control projects such as those administered by ADC are actually responsible for the creation of a "super-coyote": stronger, smarter, tougher; more apt to succeed in bringing down domestic livestock when it suits their purposes to do so.[15]

Biologist Bob Crabtree has studied coyotes in regions of Washington state that are off-limits to trappers. He believes that predator-control programs actually foster livestock attacks. Stable, undisturbed populations of coyotes tend to live in packs and forage cooperatively for food. Extensive culling throws their social equilibrium out of whack, enabling younger, more aggressive leaders to emerge. Not knowing how to forage efficiently, these new leaders are more likely to take out after livestock. The irony, says biologist Mollie Matteson, is painfully obvious: " Most ADC programs treat the symptoms, not the problems. Ultimately, they exacerbate the very problem they hope to solve."[16]

So what is the killing all about? Department of Agriculture statistics indicate that in 1991 predators killed 106,000 head of cattle and calves, or 2.4 percent of the total. The previous year, 37 percent of the 490,000 sheep and goats that died were lost to predators. Despite the availability of nonlethal methods of control, a 1990 Government Accounting Office (GAO) report indicates that, although the ADC policy manual clearly states that nonlethal methods be given first consideration, "little evidence exists of state ADC program personnel employing such methods. Rather... killing offending animals was used predominantly to control predation on livestock." Selective killing of predators was preferred by ADC personnel, says the report, "because it is the fastest and most cost-effective way to solve livestock predation problems."[17]

Nineteen eighty-eight was the last year when ADC issued an annual predator body count. (The total was 140,246 mammals, with coyotes topping the list at 76,033 and beaver next at 9,143; add birds to the total, especially urban pests such as starlings, grackles, and blackbirds, and the aggregate soars to 4.6 million.) Since 1988 the agency, under fire from media and conservationists, has developed a siege mentality. Information is difficult to come by. How difficult is illustrated by Rio Grande Sierra Club official Patricia Wolff's struggle to obtain a copy of ADC's 1992 report. She was told by the agency's New Mexico director to file a Freedom of Information request — a lengthy ordeal that frequently requires a lawyer's assistance and can take months of red tape and paperwork to process.[18]

Within the enclosed circle of ADC wagons, some pretty serious soul-searching seems to be going on. According to an internal memorandum dated April 24, 1992, entitled "A Futuring Document for Animal Damage Control," the agency is gearing itself up for a thorough overhaul of its public image and a concerted revision of its in-house management system. Included among the recommendations are 1) a more generic name-change ("Wildlife Services"); 2) a "pro-active" approach to explaining the agency's role to the public; 3) a more ethnically diverse work force (to counter the redneck, good-ole-boy stereotype); 4) increased reliance upon a philosophy known as Integrated Wildlife Damage Management, which emphasizes nonlethal predator controls and keener employee sensitivity to animal welfare issues.[19]

All of which is pure bunk, says Dick Randall. Randall is a former ADC hunter and trapper, who, by his own admission, killed thousands of predators up to 1972, when he experienced a change of heart and quit. Since then his articles and interviews, from the perspective of an insider-turned-heretic, have caused ADC a great deal of embarrassment. In 1990, says Randall, "nearly $30 million was spent to control so-called varmints. Of that amount, 87 percent was allocated for use in western states, and 90 percent of the western funds were allotted for lethal control. As it always has, ADC is relying on more killing as its principal way of solving reported wildlife depredation problems."[20]

Not so, says Stuart McDonald, public relations specialist for ADC. Things have changed since 1985 when the agency was transferred from the Department of Interior to the Animal and Plant Health Inspection Service (APHIS) of the Department of Agriculture. "The management has changed, the entire tone of the operation has changed as well," McDonald insists. "Today, we're advocating a whole new method, the 'zero-zero' approach, which means zero losses in agriculture to zero losses in wildlife. We don't want to kill an animal if we don't have to. Twenty-five percent of our current budget goes to the Denver Wildlife Research Center to develop nonlethal methods of predator control. The old days of the trigger-happy agent are gone. Our current program managers are mostly wildlife biologists. They're better educated than our managers in the past; they're more professional and career-oriented. While there's still a few old-line guys who believe in disposing of predators in wholesale numbers, the ratio is changing and will continue to change. ADC is a more accountable agency today. The image it once projected simply is not relevant any more."[21]

A prime example of a new, nonlethal device is the electronic guard. Battery-powered, equipped with an electronic sensor, the device is hung in a tree and activated at dusk. Intermittently, for intervals of 6 to 11 seconds, a strobe light beams and a siren blares. In every reported field test, the device has been 100 percent effective. (The device has been particularly effective during lambing and when sheep are being herded across country.)

Those who want to get rid of ADC need to think carefully about what they are saying, McDonald cautions. "In the past few years we have reduced nontarget wildlife deaths. A 1990 GAO report declared not only that the number of predators killed by ADC is small in comparison with overall predator populations, but that the number is also small in comparison with those killed by private hunters and trappers. Nature is cruel and unrelenting. Doing away with ADC is not going to do away with wildlife depredations of livestock. Actually, it might even get worse. Remove ADC from the equation, and you leave predator control in the hands of the individual ranchers. Who knows what kind of vigilante mentality might develop out of a situation like that?"[22]

ADC may project a negative image, it may kill animals unnecessarily at times, it may be vilified by wildlife lovers, but the fact that it continues to function can be attributed to the consequences of an even bigger problem of which the agency is but a symptom, a problem that has plagued the West since the last wild buffalo was shot to make way for the first tame cow: overgrazing on public lands. The real culprits of the predator-control problem are the ranchers and sheepherders who for generations have benefitted from a special kind of government largess, which not only grants them the use of public land at minimal rates, but has provided them with their own private constabulary to protect the health of their inventory.

The system works this way. There are approximately 307 million acres of public domain land in the U.S. — -an area roughly ten times the size of New York State. Over 130 million acres are administered by the Forest Service and the National Park Service, with the military holding a few sizeable chunks. The Bureau of Land Management (BLM), a division of the Department of Interior, currently oversees the remaining 174 million acres. Theoretically, this land belongs to the American people. Theoretically, it forms a major component of the historical legacy that is every American's birthright to enjoy.

Next time you're out west, driving along a scenic road, take a close look at the land you help maintain with your tax dollars. Notice how sparse and arid it is. Notice how the ground cover has been chewed up and destroyed. Notice how the banks of the creeks and rivers have been beaten down and trampled. Notice the presence everywhere, from fence line to ridge top, of unsightly cowpiles. A region that once nourished a tough, nutritious grass ("a sort of hay on the stem," says one naturalist) has been blighted by the effects of overgrazing.

The deluge began after the Civil War. Even before the Indians were removed, the beeves were herded in. In 1870, there were 4.6 million cattle cropping the range in 17 western states; by 1890, that number had nearly quintupled, to 26.6 million. With the Indians neutralized, the only danger to the industry was from rustlers and predators. The rustlers were hanged, and we know what happened to predators like the wolf. Unchecked by natural enemies, the beeves and sheep were free to wander and proliferate; the result has been a massive degradation of the landscape, a process akin to desertification. Not the creation of Sahara-type dunes, but something more subtle and insidious: the desiccation of a complex ecosystem through the loss of species diversity. (A UC-Irvine chemist named F. Sherwood Rowland claims that currently there are so many cows on Earth that methane belched from the depths of their four-chambered stomachs has contributed significantly to global warming!)[23]

The Taylor Grazing Act of 1934 put an end to open range foraging. Parcels were carved out to which individual ranchers could obtain exclusive permits. To qualify for a parcel today, a rancher has to own a piece of base property big enough to accommodate a ranch house. Sixty-plus years after the *Taylor Grazing Act* was signed into law by President Franklin Roosevelt, permit holders pay only $1.35 per animal unit month (AUM) to graze a cow and calf on national grasslands and

U.S. Forest Service land; twenty years ago, the fee was around $4.00.[24] But this isn't all. Though a rancher can't sell his permit, he can sell his own private holding at the "permit value" — in other words, at the enhanced value of the property to which his grazing permit is attached. And not only that: the rancher can use this inflated "permit value" as borrowing collateral at the bank.

A tasty little deal, eh? The ranchers like it. The local bankers like it. The western politicians who sit on crucial land-management committees in state and federal legislatures like it. The only people who shouldn't like it are the legions of ordinary taxpayers who know nothing about it.

Remember that only 2 percent of the marketable beef in this country is produced by the 23,000 permittees who graze their cattle on public land in 11 western states. So where does the other 98 percent come from? Primarily from private land in the Midwest and South, where it takes an average of 18 acres to support one cow, as opposed to 160 acres in the West. "Public land ranchers simply do not feed the world like they say they do," says Steve Johnson, a wildlife consultant for Native Ecosystems in Tucson. "Cattle don't belong out there. They aren't suited to the dry conditions. One cow in one month can eat its own weight. The land can't support that. By contrast, a desert tortoise eats between 7 and 23 pounds of plants a year. A cow will eat that much in one day."[25]

The cowboy myth needs to be buried, gently but firmly, with all appropriate ceremonies, six feet under the dry, flaky earth of Big Sky country. The contemporary rancher who feeds off the public lands permit system perpetuates a shibboleth of rugged individualism that's as phony as the set to a western movie. Disgusted by the sight of cow burnt wastelands, Ed Abbey, in a 1986 *Harper's* article, declared, "The rancher, after all, is only a farmer, cropping the public rangeland with his four-legged lawnmowers, stashing our grass into his bank account."[26] It's time for a new ethic that combines responsible grazing practices with proper stewardship for the land and proper recognition of the biological needs of indigenous predators."

An important step toward achieving this goal would be the formulation of a comprehensive national strategy for the treatment of predatory animals, similar in scope and detail to the Endangered Species Act of 1973. Such a strategy would help rectify numerous inadequacies in the present system. It would open the issue to public scrutiny, thus doing away with the clandestine atmosphere of the present operation. It would require that ADC produce some meaningful biological research before killing animals, thus avoiding counterproductive consequences such as the fostering of the "super-coyote." It would emphasize as its primary line of predator defense such nonlethal methods as guard dogs, strobe lights, sound alarms, and improved fencing. It would upgrade the image of ADC from a pack of trigger-happy mercenaries to a team of professional educators, who, instead of cocking their guns at the behest of the livestock owners, would be more inclined to teach them better animal husbandry.

Ranchers. A memorable breed. Benefactors of some of the finest whitewashing ever applied to the national gallery of American stereotypes. In truth, the image barely jibes with the reality. These self-proclaimed Jeffersonian idealists are among the most egregious abusers of agricultural welfare. There's a mindset among some ranchers that tends to reduce politics to a set of Manichean absolutes: obstructive federalism versus grass-roots individuality, effete liberalism to manly self-reliance. It's certainly been the source for some startling proclamations. Dick Strom, a Wyoming sheep rancher on whose land illegal toxicants were allegedly sold in 1991 to undercover Fish and Wildlife agents, has been quoted as saying, "The Supreme Court has protected a lot of habitual criminals... rapists, murderers, arsonists. As far as I'm concerned, the *Endangered Species Act* protects the predators and wildlife at the cost of other wildlife and domestic livestock."[27]

Not all ranchers are this way. Tom Lasater operates his 25,000-acre Colorado ranch as a protected wildlife area. He forbids the use of pesticides and herbicides; he observes a rigid no-kill policy toward predators; he favors prairie-dog colonies because they aerate the root system of native grasses; he allows no exotics to flourish among the strains of blue gramma and western wheatgrass; he lets his cows fend for themselves. "I like to sit back and let nature do the work," he's been quoted as saying. "She's a hell of a lot smarter than we are."[28]

Between the two attitudes yawns an abyss that a new, revamped, revitalized ADC can do a lot toward bridging. Despite the objections of its critics, ADC is here to stay. Stock growers need help dealing with wildlife problems (as do cities and airports). They need advice about the latest nonlethal techniques such as guard dogs, barrier fences, and repellants; they do not need their own private SWAT teams choppering over the range gunning for suspicious animals.

A prototype for improving ADC's methods might be the Kansas example. There are no ADC trappers or hunters in Kansas. A single state agent with a budget of $750,000 focuses primarily on nonlethal methods of predator control. Once all the nonlethal methods have been exhausted, if a Kansas stockgrower still wants to kill coyotes, he or she has to pay for the cost of the agent's services out of his or her own pocket.[29]

Meanwhile, pressure is being applied to ADC from below by grassroots activists such as Wildlife Damage Review in Tucson and the Predator Project of Bozeman, Montana. The thrust of their efforts has been to inject meaningful outside involvement into projects that traditionally have been protected by a combination of public ignorance and layers of bureaucratic swaddling. In the last two years, national forests across the West have initiated environmental analyses (EA) of proposed ADC activities. These EAs detail site-specific activities targeted by ADC and provide for public monitoring and critiquing of controversial programs. Pressure from below, coupled with pressure from above — applied by enlightened policymakers in the form of a future comprehensive national strategy for the treatment of predatory animals — can help everyone, ADC and its critics, implement a more humane attitude toward wildlife.

Logging Puts Songbirds at Risk of Predators
"There are fewer ground-nesting songbirds in the southeastern United States now than in the past. The blame may lie with logging, and not only because the songbirds' habitat is destroyed. Forest fragmentation can create avenues in which predators are funneled directly to nesting sites."

Athens, Georgia, October 13, 1998 (ENS)

Only the Rich and Powerful Are Allowed to Endanger Species
"Endangered plants and animals contribute neither hard nor soft money to political campaigns. People who care about nature are more numerous but less focused than people who want to log, mine, dam, drain, spray, pave or otherwise impose their will upon the few remaining patches of undisturbed nature. So you can imagine what the latest rewrite of the Endangered Species Act looks like."

Donella Meadows, *The Global Citizen*

A 'Dead Zone' grows in the Gulf of Mexico
"It can stretch for 7,000 square miles off the coast of Louisiana, a vast expanse of ocean devoid of the region's usual rich bounty of fish and shrimp... This is the Gulf of Mexico's 'dead zone,' which last summer reached the size of the state of New Jersey. The trouble with the dead zone is that it lacks oxygen, scientists say, apparently because of pollution in the form of excess nutrients (farm fertilizers and livestock manure) flowing into the gulf from the Mississippi River. Animals in this smothering layer of water near the bottom of the sea must flee or perish."

***The New York Times* on the Web**
January 20, 1998

Climate Change Starving Arctic Caribou
"Peary caribou are a High Arctic subspecies of caribou whose range is limited to a few Arctic islands. Since 1961 populations in the Western Arctic Islands as a whole have plummeted from a total of 24,320 animals to 1,100 in 1997. The Gunn study warns that this die-off may be the beginning of a vast Arctic wildlife decline driven by rising temperatures and precipitation."

October, 1998 (ENS)
http://www.ens-news.com

AMERICAN FORESTS
Douglas B. Trent

"Cruelty to nature is a new concept for many people. It is customary rather to speak of the conquest of nature, the victory over the forces of the elements, the conquering of the wilderness. This process of conquest has varied from sawing down in a few minutes trees which took years to grow to tearing up and destroying the sod with a plow, a sod centuries in development that ought never to have been broken."
Dr. Karl Menninger, *Whatever Became of Sin?*

Barren once-forested hills, loss of biological diversity, loss of top soil and then rivers due to erosion, watershed destruction, flooding and drought, dying fisheries, global warming...These images of tropical deforestation have stirred Americans into action. Brazil and other tropical forest countries have received hundreds of thousands of letters criticizing their blatant disrespect for nature, evident in their willingness to destroy their forests for short term profits. Multi-million dollar loans were withheld for environmental reasons. As a nation, Americans were shocked to find that the world had lost 50% of its tropical forest resources. We were infuriated to learn that the Earth's richest resource was being squandered for political gain and profits for the rich.

How, then, do we respond as a nation when we learn that we have allowed the destruction of 95% of *our own* temperate rainforests? What letters do we write when we learn that our old growth forests in the northwest are disappearing at the rate of 8 square miles per month,[1] or that our nation's native forests are logged at the rate of two football fields every minute?[2] Around 80% of Brazil's rainforest still stands, and we shudder to hear of another tree falling. At the same time, less than 5% of North America's original forests remain. Only 1% of these ancient forests are protected. Much of the rest lies in our National Forest system, and that is being logged at a rate of nearly two million acres each year — twice as fast as Brazil's rainforest. At current rates, we can expect all but that protected 1% to be gone or totally fragmented by the year 2000.[3]

Knowledge is the basis for action, and action is power. Let's take a look at the broader picture.

Historical Perspective

By the mid-1800's, much of New England's forests had been cleared for farming. Much of what wasn't deforested in New Hampshire supported a bustling tourism industry, which soon became the growing nation's favorite getaway. Grand hotels and small artist communities drew people from all around. Even then, however, shrewd timber investors were buying up what was left, and unsettled land was being given away in many eastern states. In 1867, timber interests purchased 172,000 acres of pristine forests near Mount Washington in New Hampshire for $25,000. Vast holdings were starting to be compiled.

Without modern mechanization, logging was a slow process. No one felt any cause for alarm. Then, in the late 19th century, the invention of low-geared logging trains, wood-pulp paper processing and portable steam-powered saw mills caused the cut level to rocket. Nearly 700 timber companies were stripping the White Mountains in 1885. Forests standing since the last Ice Age were turned into denuded hillsides. Unspoiled streams choked with runoff. Forest fires of previously unseen intensity fed on the waste before attacking the standing trees.

Spectacular vistas disappeared. One at a time, the hotels either went out of business or were destroyed by fire. Textile industries, dependent on the constant flow of mountain streams to turn their wheels, could only watch as the erosion-filled streams flooded and then went dry.

Elsewhere, public stewardship was gaining a foothold. New Yorkers established the Adirondack Park and Congress created Yellowstone in the 1880's. Gifford Pinchot, later to become the first Forest Service director, tried unsuccessfully to establish a forest reserve in the southern Appalachians. In 1891, vast tracts of western land were set aside by the new federal Forest Reserve Act, and this land later became national forests.

But back in New Hampshire, the problem was getting worse. The New Hampshire Land Company had accumulated most of the remaining forests by then, and refused to sell land to farmers, causing a local food shortage. The remaining hotels were unable to secure small plots of forest for their guests. With farms, homes and industry threatened, the people united with advocates for a southern Appalachian reserve. They gained the support of a former governor, and formed the Society for the Protection of New Hampshire Forests. As all land set aside for forest reserves to date had been publicly held, they reasoned that the only solution was a federal buyout.

To their disappointment, bill after bill was defeated in Congress. Not until an erosion dam high in the mountains collapsed did Congress start to come around. A gushing torrent of water rushed down the mountain valleys, took out the Amoskeag mill on the Merrimack River in Manchester. Then the Connecticut River in Hartford all but dried up. The disasters were blamed on logging steep slopes, and over 100 mill owners joined the call for action. Both houses of Congress passed the *Weeks Act* of 1911 in February of that year. This act gave the government the right to purchase land. The new Forest Service, formed out of the *Forest Reserve Act*,[4] moved quickly to establish the White Mountain National Forest.[5]

The *Weeks Act* enabled the creation of 49 other national forests, but all was not resolved. In the following decades, congressmen from the western states put pressure on the Forest Service to open national forests for logging. Low timber prices, however, kept the level of pressure relatively low until after World War II when prices soared. Lawmakers found that national forest timber was a good source of pork barrel. The

Forest Service, which had started with the forest conservationist Gifford Pinchot as its chief, had changed hands by then. When Pinchot died in 1946, the large-scale sell-out of our national forests to timber interests began.

The Forest Service aligned with western senators and representatives, who backed the timber industry which employed the voters. Congress gave the agency huge timber budgets, which directed the forests to sell timber to the industry at prices that would guarantee a profit. The industry, in return, made large donations to the campaign funds of the senators and representatives. With much of the American public assuming national forests were like national parks and protected for perpetuity, this cozy arrangement persisted until the Reagan administration came to power.

Reagan's appointments of James Watt and John Crowell to high offices changed the thinking of many Forest Service employees. Shortly after that, the Forest Service installed Data General computers which revolutionized internal communication. Concerned about their orders to promote the destruction of ever-increasingly large tracts of the forest they worked in, lower level employees started communicating their concerns via the computers, and found that their experiences were not isolated.[6]

A Whistle Blows

A timber sales planner with Oregon's Willamette National Forest was one of those who was questioning his work and the Service. Many times during his 11 years with the agency Jeff DeBonis had the idea "that things weren't quite right."[7] His transfer to the Willamette, the single largest timber producing forest in the nation, convinced him to communicate his idea to others. After attending an Ancient Forest Seminar held at the University of Oregon in January, 1989, he transmitted over the agency's DG computer communications system a two page report on the opinions expressed by the speakers, adding one paragraph of his own: "One final note. We, as an agency, are perceived by the conservation community as being an advocate of the timber industry's agenda. Based on my ten years with the Forest Service, I believe this charge is true. I also believe, along with many others, that this agency needs to re-take the moral "high-ground" i.e. we need to be advocates for many of the policies, goals, and solutions proposed by the conservation community".[8]

A copy of this letter made it into the hands of Mr. Troy Reinhart of Douglas Timber Operators, who wrote a letter to DeBonis' boss, stating "Disciplinary actions should be taken against this employee as well as others who undertake this action...a formal reprimand is most surely necessary."[9] Hank Kashdan, the acting Ranger for the Blue River District which includes the Willamette, refused to take any action against DeBonis. And the cat was out of the bag.

Selling Public Assets at a Loss

One of the important things to know in understanding the complicated issues that surround logging in national forests is how decisions are reached concerning the amount of land to be allowed to be deforested. While the Forest Service employs many biologists and botanists to study the areas proposed for cutting, it is Congress, not biologists, that sets the timber quotas. The decision is based on politics, not biology.

In addition, Congress bases the agency's budget appropriations on its ability to meet the quotas. The more the Service is able to clearcut, the greater its budget.[10] The Forest Service also gets to keep most of the receipts from timber sales. And the Forest Service will go to great lengths to get the cut out.

Our national forests, under the agency's control, have assets estimated to be worth in excess of $50 billion. Forest Service revenues in 1990 were around $1.4 billion. But according to the respected forest economist, Randal O'Toole, timber sales that year cost American taxpayers close to $400 million.[11]

How can this happen? With the exception of the Pacific Northwest and Deep South, the majority of our national forests are not economical to harvest. In the Rocky Mountains, Alaska, the Appalachian and Ozark Mountains, the Midwest and New England, timber sales by the Forest Service usually earn far less than they cost to arrange.[12]

The *Knutson-Vandenberg Act* of 1930 was passed to secure funding for reforestation. This is the Act that allows the agency to keep the vast majority of gross revenues from timber sales, and frees the agency from having to return the full economic costs of sales to the Treasury. At the time the Act was passed, the taxpayers paid an average of 50 cents per thousand board feet to arrange a sale, and the Forest Service established rules requiring that at least that amount needed to be returned to the US Treasury. Inflation, however, has driven the cost of arranging sales to $50 per thousand board feet. The Forest Service rules still require that only 50 cents per thousand board feet be returned to the treasury — 1% of the total received from the sale.[13]

Take a look at southeastern Idaho's Caribou National Forest. Taxpayers, via the Forest Service, spent over $300,000 arranging timber sales, and more than $100,000 building roads to the timber sites. Timber interests paid $814,000 to the Caribou forest managers. The Forest Service returned $757 to taxpayers via the US Treasury.[14] A significant amount of the money the Forest Service keeps is paid to counties and local governments, but the taxpayer money put into the agency's timber program is far more than what comes back out. The Forest Service receives around $3.5 billion each year from the treasury, yet receives only around $1 billion a year from timber sales. [15]

Subsidizing Private Infrastructure

The President's Office of Management and Budget has shown that in 1985, the Forest Service spent $600 million more building logging roads and administering timber sales than was received by timber sales.[16] In fact, to make national forest timber saleable, the Forest Service has used taxpayer money to become the largest socialized road building company in the world, but to whose benefit, the public or private interests? The

50,000 miles of the US Interstate Highway System seems small compared to the more than 360,000 miles of logging roads that have been forced into some of the most beautiful and biologically rich areas known.[17] The Forest Service is planning on building another 262,000 miles of new logging roads and rebuilding 319,000 miles of existing roads. They estimate the taxpayers will need to fork over another $15 billion for that. When completed, the total mileage would go to the moon and back and then circle the earth four times.[18] And this massive road system is for the exclusive use of the timber industry.

Road construction costs are not factored into the cost of the timber sold to the timber industry. In other words, taxpayers subsidize logging by private companies by providing roads for free access. As the more accessible and timber rich areas are clearcut, taxpayers pay for increasingly expensive roads into remote areas and on steep mountain slopes where the timber is increasingly poorer in quality.[19]

Timber interests and misinformation

The Forest Service is spending over $2 billion a year on logging road construction, burning slash, often ineffective erosion control and other related activities. Revenues from timber sales amount to less than $400 million. Each year, taxpayers provide over $1.5 billion of free services to the timber industry. In addition, timber companies clearcutting national forests avoid paying property taxes as they are cutting on public land.[20]

The timber industry has plenty of financial incentive to keep the system as it is, and will seemingly go to any ends to achieve that goal. There are cases and cases of intimidation, misrepresentation, retaliation and deception. The industry spends over $350 million/year on convincing the American public that things would be fine if conservation organizations would just stay out of the issue. The total budget of the top 10 conservation organizations in the country amounts to less than that, and only a fraction of that money is spent on education.[21] The industry's far reaching propaganda machine has touched most Americans, and we are unwittingly falling for their line. Let's take a look at some of their misinformation.

One ad is titled "The Truth About America's Forests." It starts out stating that "Today America has over 20% On more trees than it had just twenty years ago. And the numbers are growing daily, with trees being replenished faster than they are harvested in every region of the country." But no one is asking them not to cut lumber from their replanted tree farms. What they don't point out is that they count an 800 year old tree with an 8 foot diameter and a two month old sapling each as one tree. They don't want you to know that big timber interests have cut millions of acres of native forests in the last 20 years. They hope you don't realize that monoculture tree farms, with all trees the same age planted in rows like corn don't have a fraction of the biodiversity of an old growth forest. The truth about America's forests would be better revealed counting trees with a ten foot diameter or greater, but don't wait for that in a timber industry ad.

The ad goes on to say "Thanks, in part, to private landowners and America's forests products companies, who plant over 6,000,000 trees a day, reseed entire forests, and use other forest management techniques to promote natural regrowth." What they don't say speaks louder. The industry states that they plant six seedlings for every tree they cut. They don't say they are only counting the tree they take to market. Dozens of non-marketable species and smaller trees are cut, bulldozed and burned for each tree that is marketable. They imply that a handful of seedlings replace not only the 250-1,000 year old tree taken, but also the dozens of other trees taken down during the harvest. One doesn't need to be a rocket scientist to realize that plants and animals that flourish in old growth forest can not survive a clearcut with six inch saplings.

Furthermore, they don't indicate the survival rates for their 6,000,000 daily samplings. Without the shade of a mature forest canopy and a cool, moist forest floor, seedlings often die. Part of the reason the industry plants so many trees a day is that they have planted the same areas time and time again without success. Nor do they mention that while biologically diverse old growth forests survive , replanted single species forests often die from insect infestation or disease.

The ad continues. "We're determined to keep up with the growing demand for wood and paper products. And to make sure our forests are a continuing source of joy for every American." The industry would apparently like us to forget that the national forests already belong to the public. We certainly don't need the timber industry to be responsible for taking care of our forests. The industry uses short-term economics and is motivated by short-term profit, but it would rather the public think they are looking out for our forests for us. The industry sees no value in a forest beyond timber, while forests provide clean water, healthy fisheries, purify the air, modify the climate, provide habitat for wildlife, host a diverse variety of plants and invertebrates and support a massive recreation and recreation equipment industry.[22]

What the industry wants us to believe with this ad is that everything is OK, there is no need to worry, and that we should trust them. If those assertions were true we can be sure they wouldn't be spending so much money to try and convince us.

Misinformation and the Spotted Owl

The endangered northern spotted owl has brought the conflict between timber interests and citizen's groups to the front page. Misinformation from the industry hit the mainstream recently in a story published by the *Reader's Digest*. The magazine professes that "accuracy is the bedrock of our magazine,"[23] but accuracy editors must have forgotten to check this story with heavy timber industry influence. "The Great Spotted Owl War" was noted as authored by Randy Fitzgerald, a known timber industry spokesperson. According to the Forest Conservation Council, a very similar article appeared as an editorial in the Wall Street

Journal several months prior to the November 1992 *Reader's Digest* story. The author at that time was listed as Donald Walker, around whom the story in the *Digest* centered. The Council says both articles were in fact written by Wise Use, an industry front that distorts a whole range of issues.[24]

The article starts by telling the story of the Don Walker family and how they were caught up in the spotted owl controversy when the Forest Conservation Council threatened to sue if they cut one tree on their land. according to the author, Don was laid off from his lumber mill job in 1989. The Walkers, in financial straights, were cutting a few trees a year from a 200 acre chunk of Oregon "timberland."[25]

Referring to some of the world's richest and rarest temperate rainforest as "timberland" gives us our first clue as to this author's bias. Stating that Don Walker was interested in cutting a few trees a year seems to be a bit misleading as well. The Walkers intended to cut all their forest, and by the time you read this may have done so.

The threatened lawsuit against the Walkers was a six day notice of intent to sue if the Walkers did not apply for an incidental take permit from the Fish and Wildlife Service as required by law. The Council was using the notices to force the state government into making state regulations comply with the *Endangered Species Act*, and never did sue the Walkers or anyone else as a result of the notices sent out.

The author alleges that the trouble started when environmental groups went to work to get the owl listed under the *Endangered Species Act*.[23] But in fact the trouble started long before that. The owl's endangerment signaled the real problem: massive, uncontrolled deforestation of already rare temperate rainforest. Furthermore, the listing of the owl as an endangered species only affected timber sales on Bureau of Land Management lands; the listing did not bring about a lot of injunctions as implied by the author. Long before the owl was even an issue, lawsuits enforcing the National Environmental Policy Act (NEPA) and National Forestry Management Act (NFMA) stopped illegal timber harvest.[26] Environmental groups did not cause the owl to become nearly extinct — logging did. Environmental groups did not write the laws (NEPA and NFMA) that protected threatened forests but merely forced the Forest Service to comply with those laws.

The Owls vs. Jobs debate is the result of desperate timber industry propaganda designed to inflame public opinion. Unable to push their cause through the courts, they rely on scare tactics to move public opinion to allow them to cash in our heritage for their profit. Loggers are suffering not because of the owl or from laws restricting the cutting of old growth forest, but instead from overharvesting for decades without replanting, new technology which makes timber harvest and mills more efficient, a dramatic increase in the export of whole logs, and the shifting of capital from the northwest to the south. Timber industry claims that northern spotted owls have been found nesting in second growth forests or even in discount store parking lots only serve to cloud the issue. By concentrating on the endangered owl, timber industry propaganda aims to divert the public attention from the real causes.

The article, following this line of propaganda, then delivers some statistics about how much forest is left, how much of that is protected, how many owls were originally found, new estimates of how many owls exist, etc.[27] A lengthy rebuttal to some of these twisted statistics combined with an explanation for the new owl sightings might counter industry arguments, but it would contribute to diverting our attention from the central issue — that the vast majority of jobs in the timber industry are not lost because of protections put in place for the owl.

Overharvesting

Douglas Fir is the most valuable tree in world timber commerce.[28] Without doubt, there is a shortage of Douglas Fir in the Pacific Northwest, but protecting a portion of what is left is not what caused the shortage. Private timber companies logging private land have been doing so for years at harvest levels well above the combined harvest on Forest Service and Bureau of Land Management lands. These agencies own 40% of the forests in the region, compared with private companies which own 25%. The other 35% is owned by individuals and states. In the last decade, private companies have cleared 52% of their forests, while 37% of federal forests were cut.[29]

In Oregon, more trees were cut in the last 40 years than in the first 120 years of the timber industry's existance.[30] The peak of logging occurred in 1952. It leveled off in the 60's, and by the 1980s had slowed down to the extent that dozens of mills closed. Replanting didn't start until the 1960's, and a Douglas Fir takes an average of 55 years to reach marketable size. The industry continues to cut trees faster than they can grow back to market size. According to the Native Forest Council "The amount of timber taken from Northwest forests each year is equal to a line of log trucks 20,000 miles long!"[31] It is inevitable that more jobs will be lost as the shortage becomes more severe. Richard Haynes, chief economist of the US Forest Service, has predicted that the cut on private lands in the region will decline to the point of closing 50 mills, costing 15,000 jobs. "We are living with the consequences of what happened five decades ago," Haynes noted.[32] Clearly the loss of jobs due to overcutting has nothing to do with the northern spotted owl.

Efficient Mills Require Fewer Workers

Just as technology has made other industries more efficient, the timber industry has benefited from mechanization and automation of timber mills as well as improved logging equipment. In a typical mill outfitted with state of the art technology, the need for employees is reduced by 50% to 80%.[33] According to the Oregon Department of Employment, from 1977 to 1978 more than 12,000 logging and wood processing jobs were lost. But at the same time, there was a 10% increase in wood taken from national forests.[34] From 1961 to 1977, Forest Service data shows that timber production in Oregon and Washington climbed 29%. Timber employment, however, fell by 3% in the same time period.[35]

Exporting Logs = Exporting Jobs

Not all of the jobs lost in timber mills and the wood processing industry fell to automation. Big companies like Weyerhaeuser can sell raw logs for export and get three times the price they'd get domestically. Once Weyerhaeuser, Champion International, Georgia-Pacific and other timber giants cut all of their own old growth forests (much of it for export), they began demanding timber from federal forests to keep their mills open. Exported raw logs require little to no processing before being shipped. Lucrative processing jobs go with exported timber to countries in Europe and the Pacific Rim. Smaller private timber companies often have little or no timber of their own, and depend on federal lands for logs. The increase in demand increases prices, making less efficient mills unable to compete. The economic realities of competition are likely to close these mills in the next couple of decades regardless of the outcome of lawsuits over the northern spotted owl.[36]

In the 40 years stretching between 1950 and 1990, Weyerhaeuser cleared 230,000 acres of old growth forest near Coos Bay, Oregon. During the last several years the logs were exported, and in the spring of 1989 they announced the closing of a sawmill in North Bend. The Georgia-Pacific Corp. harvested more than 2 billion board feet of timber from another forest in the same area between 1956 and 1960. Four mills were closed or sold after that forest was cleared, and company loggers were fired. Georgia-Pacific, citing a log shortage in 1989, closed a plywood plant in Springfield, firing 250 workers. At the same time, the company was exporting second growth timber from a nearby county. The story is not much different in Washington to the north, where the majority of trees are exported as logs rather than as processed lumber.[37]

Over the last 10 years, the Coos Bay local of the International Woodworkers of America watched as its membership dropped from 3,500 to 500. For the first time in 130 years, all sawmills on Coos Bay were silent. Roy Keene, a Cottage Grove forest industry consultant commented that "Tree growing companies have degenerated into export log brokerages — shamelessly closing their domestic mills, while selling logs to foreign buyers."[38]

The Industry Moves South

Another major factor in the loss of jobs in the northwest is the shift by the industry from the northwest to the southeast. A Wilderness Society study found that from 1962 to the present time, capital investment in the northwest by the seven largest timber companies in the country dropped 60%. In the same time period, mill capacity in the south, particularly in the regions from Arkansas to North Carolina, jumped by 121%. Timber companies have found that they are able to harvest more trees and pay loggers less in the south, and more commercially usable tree species occur as well.[39]

Seven major timber companies reduced their northwest productions capacity by half while increasing their productions in the south by 155% between 1978 and 1989. Both Boise Cascade and International Paper left the west side entirely while increasing their southern operations by 500% and 144% respectively. Weyerhaeuser moved about 40% of its production south, but also owns as much mill capacity in Canada as it now holds in Washington state. After closing a pulp mill and laying off 285 workers in Everett, Washington in March 1992, the company waited six months and announced plans to spend $600 million to acquire sawmills, pulp mills and timberland in Alberta, Canada and Georgia. ITT-Rayonier laid off 650 workers at one of its mills in Hoquiam, Washington, while at the same time purchasing thousands of acres of timber in New Zealand. [40]

This shift in logging has taken its toll on jobs in the northwest, and the owl has nothing to do with it.

The timber industry emphasizes again and again that the listing of the spotted owl has cost thousands of jobs. It has not been able to prove, however, how many jobs are lost because of the listing of the owl rather than because of overharvesting, automation, exportation and the industry shift to the south and other countries. Some jobs will be lost as a result of preserving the habitat necessary to stop the owl and other threatened species from slipping into extinction. But the injunctions stopping logging because of the owl affect primarily federally owned land, and 75% of the timber supply in Oregon and Washington comes from private land.[41] The overwhelming majority of jobs lost have nothing to do with the owl.

There is little we can do about overharvested areas except plant and wait for the tress to mature. Automation and the industry's use of new labor-saving technology, along with the shift of capital from the northwest to the south are economic realities. What we can act on is the export of logs. Congress could pass a law making it illegal to export logs from both public and private land. The timber industry would fight this tooth and nail, but those interested in saving American jobs should support such a move.

The *Reader's Digest* story goes on to tell how laid off workers were forced to desert their families to find work in Alaska. One feels sorry reading the graphic descriptions of the problems both the loggers and their families face. What doesn't get mentioned is the situation in Alaska.

Alaska's Tongass National Forest

The Tongass in Alaska is America's largest national forest. It is larger than the state of West Virginia, covers 80% of Southeast Alaska and is the last largely intact temperate latitude rainforest left in the world. It is 57% forested, and spreads from the mainland to several islands off the coast. Trees 800 years old spire as tall as 200 feet. The Tongass hosts the greatest concentrations of bald eagles and grizzly bears in the US, and provides streams and spawning grounds for the salmon fishery that is vital to the

Alaskan economy. The growing number of tourists who make their way to this spectacular wilderness is rapidly becoming one of the major contributors to the state's economy. But clearcutting and tax funded Forest Service roads are taking their toll, and once again taxpayers are getting taken.

According to a Forest Service accounting of timber receipts and expenditures on the Tongass from 1977 to 1984, we are losing money. The total timber receipts for that time period total $133,475,000. Expenditures for the same time period are broken down into operating costs and amortized capital costs, and total $394,703,000. The apparent cost to taxpayers is $261,228,000, but the cost is actually higher. The amortized capital costs, such as roads, are amortized over a 20 year period. Only a portion of the cost of roads is calculated into these figures. In 1984, according to these figures, the Tongass timber program lost 93 cents on every taxpayer dollar spent.[42]

The last page of the *Reader's Digest* article contains one of its most questionable assertions, which may sum up how far the industry will go to keep tax dollars flowing into its pockets:

> We know of hundreds of cases where owl habitat was created by accident as a result of management practices. Surely, then, we can do it by design.[43]

Northern spotted owl habitat is old growth temperate rainforest, complete with its 10 foot diameter trees and full biodiversity. We have yet to show we can replace this forest once cut; assuredly, we are not creating any by accident. The statement is very vague, and it is hard to say what they are referring to. Perhaps they consider that since some owls have been spotted in selectively logged and older second growth forest that these habitats are new owl habitats. No reputable studies support the long term sustainability of owl survival in these habitats. And we can't be sure they are not owls forced into these areas as their real habitat was destroyed.

The Value of a Standing Forest

Timber is not the only valuable commodity supplied by our national forests, and given the cost of preparing land for timber harvest it is certainly not the most lucrative. Fisheries, recreational industries and tourism facilities are but a few of the industries that derive benefits from our forests that are quantifiable. Environmental benefits are just as important but harder to put a dollar value on. The timber industry and a good portion of the Forest Service would rather we think of the remaining old growth in our national forests as trees to be harvested. Seeing the whole picture requires some knowledge of other values.

Recreation industries

Industries such as outdoor sports and camping supply manufacturers and the sales outlets, including stores and catalogue sales departments, provide many jobs across the country. Cross country skis, backpacks, tents, camp stoves, hunting supplies, mountain bikes, binoculars, cameras, altimeters, bird watching handbooks, and thousands of other products support an active and growing industry that uses old growth forests. Several ecotourism businesses including birdwatching and hiking companies use old growth forests for their tours, and numerous bed and breakfast lodges and hotels derive a portion if not all of their business from people wanting to visit the spectacular old growth forests. Total revenues from such industries in 1990 came to over $122 billion, while Forest Service figures for the same year amounted to only $13 million. It's not hard to see the connection between biologically rich old growth forests and tourism. Nor is it hard to imagine the impact on these businesses when ruined forests, clearcuts and tree farms take their place.[44]

Fisheries

Old growth forests are essential to the majority of salmon stocks in the Pacific northwest. A $1 billion sport and commercial fishing industry employing 60,000 people is threatened by logging and even the removal of fallen logs.

The loss of salmon stocks up to now has been devastating, with prospects for recovery remaining slim. Four salmon species have been listed as threatened under the Endangered Species Act, and scientists are urging similar action for at least three more species. According to the American Fisheries Society, 214 salmon stocks in the region are at risk of extinction, while 106 populations of West Coast salmon are extinct already.[45]

While several factors contribute to the decline, excessive logging over the past several decades has been the most devastating. Sediment from denuded areas carried into streams has increased the sediment load by as much as 1,000 times, smothering spawning beds where the fish need clean gravel to lay their eggs. Sedimentation further disrupts the feeding grounds of young salmon. Permanent forests provide shade to keep the water cool. Branches and trees that naturally fall into streams create slow moving pools which fish use to rear their young. Logging in the old growth forests of the Pacific northwest has raised the temperature in many streams. Despite the popular idea that downed trees are of no importance and should be used, we know they provide nutrients and habitat on the forest floor, and essential habitat for fish populations as well.[46]

Medicine and the gene pool

Other non-timber forest products supporting cottage industries without serious negative effects on the ecosystems include floral greens and mushrooms. Other plant and possibly animal species with as yet unknown value are sure to exist. The Pacific yew, a tree that is used for the extraction of the anti-cancer compound taxol, has been considered waste by the timber industry for decades, and burned with other non-marketable debris.

Taxol is being used experimentally in the treatment of breast, lung, ovarian and other cancers. It is perhaps the most potent anti-cancer drug to be discovered in the last 30 years, and the Pacific yew is found primarily in old growth forests of the northwest. About 1,600 cancer patients are currently being treated with taxol, and the National Cancer Institute hopes that by mid 1993 as many as 12,000 patients will be receiving the drug. Taxol is extracted from the bark of mature trees, and it takes six 100-year-old yew trees to produce enough taxol to treat one patient. Enough was harvested from federal lands in 1991 to treat approximately 12,000 people.[47]

In 1991, the Government recognized the Pacific yew as the most valuable tree in American forests. As important as the yew tree is, however, reports from the field show that the Forest Service burns yew trees, bark and all, along with the garbage typically left over by logging operations on government lands. Both the Forest Service and Bureau of Land Management have received dozens of complaints from people who have witnessed the burning. A number of environmental groups and even congressional investigators documented instances of yew bark discarded on the ground for disposal. Forest Service documents show that 75% of the bark from yew trees is lost in traditional clear cut logging practices, while National Cancer Institute director Dr. Saul Schepartz emphasizes that "The supply is finite...in years to come, the needs of taxol will far exceed supply."[48]

In 1991, the Bush administration struck up a five year agreement with the nation's second largest pharmaceutical firm, Bristol-Myers, giving them almost exclusive access to yew trees in our national forests. But without significant competition, the tendency has been to take only the most accessible bark, leaving scraps on the ground and only taking a portion of the bark from some trees. Representative Ron Wyden of Oregon headed up the subcommittee which held hearings on the subject. "This is the only case I know of where the Government has transferred control of an entire species over to a private company," he commented. "You have the most important tree in the forest and the land agencies are running this thing by the seat of their pants."[49]

The American Cancer Society reports that in a typical year about 12,500 women will die of ovarian cancer, another 45,000 from breast cancer, and 143,000 Americans will die of lung cancer. The tree that holds the most promising treatment is in short supply due to decades of over-harvesting and, in its typical twisted propaganda, the timber industry wants us to believe the spotted owl is the culprit as logging and the removal of yew trees are forbidden in certain areas.[50] Neither the owl nor the yew tree would be in short supply if it wasn't for the short-sightedness of our timber policies.

For years, environmental groups have pointed to the rich bio-diversity of old growth forests, and the possibility of beneficial products coming from them. The Pacific yew shows us they were right, but what other beneficial products might we learn about if we had the resources? Old growth forests provide genetic "banks" for sources of genes that can help produce hybrids resistant to new diseases, climatic changes or, as the yew shows us, new solutions to existing problems.

Watersheds and our water supply

Streams that begin in our national forests provide the water for nearly 3 million residents of Washington and Oregon. The Forest Service is responsible for 211 watersheds on nearly 4 million acres of national forests. Most of the region's cities, towns, resorts and campgrounds get their water from those river systems. Portland and 53 other communities depend solely on the water from Mount Hood National Forest. But many in the region feel that the Forest Service manages the area more for timber than water. Numerous communities are either threatened or have suffered from timber harvests that have ruined their water supply.[51]

Environmental benefits

A number of known environmental benefits come from our national forests. While it is difficult to put a dollar value on these benefits, the calculable costs of recovering from a flood, respiratory illness, pest outbreaks, fires out of control and other disasters are enormous.Standing old growth forests absorb water from rain and snow, decreasing the likelihood of spring floods and providing greater flows in the summer dry season. Trees consume carbon, release oxygen, and the needles of mature evergreen trees have a filtering effect that cleans our air. Biologically intact forests host birds, bats and other animals that naturally control harmful insect populations. In many parts of the world we've seen that reliance on dangerous insecticides for pest control often kills the natural enemies of pests. Insect populations later return in full force, immune to our pesticides, and with few or no natural enemies. Moist old growth forests also provide a much more effective fire break than young plantation tree farms. Numerous small towns, isolated homes, communities, resorts, and the entire tourism industry in the region benefit from controlled fires.

Aesthetic and spiritual values.

Perhaps the hardest values to assess in terms of dollars, aesthetic and spiritual values, are nonetheless real. An ever growing number of people seek out nature for relaxation. Homes built on wooded lots have a higher market value, and a system of botanical gardens and arboretums stretch across the country. While environmentalists are sometimes pictured as mushy, hippy tree huggers, millions of people around the country vacation in our national parks and forests. The Minnesota Landscape Arboretum just outside the Twin Cities has a membership of over 13,000, but reports that more than 200,000 people visit a year from all socio-economic backgrounds and all ages.[52]

As Michael Frome put it:

> Our public lands, wilderness in particular, comprise the most priceless possessions we Americans share as a people. No other country is so enriched by its parks, forests, wildlife refuges and other reserves administered by towns, cities, counties, states and the federal government. Land is wealth, and we the people ought to hold onto every acre of it in

the common interest. Public lands provide roving room, a sense of freedom and release from urbanized high-tech super-civilization. Without public lands there would be no place of substance left for wildlife, which has shared our heritage since time immemorial."[53]

Smokey the Bear Has a Chainsaw

The arguments for preserving our national forests and stopping the logging on public lands are based on solid economic statistics and common sense reasoning. Yet the Forest Service more than just follows congressional timber quotas. Even if its mandate concerns the harvesting of timber, surely the Forest Service, as a federal agency, should ground its actions in the context of a broad-based public interest and stewardship of a valuable perishable resource rather than aligning its policies with the interests of the timber industry. However, the following instances clearly indicate its operative biases.

Environmental Impact Statements (EIS) required by law are often times effective in preventing the extinction of species, and for that reason are often the bane of timber interests and developers. When the Forest Service wants to approve a timber sale, it helps if the EIS doesn't find any endangered species, plant or animal.

In 1989, Karen Heiman was hired as the first botanist to study the national forests of North Carolina, one of the top ten states for hosting candidate species for the federal Threatened and Endangered Species List. The Southern Appalachians served as a refuge during glacial periods. The variety in elevation, aspects, rainfall and geology found in the range increased the number of different habitats available. It is not surprising that the Southern Appalachians are a center of diversity for many groups of species. Endemic salamanders, mosses, liverworts, fish and fresh water mussels inhabit areas as different as dry, fire-adapted ridge-top communities and bogs with lush green ferns. In addition to the endemics, we find plants that are more common in the circumpolar region in some areas, with plants more typical of the South American tropics in others. Square meter plots in long leaf pine savannas have been found to contain fifty plant species. The region holds world class records for temperate zones with 100 to 150 plant species per quarter acre in these ecosystems.[54]

Karen was hired by the Forest Service to carry out surveys for areas scheduled for roads and subsequent sale to timber interest for clearcutting. Until then, few if any surveys for rare species had been performed. Karen realized the weight of her survey on the timber industry as well as the likelihood of finding one or more species that could stop timber sales in the forests where she worked. But she soon learned how the agency, strongly allied with the industry, would handle the problem. She was directed to carry out her survey in February when the area was covered with snow! When she complained, she was fired.[55]

In Utah, the Friends of the Dixie National Forest is an energetic citizen activist group dedicated to advocating changes in the management of the forest. Since the group's establishment in 1988, the Forest Service has treated them more like an enemy than a friend. The callous attitude toward the "Friends" and the public in general is reflected in policy, statements and a recent highly controversial action.

The Dixie National Forest's official policy towards the "Friends" was indicated by Forest Supervisor Hugh Thompson, who in July 1991. admitted to a group he was speaking to that any information desired by the "Friends" would only be supplied in response to an expensive and time consuming Freedom of Information Act request. Public interest organizations usually receive information for free.

In another instance, citizens including members of the "Friends" challenged a Dixie NF decision to initiate an aerial predator hunt. The Chief of the Forest Service agreed and called for a reanalysis of the proposal. The Dixie Supervisor's Office called it "a hollow victory for environmentalists" in the local press. This statement is a curious characterization of the agency's top official, and well as an inappropriate remark that helped further polarize all involved.

But the most telling incident occurred September 25, 1991. Brandon Fowler, the Vice President of the "Friends", submitted a stay request on a controversial timber sale that the group had been contesting for two years. Less than three hours later, two special agents from the Forest Service and a county sheriff appeared unannounced at his job in Nevada, accusing him of an alleged vandalism of Forest Service property incident. He was barred from his office and his fellow employees were interviewed at length. It was a disrupting and embarrassing scene that lasted for two days and could damage anyone's reputation. The incident had occurred two and a half months earlier, but the Forest Service claimed they received an anonymous tip and said the timing of the arrest relative to the stay request was purely coincidental. The Friends rightfully question that. They note that a relatively minor months old vandalism incident normally doesn't bring special agents "swarming" on the work place of a suspect, and that the agents were fully aware the suspect had just been critical of Forest Service policy. Fowler was released after questioning determined he was not involved in the vandalism of Forest Service equipment.[56]

Smokey the Bear Breaks the Law

The Forest Service has reason to be nervous when taken to court. Agency supporters have stated that "we need to be reminded that the Forest Service is a Federal agency which follows the laws and direction of Congress."[57] The problem with this statement is that the agency is purposely and systematically breaking the law to get the timber cut out. In regards to the spotted owl controversy, US District Judge William L. Dwyer of Seattle stated in a 1991 decision that the record shows the Reagan and Bush administrations have engaged in "a remarkable series of violations of the environmental laws" and a "deliberate and systematic refusal...to comply with the laws protecting wildlife," not at the working levels of the relevant agencies, but higher up, for political reasons.[58]

In a letter to Forest Service Chief Dale Robertson, then agency timber planner Jeff Debonis wrote:

> Unfortunately, we ally ourselves with the timber industry and think that the environmentalists are somehow obstructing us with their numerous appeals and lawsuits. Industry's disinformation campaign has the public believing that these appeals are frivolous and counterproductive. The fact is, environmentalists are winning appeals and court cases because *we* have broken the law. The only frivolous action going on is our agency's disrespect for environmental preservation and ecological diversity. We are the obstructionists, in our insistence on promoting the greedy, insatiable appetite of the large corporate timber industry we serve so well."

> An even more poignant example of our bias towards timber industries agenda concerns how we react after environmentalists do win their lawsuits against us. In many instances, environmental organizations have won on principal in court, but end up losing "on the ground", often with our compliance and help. Instead of accepting the obvious merits of their case and rethinking our action or attitudes, we find ways to circumvent the rulings and continue our business as usual activities. Examples include the National Wildlife Federation and Sierra Club Legal Defense Fund's case against the Mapleton District of the Siuslaw National Forest in which they won in court, only to have a congressional rider attached to allow buy-back, and eventually new sales to continue to be sold and logged as usual, allowing us to essentially disregard the court injunction.[59]

Back in Brazil

While much of the American public has been duped into believing the timber industry propaganda about endangered species vs jobs, the rate of destruction of our forests, the success rate of industry reforestation programs and other issues, we have become aware of the plight of tropical rainforests. Through educational campaigns about tropical rainforests, we have learned about biodiversity and its economic importance. We've learned that after slash and burn agriculture, logging is the next major cause of destruction. It is common knowledge in many parts of the US that roads such as the Transamazon Highway have opened up forests to be destroyed at a much faster rate, and that many of these roads were financed with American tax dollars. We've heard that standing forests are necessary not just for the wildlife and plants, but are essential in keeping rivers healthy and regulating the environment. The American people have embraced the cause of protecting rainforests in distant lands.[60]

How can we be taken seriously, however, if all of these valid reasons for saving old growth rainforests *in the U.S.* are ignored, and we allow our forests to be decimated? This point was raised in a letter to President Bush from the then Secretary of the Environment for Brazil. Jose Lutzenberger wrote on October 15, 1991:

"Dear Mr. President:

...As you well know, here in Brazil, we have succeeded in considerably reducing the devastation of tropical forests in Amazonia...The reduction was accomplished by eliminating the subsidies and tax incentives that in the past promoted large scale clearing for cattle ranches.

As an ecologist with a holistic view of the world, my concerns and the concerns of our Government go beyond Amazonia. So, we are also very much concerned with the fate of the last remaining old stands of temperate and boreal forests of North America in Alaska, British Columbia, Washington, Oregon and a few remains in California. Subsidized clearcutting there, followed by monocultures of exotics may be sustainable on a very long term for the production of timber only, but it is devastating for the unique natural ecosystems with their biological diversity. I was also privileged to see these forests with my own eyes — they are among the most majestic and awe inspiring forest(s) in the World. What is left after clearcutting is one of the saddest spectacles I have ever seen, never mind the narrow strip that is left along the roads to prevent travelers from seeing the ravage behind.

At the present rate of clearcutting practices for pulp and export logs, it will all be finished in about fifteen years. An irreparable loss for your country, a shame for Mankind and a very bad example for the Third World. How can we argue against the criminal devastation to tropical forests in Indochina, Malaysia, the Philippines, Indonesia, New Guinea and Africa, as well as here in South America? The powerful and rich US can certainly afford to subsidize a few thousand jobs in a less destructive way.

The same applies to North American wetlands."[61]

To preach to and condemn countries much poorer than our own for allowing their forests to be destroyed is extremely hypocritical. To be called on it by Brazil's Secretary of the Environment is embarrassing.That the Forest Service allows and is even encouraged to blow the trumpet of the timber industry reflects serious problems in our government. It is not a solely a Republican or Democrat conspiracy...both parties are to blame. It is time we take care of our problems at home if we are to continue to criticize other nations for their forest policies. To change the practice of deforesting our last old growth forests is in our best interest economically and otherwise; to let them be destroyed is ignoring our responsibility to our children, throwing money away and destroying our credibility worldwide.

Let's lead the world by example, not rhetoric.

TAKING ACTION

1. Compared to trying to change the forest policies of a distant country, it is relatively simple to work on our forest problems at home. A number of conservation groups take an interest in the issue, and the directions they choose offer you a variety of options. They can keep you

aware of what is currently going on (some better than others), and help you join your voice with others. A few of these groups are:

Native Forest Council,
PO Box 2171,
Eugene, OR 97402.
(503) 688-2600

Led by Timothy Hermach, this is a one issue organization that has my respect. It calls for no more cutting old growth forest on any of our Federal lands, and has put forth *The National Forest Protection Act* which seeks to:
· Protect all of the remaining 5% of native forests on public lands, and confine logging to lands already logged;
· Keep jobs at home by banning exports of minimally processed wood products.
· Restore our public forests to their native biodiversity.
· Provide economic assistance to affected timber workers and timber dependent communities.

This plan has been endorsed by Greenpeace and 78 other conservation organizations, including several state and local chapters of the Sierra Club and National Audubon Society. The NFC is criticized for its uncompromising stance, which most of the big conservation organizations say will not fly in Congress and end in not gaining any ground. It responds by comparing the situation to a medical patient who has lost 95% of his blood. The time for compromise was when we still had most of our old growth forest. We've lost too much to let anymore go. Personally, my support is going to this group.

The Wilderness Society,
900 17th St. NW,
PO Box 96750,
Washington DC.
(202) 833-2300.

I found this well-known organization to have knowledgeable and dedicated staff, but after a year, the majority of correspondence I received as a member spoke in broad, desperate terms, asking for more money. Only after speaking to several staff members did I come to respect their knowledge of the issues. They take a compromising stance, but also can claim many victories in saving some forest. One clever service they offer is to write three personalized letters for a $10 fee, sending them to you to sign and mail in a pre-addressed envelope. This allows busy individuals to communicate their wishes to their congressmen with a minimal effort.

The National Audubon Society,
Washington State Office,
PO Box 462,
Olympia, WA 98507.
(206) 786-8020

This branch of the national organization is extremely knowledgeable about the situation in the northwest, and publishes some very informative and action based literature.

**Association of Forest Service Employees
for Environmental Ethics,
PO Box 11615,
Eugene, OR 97440.
(503) 484-2692**

This is the organization started by Jeff Debonis, a forester who is working to change the system from the inside. The informative *Inner Voice* newsletter is interesting, and they have an intriguing angle on the problem. About a third of the members are federal employees, most with the Forest Service.

**The Environmental Defense Fund,
257 Park Avenue South,
New York, NY 10010.**

I've found knowledgeable staff members in this organization, and they are also active in a range of other issues.

Other groups worth looking into include the Sierra Club, the Sierra Club Legal Defense Fund (different from the Sierra Club) and National Wildlife Federation. There are many other organizations working on this issue as well.

2. Write or call your congressional representatives. Call one of the numbers below, based on your time zone, and press "5" to find out your Congressmen's names, addresses, and phone numbers:
Eastern: 800-347-1997
Mountain: 800-359-3997
Central: 800-366-2998
Pacific: 800-726-4995
Tell them to "support the strongest possible ancient forest protection legislation and reject the short-sighted policies of the timber industry and government." Tell them you want all log exports banned, and ask them to tell you their position on old growth forest protection. Be brief and courteous. Address your letter:

Senator_____, c/o US Senate, Washington DC 20510
Representative _____, c/o US House of Representatives, Washington DC 20515.
Those in the non-timber industries, as well as the client base of those industries, need to be just as vocal as the timber industry and disgruntled loggers.

3) Educate yourself. There are several good books on this issue, and there is no way all of the information on it can be presented in this chapter. Go to your local library and seek out these books. One in particular is titled *Beyond the Beauty Strip: Saving What's Left of Our Forests* by Mitch Lansky (Tilbury House Publishers). The "beauty strip" in the title refers to the strips of untouched forests along roads, lakefronts and riverbanks to hide deforestation.

League of Conservation Voters Congressional Voting Scores
"Congress avoided most meaningful,pro-active legislation, choosing instead to advance environmentally detrimental initiatives under the radar screen of public attention to avoid backlash," the League said. According to League President Deb Callahan, "...rather than advancing proactive environmental policies that benefit the health and well-being of our families, our communities, our natural resources or our Earth, these Congressional leaders instead attempted to benefit a narrow set of special interests who believe that their concerns about ever-larger profits outweigh public health and conservation interests."
                                        Washington, DC, October 14, 1998 (ENS)

Antarctic Ozone Hole Largest Ever
"The latest satellite observations show that the Antarctic ozone thinning covers the largest expanse of territory since the depletion developed in the early 1980s. "This is the largest Antarctic ozone hole we've ever observed, and it's nearly the deepest,"said Dr. Richard McPeters, principal investigator for Earth Probe TOMS. Preliminary data from the satellites show that this year, ozone depletion reached a record size of 10.5 million square miles on September 19."
                                        Washington, DC, October 7, 1998 (ENS)

CASH REGISTER RIVERS
Waterways Lost to Private Profit

Ann Vileisis

"Water, gentlemen, is the one substance from which the earth can conceal nothing. It sucks out its innermost secrets and brings them to our very lips."

Jean Giraudoux

We are a nation of river lovers. Whose family photo album doesn't boast a snapshot of some relative wearing hip waders and a floppy, lure-festooned hat, proudly cradling a giant fish? In fact, forty million Americans fish for sport on rivers nationwide. Beyond anglers, hundreds of thousands of people enjoy boating on rivers in canoes, drift boats, rafts, motor boats and kayaks. Each summer, more than 150,000 people float the whitewater on Pennsylvania's Youghiogheny River alone. In addition to boaters, other people relish simply walking or biking on trails along rivers, where they may spot a heron quietly fishing or an otter playing in the current. Still others delight in the mesmerizing flow of water over rocks or the glimmering of golden light on a lazy, wide river at sunset. And what kid doesn't love to skip stones or to swing from a knotted rope into a river swimming hole? Nearly everybody enjoys rivers.

We are also a nation of river users. Even people who do not delight in the natural qualities of rivers depend on benefits derived from their use every day. For example, many products not associated with rivers, such as the living room carpet and the daily newspaper, are made with water from rivers. Aside from supplying water for manufacturing, rivers also generate electricity, dilute wastes, provide transportation and furnish water for producing food. Most significant, rivers provide drinking water for people who live in towns and cities.

While many of the ways we use rivers are necessary and beneficial, intense use has also lead to abuse, leaving most rivers choked, de-watered, straightened, dammed, impounded, and polluted. One recent federal survey determined that only two percent of our nation's rivers still hold natural values that could qualify them for the special conservation designation of "wild and scenic."[1] Another government survey found 37 percent of our rivers too polluted for people to swim or fish in.[2]

The most disturbing part of this story is that many miles of rivers have been abused and severely degraded solely for private gain. Though as a society we must always make tradeoffs when we use natural resources, in many cases rivers have been manipulated, used, and ruined solely to serve the most blatant of corporate profit motives. While a few people have amassed profits, time and time again the public has been left with ruined streams. And though this degradation has created tremendous public costs — in terms of both lost natural values and clean-up expenditures— these costs have never been accounted for. Moreover, those who have garnered private profits from destroying public values have also managed to garner sufficient political clout to maintain their iniquitous stranglehold on our nation's rivers.

People have long used rivers to reap private profits, and there is a long record of public consequences. For example, on New England rivers, dams built to power grist mills and then factories in the 1800s blocked runs of Atlantic salmon and shad. On rivers such as the Concord, the natural and public wealth of the region's fisheries were short-sightedly traded away for the private profits to be made in manufacturing.

Not until 1913, when San Francisco developers sought to dam and divert water from the Tuolumne River—right in the heart of Yosemite National Park—did citizens launch a major campaign to protect the public values of a river. Throughout the country, people argued that the national park was already set aside to preserve an American scenic treasure. Although other water development options existed, San Francisco politicians and entrepreneurs pushed to wrest water from Yosemite. While the city of San Francisco grew larger, the public was left with a flooded national park and a reach of drowned river.

Although in more recent times we've come to learn more about the important public values of rivers and riparian corridors for fish and wildlife habitat, water supply, flood control, and recreation, the private-profit impelled destruction of these critical arteries of the land has continued.

Even agencies of the federal government have involved themselves in the river rip-off, spending huge amounts of taxpayer moneys to manipulate waterways for the profit of few people. One egregious example is the mammoth Tennessee-Tombigbee Waterway, authorized in 1946 and finally completed in the 1980s. At a cost of roughly $4 billion to taxpayers, the Army Corps of Engineers built a 232 mile long canal, linking the Tennessee and Tombigbee rivers to create a navigational shortcut for an already-existing barge route from Tennessee to the Gulf of Mexico. By offering cheaper shipping rates than old barge routes and railroad lines, the new waterway essentially subsidized the oil and coal companies that make up 70 percent of the new canal's traffic.[3]

Beyond taxpayer costs, the social and ecological costs of the canal were monstrous in the biologically-rich southern Appalachian region. All in all, construction destroyed 100,000 acres of useful forests and agricultural lands. Tearing up natural vegetation and straightening meanders, builders dredged and excavated more than two times the material removed to build the Panama Canal, causing considerable erosion. Then loose sediment washed into the river, choking air from the water and creating a turbid and suffocating environment that dozens of species of native fish and other aquatic organisms could not endure.[4]

But this irreparable damage to the rivers was never accounted for. In calculating the costs and benefits of the massive canal project, Army Corps economists inflated profits to big industrial shippers. The high costs of destroying the public values of the natural river system were not considered.

In the case of the Concord, the Tuolumne, and more recently, the Tennessee and the Tombigbee, lawmakers shunned their responsibility to safeguard the many natural and public qualities of rivers. In the absence of strong resource protection policies, a window was left wide open allowing opportunists to abuse rivers for their own profit.

To gain perspective on how private profit motives have dictated the management of America's rivers, it is instructive to examine how and why some of the destruction has occurred. Although the health of our nation's rivers continues to be jeopardized on many fronts, three threats loom paramount. Rivers remain most endangered by water pollution, by the reckless generation of hydroelectricity, and by irresponsible irrigation practices.

Contaminated Rivers

We all like to imagine that the clear, cold water plunging down rocky streams is pure — that we could cup our hands and drink from inviting pools. Unfortunately, this is fantasy; in reality, few rivers are safe to drink from. In too many places, streams flowing through town parks are posted with orange signs warning against play in the polluted water. Urban rivers that could provide city-dwellers with a refreshing after-work splash are off-limits. And local health departments have issued thousands of advisories against eating fish caught in the many polluted rivers. When the Environmental Protection Agency (EPA) examined water quality for a sampling of 647,000 miles of rivers throughout the nation in 1990, they discovered that 37 percent did not meet clean water standards. For Americans in the 1990s, contaminated rivers are a way of life.[5]

Polluting is one of the oldest ways that people have abused our nation's rivers for private gain. By the late-nineteenth century, factories were already dumping industrial volumes of waste into streams. Regarding rivers as free sewers, the textile mills and shoe factories in Lowell and Lawrence, Massachusetts poured their dyes and chemicals directly into the Merrimack River. As industrial technologies advanced, the wastes generated became more harmful toxic chemicals. But river dumping continued nevertheless.

For example, between 1910 and 1960, Rockwell International manufactured universal joints for automobiles and then discarded its wastes directly into Michigan's Kalamazoo River. The company continued dumping its pollutants into two ponds near the river until 1972 when monitoring finally revealed high concentrations of lead, arsenic, cyanide, and other toxins in the river. Eventually the EPA identified the Rockwell plant as a Superfund site — one of many hazardous areas posing threats to public health and requiring federal dollars for containment and cleanup.[6]

From 1946 to 1976, two General Electric factories dumped 2 million pounds of cancer-causing PCBs (polychlorinated biphenols) directly into

New York's Hudson River. The PCBs settled and befouled the riverbed. Although 300,000 pounds of the contaminated sediment have been dredged up and removed, twice that amount may remain. These persistent PCBs have made their way up the food chain with deleterious effect. Recent studies in 1997 still found alarming levels of PCBs in tree swallows, herons, and terns. And Hudson River striped bass contain so many PCBs that public health officials advise people not to eat the fish.[7]

While Rockwell, General Electric and countless other companies gained profits at one time, they never paid for the costs of their pollution. That price has been shouldered by people who live near polluted rivers, and by people who like to fish, boat, and swim in the rivers. Still more has been paid by U. S. taxpayers who foot large portions of the giant clean-up bills.

In 1972, the Federal Water Pollution Control Act put an official end to the most flagrant forms of water pollution. The law — known now as the Clean Water Act — recognized that contaminated water posed serious public health problems and threatened many ecosystems, fisheries, and recreational opportunities. To clean up water pollution, the EPA first cracked down on industrial and municipal polluters, requiring them to treat waste water before returning it to streams. When polluters resisted the change, local citizen groups played essential roles in pressuring them to comply with the new law. As a result, the water quality of many rivers improved. In 1972, only 36 percent of our rivers were considered safe for swimming, and now 66 percent are. For example, on the Merrimack River, clean-up took twenty years, $600 million and a great deal of citizen activism, but now long-time local residents are delighted to see their river restored.[8]

While the *Clean Water Act* spurred many successful cleanups, industrial dumping has continued. In 1988 alone, industrial sources dumped over 300 million pounds of hazardous pollutants into our nation's waters. In 1988, public health officials identified the Kraft-process pulp mill as the source of high levels of dioxins in Maine's Androscoggin River. In Virginia, the Avtex Fibers Mill in Front Royal was shut down in 1989 for discharging PCBs into the Shenandoah River. In the early 1990s, 591 industries discharged their wastes into the lower Mississippi, earning that reach of river the ominous epithet of "Cancer Alley."[9]

Although regulations now exist to protect rivers from industrial dumping, they are often difficult to enforce. In 1990, 13 percent of industries that dumped directly into rivers still failed to comply with clean water laws, and industrial wastes still accounted for 9 percent of all river pollution. A General Accounting Office report in 1997 found that over 40 percent of Ohio's polluters regularly violate clean water laws. And toxins discharged into rivers continue to pose threats long after dumping stops. Though Congress outlawed the sale of cancer-causing PCBs in 1971, a 1992 EPA study found 91 percent of sampled fish contaminated with it. Toxic substances continue to threaten 15 percent of the nation's river miles, typically in areas where rivers flow past the greatest numbers of people.[10]

In some cases, industrial polluters have used their political clout and money to bypass the law. For example, in 1992, five platinum mines in

Montana sought exemptions from the state's water non-degradation policy so they could pour more wastes into the near-pristine East Boulder and Stillwater rivers. The Stillwater Mining Company, a Chevron subsidiary, intended to increase its disposal of chromium by 500 percent, ammonia by 400 percent, and nitrates by 714 percent.[11] While polluting industries continue to pile up profits, the public continues to lose the natural qualities of rivers and inherit the heavy burden of cleanup.

While Clean Water Act regulations have stemmed many of the worst cases of industrial pollution, it has done little to stop another more widespread type of river degradation: polluted runoff. Every time rain falls, it washes over the land picking up pollutants from city streets, cultivated fields, treated lawns, and abandoned mines, and carries them all directly into the nearest river.

Agricultural runoff is the most formidable pollution culprit, accounting for the greatest share of river pollution.[12] After fields are irrigated, water that has been saturated with sediment, fertilizers, and harmful toxic pesticides is funneled back to rivers. There, sediment blocks light making waters turbid, and fertilizers spur the growth of surface algae, which consumes oxygen and also blocks light from deeper waters. As a result, insects, plants, and fish that formed the base of river food chains die, and vital rivers become stagnant.

Pesticides and herbicides constitute the most dangerous type of polluted farm runoff. After flowing from cultivated fields into streams, the toxic compounds are absorbed by insects, which are then eaten by small fish and birds. Like industrial toxins, the harmful substances in pesticides persist in the environment for a long time, accumulating in the tissues of animals. For decades, the pesticide DDT entered the food chain in this way, causing the thinning of bird eggs. As a result, many populations of avian species, including Bald Eagles, plummeted. Although DDT was banned more than 20 years ago, a 1992 EPA study still found the cancer-causing pesticide in fish in nearly all of the 388 sites it tested nationwide.[13]

Moreover, pesticides routinely cause fish kills. In the spring of 1991, one-half million fish, including bass, blue gills, and gar were destroyed in southern Louisiana's Blind River when pesticides sprayed on sugar cane fields flowed into the river. Between 1988 and 1990, over 1,000 similar incidents occurred on rivers throughout the country, killing 26 million fish and harming other aquatic organisms.[14]

Beyond wreaking havoc for birds, fish and other animals, many toxins in pesticides and herbicides have been linked with serious human health problems, including cancer, sterility, and birth defects.[15] However, most of these widely-used compounds have still not been adequately tested for safety. A National Academy of Sciences study estimated that adequate health hazard assessment information was available for only 2 percent of the 65,000 chemical substances being marketed.[16]

Pesticide-contaminated water is a particular concern in rural areas where farm runoff is most prevalent. In 1989, an alarming U. S. Geological

Survey study detected pesticides in 90 percent of streams tested in midwestern states. A second study by the Iowa Department of Natural Resources found pesticides in 61 percent of that state's river and lake waters.[17]

While agricultural runoff from cultivated fields constitutes the most widespread form of polluted runoff, it is not the only source. Improper livestock grazing also results in polluted runoff. When cattle graze, trample, and destroy riparian vegetation that keeps banks stable, loose sediment washes into rivers. Without leafy shrubs to provide shade, rivers' temperatures rise too high for fish. Furthermore, each cow acts as a direct polluter by producing copious amounts of manure — much of it indiscriminately dropped right in the water. The small, creek-like Tillamook River in Oregon, for example, hosts 15,000 of these polluters along its banks. Without fencing, most manure ends up directly in the river, where it releases a variety of bacterial contaminants that cause human disease. When organisms in the river water attempt to break down the onslaught of wastes, they use and deplete the dissolved oxygen needed by fish. For these reasons, inadequately managed livestock grazing poses a severe threat to the water quality and river health.[18]

Poultry and hog farms are also notorious polluters. The average modern poultry house generates four tons of manure-and-wood-chip litter per day, which often contaminates runoff water. On the upper Potomac River alone, there are 900 poultry houses in operation. Hog farms can be even worse. In 1995, a holding tank at a North Carolina hog farm collapsed, spilling 25 million gallons of liquid manure directly into the New River. According to the television program "60 Minutes," a crackdown on this problem was hampered by the influence of a North Carolina Senator who had a financial stake in the state's $19 million hog industry.[19]

Agriculture remains an ongoing source of polluted runoff because fields are plowed, irrigated, and treated with chemicals each year, and livestock produces enormous amounts of waste every day. But even century-old sources of pollution continue to ruin rivers. Old mines in the Appalachian and Rocky Mountains still pollute water every time a rain storm carries acids and metals into nearby streams. On Montana's Clark Fork River, hazardous metals leach into the water from mounds of mining wastes piled along its banks. Largely unregulated mining and smelting have left the river's upper 120 miles — including the nation's largest Superfund sit e — severely polluted.[20] Although mines in Colorado's Leadville district at one time earned over $2 billion in profits, thousands of tunnels and pits continue to dump toxic manganese, cadmium, copper, and zinc right into the headwaters of the Arkansas River. One gulch alone — another Superfund site —contributes 250 tons of heavy metals to the river each year. Less than fifty miles downstream, thousands of tourists enjoy the Arkansas River's popular whitewater, and small towns have built economies dependent on recreation.[21]

Old abandoned mines remain among the most hazardous sources of water pollution, but modern mining techniques, including the cyanide leaching process used to extract gold, also damage water quality. All in all, runoff from drilling and mining contributes 14 percent of the nation's water pollution.[22]

Runoff from forests also creates water pollution problems where extensive and improper logging has occurred. Thousands of miles of poorly-built logging roads have caused massive erosion. Sometimes when the dirt roads become saturated, they fall into streams as whole slabs. Furthermore, after clearcutting, which removes the trees and understory vegetation that naturally anchors soil in place, tons of silt flow into rivers. These sediments smother fish spawning beds and bury the submerged vegetation critical for adding oxygen to river water.[23]

For example, the Little North Fork of the Clearwater River in Idaho's remote panhandle used to be a blue-ribbon cutthroat trout stream enjoyed by both local and visiting anglers. But in 1989, Burlington Northern Railroad began building access roads and clearcutting its checkerboard land grant parcels in the river's watershed. As a result, sediment literally poured down the steep hills into the river decimating the prize cutthroat. While the railroad reaped quick profits by selling trees, people downstream, including those whose livelihoods depended on the fishery, paid the costs of the polluted water.[24] This form of river degradation has been particularly tragic in the Pacific Northwest, where extensive clearcut logging has severely damaged the spawning habitat of many endangered species of Pacific salmon. Nationwide, polluted runoff from improper forestry practices accounts for 9 percent of water pollution.[25]

There are solutions to polluted runoff problems. For example, irrigation techniques that deliver the correct quantity of water to crops rather than overflow fields can minimize the amount of runoff, lessening erosion and siltation. Alternate means of pest control that apply fewer chemicals in a more careful and labor-intensive manner can also minimize the amount of toxic compounds that end up in rivers. And leaving wide buffer strips of trees growing along rivers can reduce the sedimentation that occurs with logging. The difficulty lies not in a lack of solutions but in convincing farmers, loggers and mining operations to try different methods. They often fear change and may be unable to spend money on new irrigation systems, new fences, new equipment, and more workers.

But the most significant obstacle to change is more insidious. Giant agribusiness farms, chemical manufacturers, mining and logging corporations carry tremendous political clout, and their well-funded lobbyists routinely block attempts to regulate polluted runoff, claiming that regulations would cost farmers and businesses too much. What the lobbyists mean is that safer products and techniques might cost corporations some of their enormous profits. If politicians, like the Senator from North Carolina, fail to push industries, agribusiness, mining and timber operations to take responsibility for the wastes they produce, the costs will be passed on to all people who enjoy rivers, drink water, and pay taxes. And those costs are heavy. Over the past 25 years, taxpayers and the private sector have spent more than $540 billion controlling water pollution, and the EPA estimates that expenditures will top $64 billion annually by the year 2000.[26] But the degradation of water quality will cost far more than dollars. Pollution will rob us all of the many natural qualities of rivers that we rely upon to sustain our health and well-being.

The Unexpected Price of Hydropower

Dams plug nearly every river in the nation. As many as 80,000 large dams block thousands of rivers, flooding a land area larger than Vermont and New Hampshire combined.[27] Built by the federal dam building agencies (the Army Corps of Engineers and the Bureau of Reclamation) from the 1930s to the 1970s, the largest dams control floods and store water for municipal supplies and irrigation. Many giant dams are also managed for the generation of hydroelectricity.

No matter what purpose they serve, all dams have serious ecological and social effects. The reservoirs they create have drowned richly vegetated riparian corridors that once supported bird, fish, and wildlife populations. Numerous river reaches enjoyed by whitewater enthusiasts and anglers have also been inundated. Dams have repeatedly flooded lands already in use and set aside for other purposes, including privately-owned home sites as well as national and state parks. The Department of Agriculture reported in 1974 that more than 300,000 acres of farmlands had been lost under reservoirs annually during America's big dam building era.[28]

Apart from areas flooded by reservoirs, hydroelectric dams produce drastic effects downstream. Managed for maximum electricity production, hydro dams fundamentally change rivers' natural flow regimes. Typically, river flows peak with early spring snowmelt and then decrease in the summer and autumn. Dams reverse the natural pattern, creating low flows year round with bursts of runoff when demand for power is high — on summer afternoons, for example, when people in Phoenix and Los Angeles turn up their swamp coolers and air conditioners.

Such has been the case at the Bureau of Reclamation's Glen Canyon Dam, where generating hydropower has caused many problems downstream in Grand Canyon National Park. Radical shifts in water flow, made to accommodate electricity demands, have scoured away beaches on the shores of the Colorado River. In the late 1980s, the river fluctuated as much as 12 vertical feet in a single day. Furthermore, the clear cold water released below the dam has no sediment to replace the sand it carries away into the next reservoir. As the shoreline has eroded away, riparian habitat that supported the unique desert plants and animals of Grand Canyon National Park has gone with it. Thousands of people who enjoy boating and hiking in the park now find fewer places to camp along the river. While the extreme fluctuation of flows was tempered following the passage of a reform law in 1992, problems from the Glen Canyon Dam's hydropower releases still have devastating effects on the canyon.[29]

The manipulated flows affect not only the physical structure of streams but also the life cycles of many riparian plants and animals that have evolved adaptations to natural flow regimes. In the Southwest, for example, stately cottonwood trees depend on high spring flows to reproduce. High water scours sand and cobble bars preparing them to receive the cottonwoods' wind-borne seeds. But since dams have flattened out flow fluctuations, few southwestern rivers experience spring time flooding. Now most cottonwood communities below reservoirs are composed only of old trees established before dams changed the flow pattern.[30]

In the Pacific Northwest, the salmon's life cycle has long corresponded to natural river flow regimes as well. During summer and autumn, salmon swim hundreds of miles up rivers where they spawn on gravel bars. The following spring, when young fish begin their downstream journey, the fast-moving waters carry them to the ocean in less than 15 days. After spending several years in the Pacific, the adult salmon return to spawn and then die in the same streams where they were born. A century ago, as many as 16 million salmon made the streams of the Columbia Basin their home. Their remarkable migration provided food, wealth, and spiritual meaning for the native peoples of the region who relied on the returning salmon for winter food supplies.[31]

But hydropower dams now blocking the rivers have spelled out virtual extinction for many runs of northwestern salmon. Although the eight large, federal projects on the lower Columbia and Snake rivers have fish ladders that enable salmon to swim upstream, 37 to 61 percent still die during their upriver journey.[32] In recent years, only a handful of sockeye salmon surmounted the dams to spawn in the upper reaches of the Salmon River. Furthermore, persistent fish that survive to spawn in headwater streams leave their offspring with the even greater challenge of swimming downstream again. Adhering to an internal biological clock, young salmon heading to the ocean quickly start making physical changes necessary for the saltwater phase of their lives. But because the journey through the slack-water reservoirs behind the dams takes so long, the salmon's bodily transformation occurs prematurely. Many seaworthy salmon thus die in the impounded freshwater. In addition to the slow water, unscreened hydroelectric turbines still tear apart 30 percent of the young fish migrating downstream. In all, up to 96 percent of Idaho salmon die on their way to the ocean.[33]

The large, federally-owned and operated dams produce cheap electricity, and thereby offer residential power consumers throughout the West very low utility bills. Idaho Power Company, for example, charges electric rates half of the national average. Moreover, the cheap Columbia River power has translated into enormous private profit for industrial users. The electricity-hungry aluminum industry uses one-third of all power generated by Columbia River dams. Aluminum factories, which can each consume as much electricity as a town the size of Akron, Ohio, pay bargain-basement power prices — less than half the low residential rates.[34] But the low price of the electricity fails to account for other natural river qualities lost, such as critical habitat and recreational opportunities. The pricing of power does not take into account the need to protect the salmon, which are nearing extinction.

Most disturbing of all, when citizens have tried to reform dam operations to restore natural values — the salmon runs in particular — politicians under the big-money influence of the aluminum, farm, and barging industries have repeatedly stonewalled efforts to change the status quo.[35]

Although large, federal hydroprojects such as the ones on the Columbia have had draconian effects on rivers, they account for only 10 percent of the nation's dams. The remaining 90 percent of dams, owned by public and private utility companies and small private hydroelectricity developers, also threaten our rivers.[36]

Since the Federal Power Act of 1920, non-federal hydro projects have been regulated by the Federal Energy Regulatory Commission (FERC), the government's dam licensing agency. Unlike other federal agencies that assess resources and develop policies, the FERC acts more like a court granting permission to study and then construct dams.

The FERC's court-like procedures make it difficult for citizens to protest dams slated for construction on rivers in their communities. For example, when developers apply for a preliminary permit, citizens generally have too little information to present a case against the project. But by the time the FERC grants a final construction license, the developer has already spent hundreds of thousands of dollars on studies and planning, making it even more difficult for local citizens to influence the FERC's decision. Because the FERC has traditionally regarded river development as its main agenda, the commission has nearly always granted preliminary permits without assessing power needs or environmental concerns.[37]

The greatest surge in building non-federal hydropower dams occurred after 1978, when Congress — in the face of gas lines and oil shortages— passed legislation to promote alternative forms of energy, including solar, wind, and hydropower. The Public Utilities Regulatory Policies Act (PURPA) provided economic incentives to small scale energy producers by requiring utility companies to buy electricity from small producers at a price equal to the cost avoided by not building or using their own oil-generated power facilities. Although few small hydropower projects would make economic sense without the subsidy, PURPA's mandatory purchasing stipulation secured a profitable market for private hydropower developers.

Though lawmakers thought that PURPA would primarily foster power generation at dams already existing for other purposes, the legislation spurred a rush of new dam projects as well. From 1978 to 1990, hydropower developers flooded the FERC office with 7,500 applications for hydropower licenses. The commission licensed 200 projects each year.[38]

In 1986, Congress tried to instill some order in the rampant dam craze with the Electric Consumers Protection Act. The new law eliminated PURPA's guaranteed market for projects on rivers already protected in state scenic river systems and also required the FERC to start weighing environmental concerns in its permitting process.

Even with the additional requirements, numerous developers have sought to build new hydroelectric projects. In the late 1980s, Idaho potato magnate J. R. Simplot proposed damming and diverting the North Fork Payette River to generate hydroelectricity that his corporation would sell to southern California and Nevada. The project would flood 6,700 acres of privately-owned ranchland and home sites and eliminate 24 miles of river, including a nationally renowned whitewater run. An Idaho citizens' conservation group, Idaho Rivers United, organized and fought the project for years before finally gaining support in the state legislature to protect the well-loved river.[39]

In southern Oregon, the city of Klamath Falls proposed the Salt Caves Project to dam and divert the Upper Klamath River. The reduced flows would damage a prime trout fishery and spoil a well-known whitewater run. Even after Oregon voters passed a statewide initiative to protect the Klamath in the state scenic rivers system, the FERC approved the project. Ultimately the state blocked Salt Caves on the grounds that the diversion would degrade water quality.[40]

Confrontation between the FERC and state governments is nothing unusual. The commission has routinely overridden state statues, fish and game codes, and state river protection codes. For example, when the FERC granted a license to dam and divert Rock Creek, a tributary of California's American River, it required developers to leave a minimum of only 20 cubic feet per second (cfs) flowing through a one mile bypass section where the water would be temporarily diverted to generate electricity. When the California State Water Resources Board set higher minimum flows of 60 cfs from March to June to protect the river's trout fishery, the FERC denied the request. Every other state supported California when it challenged the FERC's order, but the U. S. Supreme Court upheld federal jurisdiction over the stream flow.[41]

Although FERC licenses carry inordinate authority, they do not last forever. Drafters of the Federal Power Act of 1920 aptly recognized that our nation's values and circumstances could change, so they stipulated that FERC licenses be reevaluated, typically on a fifty-year cycle. The FERC granted thousands of licenses before environmental laws required the assessment of ecological damages. As many of these dams come up for re-licensing in the next twenty years, citizens will finally have an opportunity to influence dam operations and to thereby reclaim some of the public environmental and recreational costs lost to private hydropower development. For example, during re-licensing, citizens can push the FERC to require developers to install fish ladders at outdated projects or to require minimum flows that will help to protect river qualities, including habitat for birds, fish and wildlife, and recreation.

In some cases, citizens have even pressed the FERC to cancel hydropower licenses and consider the benefits of removing unneces-sary old projects. In the Northwest, the National Park Service, the U. S. Fish and Wildlife Service and the lower Elwha Tribe jointly sought the removal of dams on Washington's Elwha River, where two large hydroelectric dams, built in 1914 and 1927, decimated the runs of all five species of Pacific salmon. The dams generated only 19 megawatts of electricity, which in total supplies just one-third of the power used by a Japanese-owned pulp mill nearby.[42] In 1997, agreement was finally reached to dismantle one of the Elwha dams. Across the country in Maine, a broad coalition of natural resource agencies, citizen groups, and politicians convinced the FERC to order the demolition of Edwards Dam on the Kennebec River. Built illegally in 1837, the outmoded dam supplied only 0.1 percent of the state's electricity but blocked 10 species of anadromous fish, including shad, alewife, smelt, sturgeon, and Atlantic salmon. The re-licensing of dams provides many opportunities to partially correct the ill-conceived trade-offs made for inexpensive power.

While re-licensing promises to improve operations at many dam sites and the *Electric Consumers Protection Act* has slowed the flood of dam proposals, the threat of future hydroprojects remains always in the wings. When our nation looks to find new sources of energy, the popularity of hydroelectricity will likely resurge. Hydrodams seductively provide power without nuclear hazards or foul smokestacks but nevertheless exact their price from our vital streams. Only by creating a strong vision for protecting rivers — before that resurgence occurs — will we be able to prevent private developers from claiming more rivers as their own.

Irrigation and Dry Rivers

During the droughty summer of 1988, Montana residents found many of their rivers reduced to mere trickles. Where rivers had swirled and tumbled, small puddles sat stagnant. Thousands of silvery fish lay dead and dying on bare, dry river cobbles. Irrigators, who use 96 percent of all water consumed in Montana, had taken most of the rivers' water.[44]

Montana is no different from other semi-arid states that rely on irrigation for agriculture. Guzzling more river water than any other single use, irrigation accounts for 81 percent of all water removed from streams nationwide. This withdrawn river water irrigates more than 58 million acres of farmlands.[45]

In Montana, as in most Western states, irrigators don't own the rivers, but they act like they do. State water laws give irrigators unconditional control over the water they divert. Based on a "use it or lose it" ethic, the outdated laws allow only for agriculture, ignoring other important values such as habitat and recreation that require water to flow through riverbeds. If irrigators fail to divert water for cultivation, they lose their legal right to use it in the future. Under antiquated laws, conservation groups are unable to secure in-stream flows for fish and wildlife.

Because the water laws date from an era when agriculture formed a cornerstone of the western economy, they fail to reflect modern values and circumstances. For example, in Montana, tourism based on appreciation for the natural landscape now ranks second in the economy. But under outmoded water laws, fishing, hunting, and whitewater guides whose livelihoods depend on tourists attracted to the state's natural rivers, have no say about what happens to the river water. As a result, during the 1988 drought, blue-ribbon trout streams such as the Blue Hole River — worth $2.3 million in annual revenues — suffered severe damage and a decade later are only just beginning to recover.[46]

Leaving enough water in streams for fish has become a particularly pronounced issue in Idaho, where rivers supplying irrigation water are also home to numerous fish species, including the endangered sockeye salmon. After the salmon jump over hydrodam hurdles, they often face channels de-watered or polluted by irrigation. For example, the privately-owned Milner irrigation dam near Twin Falls, Idaho forces the entire volume of the enormous Snake River into ditches, leaving only a dry, rocky riverbed below, where no fish can live. Conservation and fish advocacy groups have only recently begun to challenge the outdated water laws.

It is troubling to see how the irrigation establishment has garnered such a stranglehold on rivers, but it's even more distressing to realize that its grip has tightened with the generous aid of taxpayer moneys.

The federal government has long subsidized irrigation. Starting in the mid-nineteenth century, the government first encouraged development in sparsely populated western states by giving away dry-land homesteads on the condition they be irrigated. When individuals and small cooperative companies could not accomplish this capital-intensive task on their own, the federal government stepped in again. With the Newlands Reclamation Act of 1902 came the birth of a federal irrigation-dam-building agency — the Bureau of Reclamation. Inspired by the ambitious vision of greening and settling the American desert, the agency set about bailing out the arid-land pioneers with federal irrigation projects.

Since then, the Bureau has dammed hundreds of rivers to supply a total of 10.2 million acres of farmlands with water. While irrigating semi-arid lands has enabled people to grow food and live in otherwise inhospitable climes, many irrigation schemes have proved far more costly and destructive than the benefits they produce. For example, North Dakota's Garrison Project, proposed time and time again during the 1970s and 1980s, would have actually flooded nearly as much existing farm acreage as it would bring into production.[47] Eventually, the boondoggle project was scaled back.

Nonetheless, while the very premise of reclamation law — the need to encourage settlement in the arid West — has become obsolete, the government assistance intended to help family farmers open up new lands for cultivation has never stopped. Instead, federal subsidies have grown to become an endless drain on taxpayer moneys and a multi-billion dollar incentive to ruin rivers.

Although the original legislation stipulated that all costs of water projects be returned by irrigators in the form of water payments, this repayment plan has not exactly panned out. As anyone who has bought a house knows, interest expenses far exceed principal costs. But the Bureau charges no interest over long, forty-year repayment periods. As a result, the irrigator's water payments cover only 6 to 7 percent of the government's actual costs.[48] These interest-free loans for irrigation projects translate into giant federal subsidies and put western agribusiness at the heart of the welfare state.

For example, western farmers may pay only $3.50 for an acre foot of water that sometimes costs the federal government $80 or more to deliver. In particularly dry locales, the costs soar even higher. Irrigators using water from the Bonneville Unit of the Bureau's Central Utah Project, for example, pay $17.84 for an acre foot of water that costs the Bureau $306 to convey across the searing desert. Even realizing this great deficit-producing discrepancy, the Bureau continues to propose economically unsound projects. Water delivered by the proposed Animas LaPlata Project in southwestern Colorado will cost the Bureau $673.47 per acre foot, but growers will be charged only $40.60.[49] In 1986, federal water sales generated $58 million in revenues, but taxpayers covered the additional $534 million necessary to deliver water for irrigation.

Even more deplorable is the fact that one-third of that staggering sum went to irrigate surplus crops eligible for other federal agricultural benefits, including price supports and direct cash subsidies to reduce the amount of crops grown. According to a 1988 Bureau of Reclamation study of 9.9 million acres it services in 17 western states, 3.7 million acres were planted in surplus crops eligible for subsidy payments. In California, home of the Bureau's largest projects, 43 percent of lands irrigated by federally delivered water produced surplus crops. In the Missouri River Basin, where the Bureau waters 2.17 million acres of crops in Nebraska, Montana, Wyoming, the Dakotas, and Colorado, 48 percent produced surplus crops. A Department of Agriculture study similarly estimated that the U. S. government annually paid $730 million for subsidy payments to farms that also received subsidized water.[50] These government studies indicate that many irrigators have been as successful at cultivating the federal coffers as they have been at cultivating fields.

While the overall dollar amounts of these double subsidies boggle the mind, just the federal subsidies to individual farms are enormous. In a 1987 congressional hearing, California Representative George Miller explained that "a farmer with a 960 acre farm receiving water from the Columbia Basin Project in the state of Washington [would] receive a total subsidy in the neighborhood of $4.4 million for a single farm." Regarding his own state, Miller went on to explain that "a single farm receiving water from the Central Valley Project [would] receive a total subsidy of $1.8 million."[51] With such large amounts of money at stake, it is little wonder that the irrigation establishment fights against regulations that would protect river environments.

Nowhere have the abuses of rivers and taxpayer moneys been more outrageous than in California's Central Valley, where the Bureau's Central Valley Project — the CVP — dams and diverts waters from the Sacramento and San Joaquin river systems to provide subsidized water to corporate irrigators. During times of drought, heavy irrigation diversions compound water quality problems in the valley's rivers. Below Friant Dam on the San Joaquin River, irrigation diversions totally dry up a ten-mile stretch of river. After these massive diversions and others, the low flows of the San Joaquin River can no longer supply the wetland wildlife refuges downstream. Worse still, polluted runoff has contaminated refuges, such as Kesterson, killing thousands of birds that used the area as a critical stopover point during migration. As a consequence, waterfowl populations in the valley have decreased by 50 percent since 1974, and winter salmon, delta smelt and other native fish populations have plummeted.[52] In addition to destroying habitat, one of the CVP's sixteen dams flooded a popular whitewater reach on the Stanislaus River. Before New Melones Dam, more people boated the Stanislaus than any other whitewater in the West. The Bureau's irrigation project transformed this and other vital rivers of the Central Valley into a sterile plumbing system at the expense of local river lovers and fish and wildlife dependent on river and riparian habitat.

While the social and ecological costs of the irrigation project have been monstrous, the costs to taxpayers have been enormous as well. Although reclamation law stipulated that federal projects should supply water only to resident farmers with limited holdings, big landowners stood poised to reap profits in the CVP's Westlands district. When federal water first arrived in 1968, eleven percent of the growers possessed 84 percent of the land. Some corporate farmers owned huge plots, such as Southern Pacific with 106,000 acres, Giffen Inc. with 100,000 acres, J. G. Boswell with 25,000 acres, and Standard Oil with 11,000 acres. The Bureau of Reclamation essentially ignored the acreage and residency violations.[53]

With federally subsidized water from the San Joaquin River and its associated groundwater, these rich landowners grew richer and richer. Farm products from the Westlands district increased in value from $80 to $489 million between 1964 and 1982. Land values rose from $32 per acre in 1952 to $3,000 per acre in 1982. According to a 1981 Department of Interior study, the CVP cost $79,000 each day, a tab picked up by taxpayers. The long no-interest repayment period for the Bureau's Westlands project amounted to a $1,500 per acre federal gift to the corporate landowners.[54]

Federal subsidies coupled with outdated water laws have encouraged irrigators to carelessly overuse scarce water resources. Rather than adopt water conserving irrigation methods that would reduce erosion and pollution, farms continue to use huge volumes of water. In California alone, the cultivation of rice annually loses more water to evaporation than the city of Los Angeles uses. While California's most water-consuming crops use one-third of the state's scarce water, these crops — rice, alfalfa, cotton, and irrigated pasture — provide only 1 percent of the state's economy.

In addition to wasting inexpensive federal water, the giant Central Valley farms have also collected giant double subsidies. From 1966 to 1971, J. G. Boswell, received over $25 million in crop subsidies. Even though the corporate farm received $20 million in subsidies in 1986 not to grow cotton on land served by heavily subsidized water, it was allowed to grow other crops with the cheap water. For decades, Boswell has scored both federal subsidies while the Central Valley rivers and their webs of life deteriorated.[56]

Although California's agribusiness creates pollution, unabashedly consumes the state's scarce water, and contributes only 2.5 percent to the state's economy, the industry has been long coddled by outmoded reclamation law that ignores many modern realities. Now scientists better understand that polluting and de-watering river channels for large scale irrigation wreak ecological destruction. And many more people now live in urban areas and need water for domestic purposes and rivers for parks and recreational opportunities. Furthermore, less damaging water-conserving irrigation methods are also more widely available today. Yet the giant farms have long wielded disproportionate power and thus maintained the status quo with an iron hold.

Recognizing the need for change, House Interior Committee Chair George Miller explained, "You can't let people who placed a claim on a resource forty years ago continue as they did when their claim today is so politically weak. It is not the dead hand from the grave that should dictate water policy today." Together with Bill Bradley, chair of the Senate Interior Committee, Miller pushed the Central Valley Project Reform Act in 1992. The new law reserved more than 800,000 acre feet of water for fish and wildlife, guaranteed 15 percent of the project's water supply for environmental restoration of areas badly damaged by years of destructive operation, and gave new emphasis to urban water needs while gradually increasing irrigators' fees to cover the government's costs. But the powerful Central Valley irrigators vowed to block the law, and have managed to weaken its full implementation.

Addressing the many pressing demands on rivers will require careful attention not only in California but nationwide. As scarce water is re-allocated to serve urban needs and not just irrigators, the natural qualities of rivers must also be protected to serve the needs of all citizens. People who value and enjoy the natural qualities of rivers must have a say in deciding how these public resources will be used.

The Time for Vision

The threats of pollution, hydrodams, and diversions continue to whittle away at river values. On a single river, the FERC may license a small hydroproject, a state water board may approve additional dumping from an upstream mine, and a farm may divert flows to flood a cotton field and then return its wastewater. Over time, such incremental developments cumulatively degrade entire river systems. Because various government agencies deal with each part of the river pie separately, no vision exists to protect whole-river ecosystems.

One important political vehicle for protecting rivers is wild and scenic designation. Passed by Congress in 1968, the Wild and Scenic Rivers Act guards rivers that possess "outstandingly remarkable" qualities. As of 1998, Congress has protected sections of 230 rivers with this law. But that figure represents less than one-third of one percent of the nation's rivers; only 5 yards is protected for each mile of the nation's rivers.[59] While wild and scenic designation is the strongest political protection a river can have, it protects only against new dams and not other threats to rivers, such as pollution and riverbank development. Furthermore, criteria for wild and scenic designation applies to only a small number of rivers, leaving out many community streams that lack superlative scenery but are nonetheless enjoyed by citizens.

Another tool for protecting rivers, the Clean Water Act, was intended to cope with pollution, but enforcement has always been underfunded — often because corrupt politicians have used their influence to undercut agency budgets. For example, in 1995 and 1996, the House of Representatives passed budget bills that would have cut $760 million from drinking water and clean water programs and slashed the already-minimal EPA enforcement budget by 25 percent. House members voting for the clean water cuts had each received an average of $55,000 from 212 Political Action Committees associated with lobbies seeking to undermine the Clean Water Act. Fortunately because of strong public outcry, the cuts failed to pass the more moderate Senate.[60] Only recently has the EPA begun to deal with the ubiquitous problems of polluted runoff.

Beyond the Wild and Scenic Rivers Act and the Clean Water Act, no federal policy explicitly recognizes rivers' natural values. Although President Clinton announced a new Heritage Rivers Program in 1997, it is non-regulatory. Essentially, it is a voluntary program designed to encourage citizen participation in river conservation. A new and stronger policy — perhaps also in the form of a presidential directive — needs to recognize the vital functions that natural rivers serve, to better balance different forms of river use, to integrate the work of various agencies with jurisdiction over rivers, and to take whole-river ecosystems into account. A new national policy could protect rivers of all types.

Meanwhile, the necessary reform of existing federal water projects will require pressure from both the White House and Congress. Although such progress has been difficult to mobilize, some recent efforts afford hope. For example, the 1992 laws to protect the Grand Canyon from wildly fluctuating flows and to reform California's Central Valley Project signify a new awareness about the broader importance of rivers. The pending removal of Edwards Dam on the Kennebec also indicates a new willingness to consider more carefully the natural and public values of rivers.

Most promising of all, numerous citizen groups have formed to steward the rivers in their communities. When the national river conservation group River Network conducted a survey in 1990, it found roughly 500 groups working on river conservation; in a 1996 survey, it found 3,000 active groups! For example, the Merrimack River Watershed Council worked to clean up their heavily-polluted river. In California, the

South Yuba River Citizens' League successfully defeated a proposed hydropower dam on their community's river in 1995 and fended off other destructive projects in 1998. Citizen groups such as these can routinely monitor all aspects of their river's well-being, including water quality, flows, fisheries, dam re-licensing opportunities, new hydropower threats, and other developments. These groups can also pressure river abusers to comply with regulations and push for more stringent river protection laws. According to river conservationist and author Tim Palmer, "When every river has a steward, we'll see much better care of our waterways."

Although rivers belong to all people, opportunists still exploit these public resources. And lawmakers still fail to balance river uses to protect the natural qualities of rivers that most people value. If our nation continues to view rivers as cash registers, the consequences will befall all citizens. Our rivers will continue to be poisoned with chemicals, blocked and flooded by dams, and de-watered by diversions. The birds, fish, and wildlife dependent on healthy riparian habitats will perish. Our drinking water will be contaminated and require more costly treatment. The people who enjoy fishing, swimming, boating, and hiking along rivers will have far fewer places to refresh their spirits.

A vision must exist to protect the rivers that threads our nation's landscape with blue-green corridors of life. In the long term, clean, healthy rivers will prove a far greater economic asset than any profits earned now at their expense. With only two percent of our rivers still holding their finest qualities, the plea of ignorance no longer suffices. It is time for the river lovers to stand up and be counted.

Very Troubled Waters

"Four of the most prevalent herbicides—atrazine, simazine, alachlor, and metolachlor—are applied nationwide and grain belt states receive large shares of the estimated 135 million pounds that is used annually. Agricultural pesticide use jumped 10 percent between 1993 and 1995, and the U.S. Geological Survey increasingly finds the chemicals in river water. Several studies by the Environmental Working Group, which advocate tougher environmental laws, have found that 14 million Americans drink public water that contains the four major herbicides found in rivers."

**Penny Loeb, *U.S. News & World Report*
September 28, 1998, pp 39-41.**

"[I]n the U.S. and many other parts of the world, a strong conservation ethic means that nature may be starting to gain a stronger voice in water quality and allocation decisions. Hundreds of grassroots groups are defending rivers and lakes in every corner of the globe—and they've been winning significant victories."

**Elaine Robbins, "Water, Water Everywhere"
E, The Environmental Magazine, October 1998**

THE MINING LAW OF 1872
A Law Whose Time Has Gone

Johanna H. Wald & Susannah French

"The more we get out of the world the less we leave, and in the long run we shall have to pay our debts at a time that may be very inconvenient for our own survival."

Norbert Wiener

Introduction

When Congress passed the Mining Law of 1872, the telephone had not been invented. Alaska had only recently been purchased from the Russians, women could not vote, and Ulysses S. Grant was President of the United States. In 1872, miners rode mules, and mining was done with picks and shovels. The West was unexplored and, in many areas, unpopulated. The country's vast resources were seen as unlimited.

The Mining Law of 1872 established the rules miners had to follow to claim mineral deposits they discovered and protect their claims from other prospectors. To encourage exploration of the nation's mineral resources, the Law provided that prospectors could purchase, or "patent," valuable claims at modest cost. Unlike later laws for commodities like oil and gas, the 1872 Law charged no royalty on the minerals removed from the public lands. The Law imposed no environmental controls, but of course at that time the extent of environmental damage was limited in most cases to what prospectors with hand tools could accomplish.

Much has changed in the last 125 years. Growing public concern over abuse of the environment has led to sweeping reforms in the way government regulates most industries. Today mining is done with giant earth movers which create pits almost as deep as the world's tallest building is high. Poisonous cyanide is sprayed over gigantic piles of unearthed rock and ore to leach out minute concentrations of gold. Hundreds of acres of land are covered with heaps of waste rock and toxic mine tailings. Speculators can buy up mineral claims which they then turn around and sell to other speculators for hundreds of times more than they paid for it. In many of the West's wildlands, squatters are living off their claims rent-free, with no intention of ever mining them. Yet, despite these abuses and the changes in mining technology that have occurred, all these activities — and more — are taking place pursuant to the Mining Law of 1872.

Despite passage of a century and a quarter, the Mining Law of 1872 still remains the primary law governing hardrock mining on federal land today. As a result of inadequate federal control of mining activities, the public's lands are suffering extensive and unnecessary damage, and the public is being saddled with enormous charges for environmental cleanup. The government officials who manage the hundreds of millions of acres open to hardrock mining have almost no authority to reject mining proposals, even where serious environmental damage will result. They are reluctant to impose environmental controls which could reduce a mine's profitability. And, they lack the funds and staff to enforce the limited controls they do impose.

The need for a new mining law has become increasingly obvious during America's latest "gold rush." From 1980 to 1995, with the growing use of cyanide heap leach processing, gold production soared from almost one million ounces per year to 10.6 million.[1] Throughout the country, a new generation of gold miners has been spurred on by the prospect of using this cheap process to make enormous profits without paying a penny of royalty or rent to the United States. The tremendous proliferation of new gold mines has inspired renewed efforts to jettison the Mining Law of 1872 and to bring hardrock mining into line with modern values and standards for environmental protection.

Efforts to reform the 1872 Law have been going on for almost as long as it has been on the books. Reports criticizing the Law began appearing as early as the 1930s. When Secretary of the Interior Stewart Udall left office in 1969, he called replacement of the Mining Law of 1872 "the most important piece of unfinished business on the Nation's natural resource agenda."[2] Presidents from Hoover to Carter and, more recently, Clinton, have called for change in the law. Over the past decade, a number of reform bills narrowly missed passage in Congress. As new mines spring up across the country, the public outcry has grown. It is now more clear than ever that, after 125 years, the Mining Law of 1872 needs to be reformed. The only questions are precisely when and how.

The Mining Law of 1872

The Mining Law of 1872 was the last major act passed by Congress to encourage settlement and exploration of the great western territories. Previous acts had spawned the two largest land giveaway programs in American history. Two hundred and eighty-five million acres were given away under the Homestead Act of 1862 which granted 160 acres in the west to any person willing to establish residence and pay a minimal filing fee. Another hundred million acres were disposed of in land grants to the railroads between 1850 and 1871.[3] In keeping with the expansionist frontier policies of the day, the Mining Act of 1872 encouraged the exploration of the nation's natural resources and the occupation and purchase of "vacant" land.

Eighteen seventy-two was a watershed year in public lands management, marking the beginning of the end of the land giveaway programs and ushering in a new era of federal retention and protection of the public lands. In 1872, Congress not only passed the Mining Law, it set aside two million acres of public land to form the country's first national park at Yellowstone. Between 1872 and 1936, Congress withdrew almost all remaining federal lands from settlement and created our national park, forest, wildlife refuge and other land systems.[4]

Early laws like the 1872 Mining Law contained no provisions for environmental protection. In the late nineteenth century, water and air pollution were unknown. The west was largely pristine wilderness. The environmental movement had barely begun. John Muir, who would become America's first popular environmentalist and found the Sierra Club in 1892, was exploring the forests and mountains of the west. The environmental consciousness which did exist was largely focused on preserving America's spectacular wilderness areas. There were no smog alerts, no Superfund sites, no ozone holes. The American public would not begin to understand the far-reaching effects of industrial pollution for another hundred years. And, much of the most serious pollution from mining operations begun in the late 19th century would not appear for decades.

Mining before 1872, in any case, was far different from the mining practices of today. It was dominated by individual prospectors with pick axes and shovels. These prospectors usually worked in unsurveyed and unexplored territory under the most primitive conditions. They ate beans, bacon, and dough cooked over an open fire and slept in tents or in hand-built cabins. Although copper, silver, and other minerals were also mined, it was gold that captured the national imagination.

New gold discoveries led to sudden migrations by men who lived from camp to camp. One writer in the 1860s described the desperate conditions in which gold miners would rush to stake their claims to new finds: "They were strung out for a quarter of a mile, some were on foot carrying a blanket and a few pounds of food on their backs, others were leading packhorses, others [rode on] horseback leading pack animals."[5] Bringing with them mining traditions from all over the world, the early prospectors developed their own codes of behavior for the mining camps. With little access to courts and judges, and no legal rights to the land or the minerals they found, they even created an informal system to identify and claim their finds.

When Congress decided to regulate the development of federal western mineral deposits, it looked to the traditions established in the early mining camps. In addition to encouraging the exploration and settlement of the West, the Mining Law of 1872 sought to create a coherent set of rules for resolving proliferating disputes over the ownership of mining claims. The 1872 *Law* codified the early mining practices regarding rights of possession, abandonment, marking of locations, recording claims, and patenting. At a time when there were few competing demands for the public's lands, the *Law* also effectively established mining as their highest and best use: "[e]xcept as otherwise provided, all valuable mineral deposits in lands belonging to the United States, both surveyed and unsurveyed, shall be free and open to exploration and purchase."

The Mining Law of 1872 applies to the exploration and extraction of "hardrock" minerals on federal lands. These minerals include gold, silver, copper, iron, lead, fluorspar, and asbestos. "Uncommon varieties" of more mundane substances are also covered, including certain varieties of sand and gravel. The Mining Law established a claim system intended to protect mineral deposits discovered by one prospector from subsequent encroachment by others. According to estimates by the U.S. Department of the Interior, there were approximately 300,000 active mining claims as of August 1995, down from about 1.2 million active claims at the end of fiscal year 1985.[6]

Today more than 430 million acres of federal lands are subject to the Mining Law of 1872.[7] Located in the western United States, they are administered principally by two federal agencies, the Bureau of Land Management (BLM) and the U.S. Forest Service. This huge acreage does not include all of our federal lands; some of them, such as Wilderness Areas established under the 1964 Wilderness Act, have been "withdrawn" from mining. Any lands which Congress has not specifically withdrawn, however, remain open to mining claims. What is more, even Wilderness Areas, National Parks and National Wildlife Refuges can be mined under claims which existed before these areas were set aside for protection by Congress.

Today mining claims are scattered through California's Death Valley National Monument, Oregon's Crater Lake National Park, and Nevada's Great Basin National Park.[8] Overall, the National Park System contains over five thousand patented and unpatented mining claims. If the holders of those claims meet the terms of the Mining Law, they are entitled to develop them no matter what other resources exist on the surrounding lands or what the environmental effects would be.

In the early days of the Law, a valid mining claim had to be based on an actual discovery of mineral deposits. Early in the 20th century, however, the courts held that a miner's rights to a claim persisted as long as he diligently continued to work the land, even if no mineral resources had been found. This doctrine, known as *pedis possessio* (or "foothold"), was intended to protect the original claimant from having a mineral deposit taken away before it could be fully explored.[9] The doctrine has led to widespread abuse. Since claimants need not make an actual mineral discovery, many claims have been located by people with no intention of ever beginning exploration, let alone mining.

Any business or American citizen can locate a claim on up to 20 acres of non-withdrawn or "open" public land. There is no limit to the number of claims that can be held by any one claimant. To gain permanent title to the land, a claimant must obtain a patent from the federal government. A patent may be obtained when the claimant has 1) found a "valuable" ore deposit, 2) performed $500 worth of development work, 3) paid an application fee, currently $250 plus $50 per claim, and 4) paid $2.50 or $5 per acre, depending on whether the claim is a concentrated vein ("lode") or a less consolidated ("placer") claim.

The federal government employs marketability and profit tests to determine whether a "valuable" deposit has been discovered. If a mineral deposit can be extracted, removed, and sold at a profit, it is considered valuable. Although the profit test is supposed to count environmental costs against the claimholder's profits, the government neither routinely identifies nor systematically quantifies such costs. And, even if the government does choose to challenge a claim, it has the burden of proving that the claim is not valuable.[10] Due to the limited resources available for claim reviews, bogus mining claims are rarely invalidated.

Only a small number of claims are ever patented. In 1985, for example, only 51 out of 1.5 million outstanding claims were patented.[11] In the early 1990s, however, the number of patent applications increased significantly, until, in 1994, Congress enacted a moratorium on patenting, which has been extended in each successive year. At the end of fiscal year 1991, the number of patent applications was at an all time high in California, with 66 patents pending on 14,915 acres.[12] As of mid-1994, patent applications involving 30 mines and minerals worth some $34 million were pending.[13]

An unpatented claim is also considered a valid property right, which can be traded or sold.[14] In practice, the owner of an unpatented claim enjoys many of the same benefits as the owner of patented land, without having to defend the value of the claimed mineral deposit or becoming subject to local laws and regulations as patented lands are -- at least as long as an annual "holding" fee of $100 is paid for each claim. (Until 1992, the Mining Law required that $100 work of work be done on each claim each year — a requirement that led to purposeless environmental damage as claimants stripped topsoil and dug trenches to demonstrate their "diligence.")

How the 1872 Law Works Today

Subsidization of the Mining Industry

The Mining Law of 1872, which may have made sense in a young country with a hand tool mining industry, is dramatically out of date today, economically as well as environmentally. The Law's economic obsolescence is strikingly illustrated by the fact that the price charged to patent land and the dollar amount charged in lieu of previously-required assessment work remain unchanged after 125 years. In 1872, the $100 annual work requirement represented approximately 25 days of labor and was meant to ensure that claimed land was actually mined rather than held for speculation.[15] By 1992, when this requirement was replaced with a $100 holding fee, it took only a few hours of digging or half an hour of bulldozing.[16] At a time when claims can be worth millions of dollars, the new $100 fee is literally "a drop in the bucket." Similarly, the patent price of $2.50 or $5 per acre reflected the approximate value of federal land for agricultural uses when the Law was passed in 1872. Because this price has never been changed, a patent holder may gain full legal title to 20 acres of land today for an investment of only $550 to $600 plus the application fee. Since 1874, almost 3.5 million acres of land — an area larger than Connecticut and the District of Columbia combined — have passed from public to private ownership via the Mining Law. Once in private ownership, patented minerals do not have to be mined. The land can be resold or used for other purposes. Hence the law permits an extensive privatization program of public lands under the guise of mineral resource management.

Unlike all other commodity users of the public's lands, mining operators still pay nothing for the valuable resources they extract. While the prices paid by other economic users of these lands — timber companies, livestock producers, oil and gas companies —- are typically heavily subsidized, no subsidy is as great as that given the mining industry. Even Interior Secretary Manuel Lujan, who oversaw BLM's mining program from 1989 through 1992, was shocked to learn that the federal government receives no rent or royalties on even the most profitable gold mines. "You mean we don't get any money?" he asked in disbelief to an assembly of reporters in 1989.[17] Given the ignorance of the former Secretary, one of the country's most powerful mining policymakers at the time, it is hardly surprising that most of the general public remains in the dark about the government's ongoing subsidization of the mining industry.

The most the federal government receives from mining claimants is their patent application fee plus $5 per acre and then only if their claims are patented. Whether or not their claims are patented, mining companies extract hardrock minerals from the public's land free of charge. The value of these extracted minerals is enormous. Hardrock mineral production from public land is worth between $2-3 billion per year.[18] In all, the government has given away more than $231 billion worth of minerals through patenting and royalty-free mining, according to the Mineral Policy Center.[19] If the government demanded the 12.5% royalty which private landowners frequently charge, it could add almost half a billion dollars to the federal treasury annually,[20] while an 8% royalty could raise roughly $1 billion over five years.[21]

Mining Claims For Non-Mining Purposes

Because there are so few restrictions on patenting a mining claim, speculation in patented claims is widespread. The list of development scams involving claims near suburban or resort areas that have been patented, never mined, and resold at enormous profits is already long and still growing. Near the popular ski resort of Keystone, Colorado, for example, 160 acres of land were patented under the Mining Law in 1983 for $2.50 an acre. Six years later, that same land was on the market for $11,000 an acre as part of a real estate development. In 1987, the government patented a granite claim located in a prime residential area of Phoenix, Arizona for $47 per acre. The General Accounting Office found that the parcel was worth $3.8 million. In the outskirts of Las Vegas, 449 acres appraised at over $2 million were patented in 1981 for a mere $1,124.[22]

Patented land doesn't have to be located in urban or suburban areas to be valuable, of course. The mineral deposits themselves are often highly valuable. The 1872 patent provision became the subject of Congressional scrutiny in 1986, for example, when 17,000 acres of oil shale land were patented for $42,500, only to be resold several weeks later to major oil companies for $37 million.[23]

Patents can be filed strategically near or within property valuable for recreational purposes as well. In an Oregon incident known popularly as "Sandscam," claimants filed a patent application covering 780 acres for an "uncommon variety" of sand in the middle of the Oregon Dunes National Recreation Area, one of the state's most popular and spectacular tourist attractions. Not surprisingly, a massive public uproar resulted. But the only way that the federal government could protect its own land against the mining claim was to offer an exchange of suitable land in another area. The claimants rejected offers of land valued at $500 to $700 per acre for land for which they had paid only $5 per acre. When alternative sites with similar kinds of sand were identified for potential trades, the company simply filed claims on those lands as well.[24]

Easy access to land under the 1872 Law also encourages abuse of unpatented mining claims. The BLM generally devotes little in the way of funds or effort to investigate illegal uses of unpatented lands. As a result, rather than being mined, many unpatented claims are being used as garbage dumps, hazardous waste disposal sites, marijuana farms, taverns, permanent residences and vacation homes. In California, people are living rent-free in cabins in the Sierras, around Lake Tahoe, and near other highly scenic and popular recreation areas. Some of the cabins do not meet local health regulations; others are more luxurious and are equipped with amenities such as satellite television dishes and greenhouses. These illegal residents often post no-trespassing signs and erect fences and barriers, effectively excluding the general public from access to public lands. Some of these unauthorized claimants even carry weapons to ward off "trespassers" on land which actually belongs to the public.[25]

Nor is abuse of mining claims restricted to lands under the jurisdiction of the BLM or Forest Service. In 1990, the National Park Service received a proposal for a major tourist development in Denali National Park in Alaska that involved lands obtained through mining claims. Any of the thousands of claims that litter our parks might give rise to similar proposals, proposals that the Park Service undoubtedly could not afford to buy out in order to prevent development.

Only a small fraction of claim locators will ever have a valuable mineral deposit; even fewer will actually mine their claims.[26] An audit of 240 claims in 1974 revealed that only three showed any evidence of mineral extraction.[27] Even lands held by legitimate mining companies are often not actively mined. The cost of holding claims — now $100 per claim each year, previously $100 worth of work — is so low that companies may hold 80 to 90% of their claims for future development.[28] When they do develop their claims, the companies frequently make huge profits — particularly if they are gold mining companies using cyanide to produce gold ore.

The Gold Boom

Cyanide has been used as a means of extracting gold and silver in the United States since the turn of the century. In the late 1960s, the Department of Interior's Bureau of Mines developed a new technology using cyanide which allows gold to be removed from very low-grade ore. This process, known as cyanide heap leaching, can extract as little as 0.02 ounces of gold from a ton of ore.[29] When gold prices rose from $35 an ounce to $350 an ounce during the 1970s, the technique became extremely profitable.

From the early 1980s to the 1990s, the United States' trade in gold went from a $6.7 billion deficit to an $8 billion surplus.[30] The 900% increase in gold production since 1980 has made the United States the world's second largest gold producer. This country produced 9.6 million ounces of gold in 1990,[31] and 10.6 million ounces in 1995.[32] Most of that gold was extracted using cyanide,[33] and 60% of it went into the manufacture of jewelry.[34]

Environmental Damage

The giveaway of the nation's minerals and the prevalence of land scams and illegal squatting highlight the 1872 Law's inadequacies as a tool to manage the public lands. From an environmental point of view, however, the Law's economic inadequacies are not the major reason why it needs to be replaced. Even more disturbing than those flaws is the Law's failure to address the environmental destruction that modern mining technology can cause. The dangers posed by the cyanide leaching method of producing gold vividly demonstrate the need for reform of the 1872 Law.

The environmental damage that results from hardrock mining begins at the exploration stage. When miners bulldoze access roads, construct drill pads for exploration, trenching and excavation activities, and process even limited amounts of ore, they can cause significant damage to wildlife and land and water resources. Even the plastic pipe miners use to stake mining claims can be hazardous; pipes have entrapped tens of thousands of small birds, mammals, and reptiles.[35]

The actual mining of hardrock minerals creates even more environmental damage. Today's gold mines use huge amounts of cyanide in solutions which can be highly toxic. Gold and other hardrock operations can create pits in the earth that are thousands of feet wide and deeper than Chicago's Sears Tower — the world's second tallest building — is high. These unbelievably huge pits are virtually never refilled once mining is completed, because the added costs could make a mining operation uneconomical, or at least less profitable. Instead, they are left as permanent monuments to the industry.

These enormous mines do more than scar the earth. They disturb and contaminate groundwater aquifers in ways which are not fully understood today. Streams and rivers may be sterilized for miles by toxic runoff and sedimentation. In fact, scientists have only recently begun to understand the scope of the environmental damage caused by mines begun over a hundred years ago, let alone come to grips with the problems caused by modern technologies.

Mining with Cyanide

The heap leach process now being used in countless mines around the country employs cyanide in solution to extract microscopic specks of gold from huge quantities of ore. To create a simple gold ring, for example, requires 2.8 tons of ore![36] To expose this ore, topsoil and waste rock are stripped away, creating huge open pits. Enormous trucks that can each carry up to 180 tons transport the ore to large piles, or "heaps," which can cover hundreds of acres and reach over 400 feet in height. A weak solution of cyanide is then sprayed or dripped over each heap. The gold binds to the cyanide solution, which runs into a holding pond of one to five acres in size. The gold-cyanide solution is then pumped to a processing plant which removes the gold, and the cyanide is recycled through the leaching process.

Higher-grade ore is processed in a similar manner, known as "milling." The ore is first crushed and then placed in a tank with cyanide solution. After the gold is extracted, the remaining materials, known as tailings, are placed in "tailings ponds" of up to 500 acres. The tailings retain large quantities of cyanide. Modern tailings ponds and holding ponds typically have synthetic liners to prevent leakage into groundwater.[37]

Cyanide used in gold mining can cause environmental degradation in several ways. The holding ponds attract wildlife which are poisoned by drinking the cyanide-laden water. Cyanide is highly toxic — as is popularly known. Fortunately, it does break down into less hazardous components on exposure to air and sunlight. It also is not generally considered to be bioaccumulative; that is, it does not build up in the body. In sufficient concentrations, however, cyanide will kill both humans and wildlife. Cyanide poisoning from contaminated tailing ponds has killed countless birds and small mammals. There were over 9,000 reported cyanide-related wildlife deaths between 1984 and 1989 in California, Arizona, and Nevada.[38] This total undoubtedly understates the number of fatalities that occurred, since reporting during that period was neither mandatory nor universal.

Most of the birds that have been killed were migratory waterfowl which are attracted to cyanide-laden ponds as they fly through arid areas. One of North America's largest gold mining companies, Echo Bay, pled guilty in 1990 to the deaths of more than 900 birds over a 12-month period at a tailing pond in Nevada; it paid a $250,000 fine. At another Nevada mine, Horizon Gold Company reported more than 200 wildlife deaths in one quarter of 1990 alone.

New tactics have been developed to discourage wildlife from drinking from the ponds, although the fact that wildlife deaths continue to occur shows that none is completely successful. For example, mine operators can cover ponds with nets or pennants or haze birds by using noise or small cannons. The safest system, however, a closed operation where the cyanide solution is not directly exposed to air or land, has rarely been implemented.

Cyanide leaks and spills pose a potentially more serious — and less understood — threat both to wildlife and people. Liners underlying the holding ponds may leak, or large storms may cause holding ponds to overflow, releasing cyanide into streams, rivers, other surface water bodies and groundwater. Cyanide spills are common in gold mining operations and their numbers are likely to increase as the number of operations increases. In 1990, 12 million gallons of cyanide solution overflowed a holding dam in Jefferson, South Carolina. The solution flowed into a nearby river, killing 11,000 fish.[39] Failed holding pond walls caused a 24,000 gallon cyanide spill in California's Mojave Desert and a 225,000 gallon spill in the Nevada desert.[40] After the Noranda Grey Eagle Mine in northern California was closed in 1987, cyanide solution leaked from its tailings dam.[41] Leaks of cyanide, sulfuric acid and toxic heavy metals began shortly after Galactic Resources Inc., a Canadian company, stared operations at the Summitville Mine in Colorado's San Juan mountains in 1986. Over the next several years, the mine eradicated aquatic life along 17 miles of the Alamosa River. In 1992, the company declared bankruptcy, leaving a cleanup job estimated to cost more than $100 million.[42]

While most spills have occurred in relatively remote areas, the threat to developed areas and to drinking water sources may be growing. Over a million cubic feet of cyanide and heavy-metal laden water escaped from a tailings dam in Whitehall, Montana in 1983. Two years later, cyanide was found in local wells and the mine operator eventually supplied local residents with an alternative drinking water supply.[43] In 1988, cyanide solution from a tailings pond in Nevada seeped into groundwater resulting in cyanide concentrations 30 times higher than drinking water standards allowed.[44]

Notwithstanding the claims of the mining industry and some government employees to the contrary, leaks from liners and pads underlying cyanide ponds and tailings frequently occur. Liners tend to break down with age, and large discharges may occur decades after their original installation.[45] In Utah, three of the six registered heap leach operations in 1989 had leaking pads or ponds which threatened local groundwater sources.[46] At the Gilt Edge Mine in South Dakota's Black Hills National Forest, 6,800 gallons of cyanide solution were leaking through the 14-acre liner each day in October 1988.[47] In fact, according to records filed with the Montana Department of Environmental Quality, every mine in the state that has used cyanide leaching has leaked or had a breach of tailings facilities

Open Pit Mines

In addition to the particular problems caused by the use of cyanide, open pit gold mines present environmental risks and impacts common to all open pit mines. They damage underground aquifers and dependent riparian areas, lead to extensive road building in undeveloped areas, destroy wildlife habitat and damage scenic resources.

Consider, for example, the Barrick Goldstrike Mine in Nevada: in a single year, 1990, the Barrick mine recovered over 500,000 ounces of gold, worth $174 million.[48] At the end of the mine's 20-year life, 15.1 million ounces of gold will have been recovered and approximately 2,189 acres of public land disturbed[49] — twice the size of Golden Gate Park in San Francisco. The heap leach pad alone will cover 142 acres. Over 900 million tons of rock and ore will be excavated[50] — more than 100 times the size of the Great Pyramid.

The open pit will be 8,000 feet long, 4,500 feet wide, and 1,800 feet deep[51] — deep enough to hold six Statues of Liberty stacked one on top of the other. The pit will eventually penetrate three groundwater aquifers. As the mine becomes deeper and deeper, the mining company will be forced to continually pump out the groundwater flowing into the giant pit. By the year 2000, the operator could be pumping as many as 44,550 gallons of groundwater per minute out of the pit and into local creeks and reservoirs — almost three times as much water per minute as was used during the peak hour on the peak day in July, 1991 in the capitol of Nevada, Carson City. Nearby springs will dry up as more and more water is drawn into the pit, leaving wildlife such as grouse, riparian songbirds, and mule deer without a source of water. According to the BLM, water levels in the pit and surrounding areas will not be close to normal levels for 100 years.[52]

The Barrick mine, like other open pit mines, will generate huge amounts of waste. When the Barrick operation is finished, piled-up waste rock will tower over the existing terrain by more than 400 feet, and will block views of the surrounding mountains.[53] In the case of many mines of this type, the waste is hazardous. Indeed, according to the Environmental Protection Agency, non-coal mining generates 800 million tons of toxic waste every year — nearly three times the amount produced by all other U.S. industries combined.[54] The mountains and lakes of waste that modern mining is creating contain tons of toxic materials like copper, arsenic, mercury, lead, asbestos and selenium, as well as radioactive materials.

The most serious harm caused by older mines stems from their production of a highly acidic runoff termed "acid mine drainage." At many sites, the rock and soil disturbed by mining operations contain high quantities of heavy metals and sulfides, often in the form of iron pyrite, or "fool's gold." Before mining begins, the sulfides and heavy metals are locked inside the earth, posing few hazards. Once disturbed, however, metal sulfide ores react with the oxygen in water and air, forming a highly acidic solution. This solution may be formed as groundwater fills abandoned mines and pits or as surface water runs over tailings and waste rock piles.[55] As the acid drainage percolates through mining shafts, pits and tailings, it leaches out heavy metals from the exposed rock. The acidic water also accelerates the growth of bacteria, which in turn further oxidize and acidify the solution.[56] Acid mine drainage can be highly toxic. It is typically 20-300 times more acidic than the acid rain which has sterilized streams and lakes in the Eastern United States.[57] Once acid mine drainage begins, it may persist for centuries.[58]

The legacy of these abandoned mines is a perpetual stream of pollution into the natural environment. The U.S. Bureau of Mines (now shut down) estimated that 12,000 miles of rivers and streams and over 180,000 acres of lakes and reservoirs have been poisoned from coal and metal mining acid waste piles.[59] In Colorado alone, acidic drainage has sterilized 300-400 miles of streams.[60] As of August, 1996, there were 66 hard rock mine sites on the Superfund list of worst hazardous waste sites.[61] In fact, the nation's largest Superfund site is a 120-mile stretch of the Clark Fork River in Montana, the product of a copper mining operation that began more than a century ago.

One of Clark Fork's most hazardous places, the Berkeley Pit, is 1.5 miles long, 1 mile wide, and about 2,000 feet deep,. Pure water from local aquifers is slowly flowing into the pit where it becomes acidified. Acid drainage (pH 2.5) is seeping out of the pit and is expected to reach groundwater aquifers in 5-10 years.[62] A nearby community has been evacuated. Local wells have been closed because their levels of arsenic exceed federal drinking-water standards. The surrounding area has consistently shown above-average health problems, including elevated levels of kidney and cardiovascular disease as well as cancer. The trout fishery has been devastated; fish densities are only 2-10% of comparable rivers in Montana. In 1989, 5,290 fish were killed from high levels of copper, cadmium, and zinc which had washed through mining tailings after a storm.[63] The EPA estimated that $1.5 billion will be needed for cleanup and continued monitoring of the Clark Fork site[64] which has been referred to as an example of an irreparable "environmental mistake," for which the only reasonable remediation response may be perpetual monitoring, or perhaps the creation of a "National Environmental Disaster Monument" that could illustrate the costs of past mistakes to future generations.[65] Many more "Environmental Disaster Monuments" may appear all across the west.

In California, the Penn Mine, first mined for copper in 1861, continues to cause regular fish kills in the Mokelumne River. To hold back the acidic runoff from the mine, the state built several dams, which have repeatedly failed during winter storms. In 1986 alone, an estimated 13 tons of copper, lead, aluminum and other heavy metals spilled into a nearby reservoir. Indeed, damming may have made the problem worse by causing increased acid production and seepage of the acidic solution into groundwater. Possible long-term solutions to the problem would be expensive and would involve frequent monitoring. State agencies have not even begun the search for alternatives due to lack of money and lack of information about how effective new "solutions" will be.[66]

There are numerous other examples of continuing mining damage. Until 1994, California's Iron Mountain Mine — the largest mine in the state — spewed thousands of pounds of iron, copper and zinc every day into the Keswick Reservoir on the Sacramento River, making its water nearly as acidic as lemon juice and capable of killing a fish in a few minutes. That year, the U.S. Environmental Protection Agency forced the mine owner to build a treatment plant that treats up to 80% of the toxic material. The rest flows into creeks and reservoirs, which can — and do — experience excessive levels of metals.[67] The Mike Horse mine in Montana, shut down over 40 years ago, is still leaching arsenic, cadmium, zinc, and other toxics into Big Blackfoot River, once famous for its trout fishing.[68] Serious environmental damage results from new mines as well. One gold, silver, and lead mine in Montana was abandoned as recently as 1982. The operator went bankrupt, leaving behind $250,000 in unpaid pollution fines and a bond which could not cover the tens of thousands of dollars worth of needed reclamation. A stream, tinted orange by heavy metals, gushes out from the mine site and into two beaver ponds.[69]

The mining industry claims that these examples result from practices and technologies which are no longer used. But current technology cannot adequately address the problems of acid mine drainage from waste rock dumps and mining tailings. Acid drainage is difficult to halt once it begins and the mitigation measures which are currently available are very expensive. There are no widely available, practical, and cost-efficient means of cleaning up contaminated groundwater or halting acid mine drainage from underground mine workings.[70] And, while the most serious current acid drainage problems have resulted from abandoned mines, a study carried out by the University of California for the California State Legislature, found that "current mining operations continue to pose environmental and public health hazards that warrant serious attention."[71]

In fact, one of the greatest concerns about today's mining boom is that its full environmental legacy may not appear for decades after activities now underway have ended or been abandoned. The California study found that current problem mines had not shown evidence of acid formation at the early stages of mining and warned that "active mines that currently comply with waste discharge requirements could well produce [acid mine drainage] in the future if their wastes have a net acid-generating potential."[72] Clearly, we are engaged in a dangerous experiment with dire consequences for human health and the environment.

Current Regulation

Mining was virtually unregulated for over a hundred years.[68] Today mines are subject to a hodgepodge of federal, state, and local regulations – – but no uniform standards designed to protect the environment.

The extent of regulation varies from state to state and county to county. Numerous gaps exist in state programs, precluding comprehensive regulation of mining impacts.[73] A review of western state mining laws and regulations revealed that they are often too vague or give regulators too much discretion. Reclamation bonds are set far too low to ensure cleanup in the event mines are closed down and their operators default. Often inadequate budgets prevent effective inspection while agencies often turn a blind eye to violations of requirements as well as to environmental problems. Even when the regulatory framework is adequate, state agencies consistently give mining companies broad variances.[74] As a result of these problems, environmental damage is occurring.[75]

Existing federal laws do not eliminate regulatory gaps or cure other deficiencies in state programs. For example, although mining operators must comply with the Clean Water Act, until recently there were no regulations to control storm runoff from mines, a significant source of water contamination.[76] Nor have federal land management agencies rushed to fill the gaps. They have interpreted the Mining Law of 1872 as prohibiting them from requiring environmental protection measures which will increase operating costs to miners.[77] As a result, they have not demanded enough from mining operators in the way of either environmentally-sound practices or reclamation of mining sites. Reclamation provides a good example of how little public regulators do to protect the public interest.

Reclamation

Reclamation is a complicated process which must be based on a thorough evaluation of the needs of each site. It may include removing buildings, tanks, and debris, as well as stabilizing pit walls, grading rock heaps to prevent landslides and erosion, replacing topsoil, replanting disturbed area(s), neutralizing, containing, or disposing of toxic waste and by-products and "backfilling" or refilling open pits with the waste rock that was removed from them.[78] The BLM has required few of these measures, preferring instead to permit miners to leave behind extensive environmental devastation which will cost taxpayers hundreds of millions of dollars to clean up -- assuming attempts are ever made to do so.

The BLM's hardrock mining regulations require the agency to "prevent unnecessary and undue degradation." From 1980, when this requirement was first imposed, until today, the BLM has looked to standard industry practice in applying this test. Since backfilling is definitely not standard industry practice, leaving open pits unfilled is considered neither unnecessary nor undue. Similarly, since re-vegetating with native species is not standard practice, it is rarely required. When waste rock is piled in heaps, steep "angle-of-repose" slopes are formed. These slopes are difficult to revegetate and subject to landslides and erosion. The BLM, however, does not require operators to recontour the rock heaps so that their sides will be less steep and can be reclaimed.[79]

Nor has the BLM forced operators to comply with the reclamation measures it has required. Even when the agency required operators to post bonds to ensure cleanup and restoration, it rarely went after the bonds when the companies defaulted.

Under the circumstances, it is hardly surprising that numerous unreclaimed mines now scar the landscape. One study found that most BLM managers had no immediate plans to follow up on abandoned operations. They could only promise to do so if and when funds became available or hope that another operator would take over and subsequently reclaim the site.[80] An Interior Department audit found that, as of the middle of fiscal year 1990, the BLM had conducted only 10 percent of the required mining claim inspections in California.[81] Without regular inspection or adequate bonds, operators who fail to comply with reclamation regulations can easily escape notice and disappear.

Unreclaimed sites are not only permanent scars, they may present serious health and environmental risks. An investigation of unreclaimed mine sites in Colorado and Nevada revealed leaking 50-gallon barrels of sulfuric acid, abandoned cyanide leaching ponds, and trenches five feet deep and twenty feet wide. On one abandoned site, 25 acres of topsoil had been stripped away.[82] Similar problems were revealed by a series of reviews the BLM conducted of state mining programs during 1991 (California), 1992 (Nevada), 1993 (Idaho).

In 1988, the General Accounting Office estimated that there were 424,049 acres of unreclaimed land resulting from hardrock mining operations in eleven western states. Two-thirds of this acreage resulted from unauthorized or abandoned sites which the U.S. General Accounting Office estimated would cost the government $284 million to reclaim.[83] These estimates probably fail to account fully for the cost of hazardous acid mine drainage cleanup. In 1991, a single state, Montana, which began inventorying abandoned sites in 1979, estimated that reclamation costs for over 20,000 identified problem sites covering 153,800 acres and 1,118 miles of streams would be $912 million. The estimates of many other states also amounted to hundreds of millions of dollars, including Arizona ($654 million), Colorado ($245 million), Idaho ($316 million), New Mexico ($332 million) and Utah ($174 million).[84] Nationally, the Mineral Policy Center has estimated that there are more than 550,000 abandoned hardrock mines and that the cost of cleaning them up is between $32 and $72 billion.[85]

Environmental Regulation

At best, even strong state and federal regulations can only mitigate environmental damage from hardrock mines. The greatest weakness of the 1872 Law is that it essentially gives mining companies "the right to mine," regardless of the damage their activities may do to other values. Some of the West's most sacred places, including our national parks — lands which most Americans assume are protected in their natural state — are threatened by mining, because the Law makes mining the highest and best use of our public lands, regardless of their non-mining values. Thus, mines are frequently proposed, and ultimately permitted, in areas which should be protected.

For example, a gold, silver and copper mining operation was proposed to be located two miles northeast of Yellowstone National Park. The area selected was in the middle of prime grizzly bear habitat and is surrounded on three sides by the Absaroka-Beartooth Wilderness.[86] Local environmental groups as well as the National Park Service became concerned that acid runoff from this project, the New World mine, could flow into the Park, threatening the Park's water purity and the survival of the endangered Yellowstone Cutthroat trout for decades.[87] They raised a sufficient hue and cry that the Clinton administration committed to paying $65 million to the project's Canadian proponents to prevent the mine from going forward.[88]

Another Canadian mining company is currently operating in the California Desert Conservation Area, which was established by Congress in 1976 to protect fragile and unique environmental values. At the time the BLM permitted this operation, known as the Viceroy mine, it was within the East Mojave National Scenic Area, which BLM itself had established to protect the area's outstanding natural and cultural values. And, it was in the vicinity of two federally designated wilderness study areas, as well as habitat for a federally-listed endangered species, the desert tortoise, and another rare desert resource, a year-round spring.

Both of these sites, along with uncounted others, are places that should be protected from the impacts of modern, large-scale mineral development. Yet, because the Mining Law of 1872 still controls development decisions, administrative or Congressional "withdrawals" are the only way to protect land from mining. Withdrawals are awkward tools for land management. They can be very effective, however, as demonstrated by the following description of a local citizen group's successful effort to protect Cave Creek Canyon in the Coronado National Forest from mining.

Citizen Action

In June 1990, Newmont Exploration Limited filed a plan of operations with the U.S. Forest Service for a gold mining claim located in Cave Creek Canyon. Cave Creek is a haven for amateur naturalists, birders, photographers, and biologists. The Southwestern Research Station of the American Museum of Natural History in New York is located in its main canyon. The area is renowned for its spectacular scenery and had changed little since 1854, when a railroad surveyor described in his journal its "upright walls rising majestically ... tapering like spires amid the clouds."[89] The Newmont plan called for the building or improving of about 6,000 feet of dirt roads and the construction of four or five pads to accommodate drilling rigs capable of taking soil and rock samples from a depth of 500 feet. Negotiations between Newmont and the Forest Service went on for several months before the plan was revealed to the public. Confronted with an immediate protest from the local population, the Forest Service responded that, under the Mining Law of 1872, it had no authority to deny the plan.

Local citizens began a two-pronged attack on the mining proposal in concert with the American Museum of Natural History, a powerful national institution. They filed an administrative appeal with the Forest Service asking that the land be withdrawn from mineral entry, and initiated a massive grassroots political campaign to encourage Congressional enactment of a law to protect Cave Creek Canyon. Newmont agreed to suspend operations until the Forest Service appeal was decided. A year later, the Forest Service agreed to seek administrative withdrawal of over 13,000 acres within the Canyon. The mining plan was canceled and the Canyon, at least temporarily, was spared while citizens work to secure enactment of permanent protective legislation by Congress.[90]

Not all groups are as successful in demanding that Congress and the public's land management agencies protect these valuable scenic resources. Usually, mining operations go forward with only as many environmental controls as citizens have been able to force the agencies to impose or the companies to accept.

In the case of the Viceroy mine described above, citizen pressure did lead to the adoption of significant mitigation measures by the company. Despite the unique and nationally recognized scenic and wildlife values of the area surrounding the Viceroy mine, the BLM was reluctant to use its authority to force the mining company to mitigate the environmental dangers the mine posed. When pressure from a coalition of national and local environmental groups failed to budge the BLM, the coalition decided to negotiate directly with the mining company. After more than two years, the coalition won a number of important concessions from the company. Rather than utilizing open ponds and heaps, the company agreed to use closed tanks which protect wildlife from cyanide poisoning and help prevent leaks and spills. It installed a monitoring system to check the effects of the operation on aquifer levels and the nearby spring. And, it set aside a $2 million fund to help restore the area.[91] At least on paper, the citizen groups succeeded where the BLM essentially abdicated its environmental management responsibilities. A stronger federal mining law would have established clear guidelines to protect the environment or, better yet, allowed the agency to say "no" to mining in this particular area. In addition, a stronger mining law would have given citizens groups even more leverage to secure responsible environmental practices from industry.

Reform

The Mining Law of 1872 has been under attack for decades. Over the years, the Public Land Law Review Commission (1970), the Council on Environmental Quality (1977), the U.S. General Accounting Office (1989), and others have called for replacement of the law's archaic provisions with new requirements that will protect the beauty and health of our public lands.[92] To date, reformers have been unsuccessful in overcoming the tremendous political clout the mining industry and its allies wield in Congress — clout that has been bolstered by the continuing unwillingness of the federal government to require miners, large and small, to change the way they have been doing business in order to protect the environment.

At times in the past, including the recent past, the BLM has been a zealous advocate for the mining industry. In 1991, for example, the agency tried to develop a "marketing plan" to sell the Mining Law of 1872. Championing the cause of the mining industry, the draft plan lashed out at the agency's own employees, charging that they "are not good ambassadors for mineral development [and] the 1872 mining law... Some, perhaps many, BLM employees want the mining law abolished or radically changed." The plan publicly criticized employees for expressing "their own personal opinions" to the media about the problems with the 1872 Law. Although the plan was scrapped after it was leaked, the agency may attempt to reach out to a less skeptical audience in the future under a pro-mining administration. One of the solutions the agency proposed to remedy its bad public image was a joint project with industry to develop informational posters for school children.[93] Nonetheless, the question of the source of BLM pro-mining policy remains, given its controversial nature among the agency's own personnel, who are well placed to measure its benefits and liabilities.

Of course, the mining industry has considerable means at its disposal, even without the BLM. Although it has fostered the image of a small-scale, entrepreneurial industry, the fact is hardrock mining is currently undergoing a spectacular boom, posting record production and profits. It is a highly-concentrated, multi-billion-dollar business, not a marginal industry. Gold production grew tenfold during the period from 1981 to 1996,[94] and by 1990, the combined value of mine production for gold, silver, copper, iron ore and lead was more than $10 billion.[95]

Nor is this an industry typified by the small, independent prospector, contrary to the claims of the American Mining Congress and others. The small prospector is as likely to be a nuisance as a benefit to the major mining companies, which reap the bulk of the profits. For example, while there are hundreds of gold mines across the country, 25 gold mines accounted for 82% of the nation's gold production in 1990.[96] The largest companies' net incomes may exceed $100 million.[97] The Bingham Gold Mine in Utah, for example, showed a profit of $160.8 million in 1990,[98] and other non-gold mines are also very profitable. In 1989, for example, both Cyprus Minerals and Magma Copper reported record earnings.[99]

The mining industry also complains that the imposition of tighter environmental controls and reclamation standards will put an end to mining. These complaints are no different from the arguments that other industries employ when protesting environmental regulation. The Clean Air Act and the Clean Water Act both engendered vehement opposition from industry groups, who claimed the stringent regulations of those laws would put them out of business. The coal industry made the same complaints prior to the passage of federal strip mining laws in 1976, but has enjoyed steady growth despite strict standards. Tighter controls can increase costs, but mining companies can learn to accommodate new environmental protection laws and regulations — as have other resource exploiting industries.

Finally, miners protest, they already give a fair return to the government in the form of income taxes. They ignore not only the fact that every business must pay taxes, but also the damage that has been done and the communities that have "boomed" and "busted" as mining companies have continued to profit.

Clearly it is time to change the Mining Law of 1872. A comprehensive overhaul of its underlying policies and assumptions about hardrock mineral mining is essential. The new law must prevent mining on land best used for other purposes, require miners to pay a fair price for the minerals they extract, ensure environmental cleanup, and grant citizens access to the courts and to information about mining abuses. The new law should incorporate the following features:

1) Federal land managers must have explicit authority to exclude or limit mineral exploration and development activities on federal lands as well as to adjust the scope of these activities to protect other public values. Federal managers must be able to protect sensitive areas — such as critical wildlife habitat, areas of exceptional scenic or recreational value, lands recommended for wilderness designation or within wilderness study areas, riparian and wetland areas, significant archeological and cultural sites, and surface and subsurface waters from damaged by mining activities. In other words, the new mining law must recognize, as do other natural resource laws, that our nation has many needs. Mining is not the best use for all public lands. While it is true, as the mining industry claims, that "minerals are where you find them," so are the irreplaceable land, water, and wildlife resources which mining too often destroys.

2) Uniform performance standards that will protect the environment and ensure full reclamation and environmental cleanup must be enacted. These new standards should force mining companies to internalize the environmental costs that mining generates. If a proposed mine is not profitable enough to be operated in an environmentally responsible manner and then properly restored, it should not be allowed to go forward. The industry must be required to post bonds which will cover all cleanup costs should the operator go out of business or fail to perform adequate reclamation. The public should not bear the cost of cleaning up the damage irresponsible companies leave behind.

3) Mining companies should pay royalties on the minerals which they extract, above and beyond the cost of restoration. The oil, coal, and gas industries all pay royalties in exchange for public resources. Hardrock companies commonly pay royalties of up to 12.5% on gross production to prospectors whose claims they mine and up to 10% to states. There is no reason the industry should not pay the American public a reasonable royalty on the minerals which they extract from public lands.

At least part of the increased monies paid by the industry should be used to fund an effective enforcement program which should include such features as mandatory regular inspections and civil penalties for failure to meet environmental standards.

4) Federal lands should remain in the public domain. By virtue of its patent provision, the 1872 Law is the only remaining mining law which allows public property to be permanently converted to private use. Miners

should be able to gain title only to mineral deposits, not the surface land. Once an area has been commercially mined, it should be restored and returned to the public domain.

5) The new law should provide the public with the right to comment and participate at all levels of the permitting and regulatory processes. In addition, it should include a citizen suit provision to allow the public to go to court to enforce the new requirements. Indeed, federal agencies are often understaffed and underfunded and do not enforce current laws and regulations — even when they have the best intentions. The federal strip mining legislation, the Clean Water Act, and numerous other laws have recognized that allowing citizens to bring suit to enforce the law is one of the most efficient ways to ensure compliance.

Enactment of these reforms would bring the mining law into the 20th century just in time for the 21st. Hardrock mining would finally be regulated in a manner consistent with other public land uses. While there will doubtless always be conflicts over the best use for a particular piece of land, public land managers would have the authority to ban mining which would destroy other valuable resources. Federal land would remain in the public domain. Uniform minimal standards for operation and cleanup would be in place across the nation and citizens would have access to the courts to enforce those standards. Finally, more information would be available to the public on mining wastes released into their environment. With these tools, public land managers and private citizens can break the mining industry's 125-year-old stranglehold on the public lands and begin a new era of balanced use and respect for some of our nation's most irreplaceable resources.

RECOMMENDED READING:
Leshy, John (1987) *The Mining Law -- A Study in Perpetual Motion.* Resources for the Future.
The Mineral Policy Center (1994) *Golden Patents, Empty Pockets.*
The Mineral Policy Center (1997) *Righting the Regs.*
*States' Rights, Miners' Wrongs — Case studies in water contamination from hardrock mining, and the failure of states to prevent it.*The Mineral Policy Center, American Fisheries Society, American Rivers and Trout Unlimited.

LIVING DANGEROUSLY
Chemical Contamination in Our Air, Water & Food*
Dr. Sandra Steingraber

"We hear of EPA standards for toxic chemicals stated in parts per billion, and we hear corporate interests say that so many parts per billion represents one drop in a gallon of water —how could that possibly be harmful. In the summer of 1997, British scientists pointed out that chemicals which act as endocrine disruptors are causing genital malformations among wildlife and are believed to cause breast, ovarian and testicular cancer among humans. The reason: 'Parts per billion doesn't sound like very much... But if you put that in the context of your own hormones, which work in parts per trillion, then these substances are there in a concentration a thousand times higher than your own chemical messengers.'"

"'Something nasty' happening to humans, wildlife."
Margaret Lowrie, CNN, June 27, 1997

"All Flesh is Grass." **Isaiah**

There are individuals who claim, as a form of dismissal, that links between cancer and environmental contamination are unproven and unprovable. There are others who believe that placing people in harm's way is wrong — whether the exact mechanisms by which this harm is inflicted can be precisely deciphered or not. At the very least, they argue, we are obliged to investigate, however imperfect our scientific tools; with the right to know comes the duty to inquire.

Consider, for example, herbicides which kill by a variety of different poisoning mechanisms. Some interfere with plant hormones; other herbicides halt the production of amino acids, from which proteins are made; still others strike directly at the process by which plants use sunlight to transform water and carbon dioxide into sugar and oxygen. These are the triazine herbicides, and they are used most often in cornfields but also in orchards and lawns, as well as cotton, sugarcane, and sorghum fields. One member of the group, atrazine, is currently one of the top two most widely used pesticides in United States agriculture.

The wisdom of broadcasting over the landscape chemicals that extinguish the miraculous fact of photosynthesis — which, after all, furnishes us our sole supply of oxygen — is, in and of itself, questionable. Applied directly to the soil, triazines are absorbed by the roots of plants

* Excerpted with permission from Dr. Sandra Steingraber, *Living Downstream: An Ecologist Looks at Cancer and the Environment,"* Addison-Wesley Publishing Company, Inc., New York, 1997.

and transported to the leaves. They poison from within. Triazines are thus water soluble. And because of their solubility, they tend to migrate to many other places.

Traces of triazine herbicides are now found in ground water, as well as about 98 percent of all midwestern surface waters. They have demonstrated a remarkable capacity to poison plankton, algae, aquatic plants, and other chloroplast-bearing organisms that form the basis of the freshwater food chain.

If photosynthetic inhibition were the only problem, the widespread use of triazine herbicides in the Corn Belt would be worrisome enough. But triazines also have effects inside our own bodies, where they gain entry as contaminants of drinking water and as residues on food. Three of the triazines — -cyanazine, simazine, and atrazine — -are classified as possible human carcinogens. Atrazine, a known endocrine disrupter, is restricted for use in Germany, the Netherlands, and several Nordic countries. However, in the United States, atrazine is used on about two-thirds of all cornfields. Simazine is used on lawns and on fruit crops, including oranges, apples, plums, olives, cherries, peaches, cranberries, blueberries, strawberries, grapes, and pears. Until 1994, it was also employed to kill algae in swimming pools and hot tubs, but these uses have since been disallowed after the EPA determined they posed "unacceptable cancer and non-cancer health risks to children and adults." In 1995, the manufacturer of cyanazine voluntarily agreed to phase out production over a four-year period in response to concerns about the herbicide's carcinogenic potential as a contaminant of food and drinking water. In the meantime, cyanazine continues to be used to control grasses and broadleaf weeds in cotton, sorghum, and, most of all, cornfields.

The triazine herbicides are under particular scrutiny for their possible role in breast and ovarian cancers. Evidence comes from both animal and human studies. Atrazine, for example, causes breast cancer and significant changes in menstrual cycling in one strain of laboratory rats, but not in another. Atrazine has also been demonstrated to cause chromosomal breakages in the tissues of hamster ovaries growing in culture. This damage occurs even at trace concentrations — that is, at concentrations comparable to those found in Illinois's public drinking waters.

We are exposed to these herbicides not only through consumption of produce — oranges and apples appear to be major sources of dietary exposure to simazine, for example— but also through meat, milk, poultry, and eggs resulting from the ubiquitous use of corn in animal feed.

In other words, more than three decades after the introduction of chloroplast-destroying chemicals into American agriculture, the cancer risks we have assumed from eating food grown in fields sprayed with them — and from drinking water that has percolated through these fields-have yet to be determined.

In 1994, the General Accounting Office, the investigative arm of Congress, reported that the federal system of food policing is fragmented, inconsistent, and fails to protect us adequately against illegal residues of pesticides of all kinds — triazines and non-triazines alike.

The FDA reports that 3.1 percent of the fruits and vegetables consumed by the public contain pesticide residues above the legal tolerance level. One research institution, Environmental Working Group, places this estimate higher. In their review of FDA monitoring data, EWG staff discovered that many more violations were detected by FDA chemists than were reported by FDA enforcement personnel. The actual violation rate, according to their reanalysis, is 5.6 percent — nearly double the FDA's official claim — and involves sixty-six different pesticides, including many banned or restricted for use. Green peas showed a violation rate of nearly 25 percent, pears 15.7 percent, apple juice 12.5 percent, blackberries 12.4 percent, and green onions 11.7 percent. In response, the FDA asserted that many of the additional violations uncovered by the group were technical in nature and "of no regulatory significance." In any case, these numbers reflected only the violations caught by FDA inspectors, who still examine a tiny fraction — less than 1 percent — of food shipments sent through interstate commerce. How representative these samples are of the entire food supply is anyone's guess. Compounding the uncertainty, the methods used for evaluating compliance are only sensitive enough to detect about half of the pesticides known to be used on food.

In 1991, the USDA itself began collecting data on pesticide residues in certain fresh fruits and vegetables. According to its latest report, the USDA's Pesticide Data Program, using more sensitive tests, found 10,329 detectable residues on 7,328 samples.

The Phenomenon of Biomagnification

It requires an ecological mind to understand why fish, as well as meat, eggs, and dairy, function as a major source of certain kinds of pesticides in our diet. The path between cause and effect can take us into the groundwater, down a river, along an air current, and through a complicated food chain involving organisms whose names we may have never heard. Ecologically speaking, a food chain consists of a series of organisms who pass chemical energy unidirectionally, through each other. Each link of the chain is officially referred to as a trophic level. At the bottom are the producers — green plants that transform sunlight into food, thereby making chemical energy available to everyone else. The primary consumers feed on the producers directly, the secondary consumers feed on the primary consumers, and so forth. About 90 percent of the energy transferred from one trophic level to the next is dissipated as heat.

The phenomenon of biomagnification flows from this fact. Because so much food energy is lost, fewer organisms can be supported at each succeeding link. In order to survive, each individual must consume many individuals at lower trophic levels.

Thus, as organisms continue to feed on each other, any contaminant that accumulates in living tissues is funneled into the smaller and smaller mass of organisms at the top of the pyramid. There are fewer of them. Like a sauce simmering slowly, the poison concentrates. The fat-soluble organochlorine pesticides are well known exemplifiers of this principle. (Happily, triazine herbicides do not appear to biomagnify.)

With pyramids and chains firmly in mind, three mysterious facts become explainable. First, we see why infants are at special risk: residues of fat-soluble pesticides contained in the food eaten by nursing mothers are distilled even further in breast milk. In essence, breast-feeding infants occupy a higher rung on the food chain than the rest of us. In many cases, human milk contains pesticide and other residues in excess of limits established for commercially marketed food.

Second, we see why a diet rich in animal products exposes us to more pesticide residues than a plant-based diet, even though the plants are directly sprayed. For the most part, the flesh of the animals we eat contains more pesticides than the grasses and grains we feed them. Indeed, the largest contributors to total adult intake of chlorinated insecticides are dairy products, meat, fish, and poultry. In 1991, the National Research Council reported that more than half the cattle tested in a sample of Colorado ranches had detectable levels of pesticides in their blood serum. A commonly found contaminant was the banned pesticide heptachlor. Similarly, the FDA's Total Diet Study, which regularly monitors the concentration of contaminants in cooked, table-ready foods, continues to find traces of DDT in many food types, but particularly those of animal origin. Third, we see why those of us born in the heyday of organochlorine pesticide use bear special risks. Passed along from one creature to the next, these persistent pesticides are especially subject to biomagnification.

Global distillation*

When certain chlorinated pesticides and other persistent pollutants are released in warmer climates, they evaporate and are carried by winds to cooler areas, where they condense and descend back to earth. These trespassers overwinter in soil, snow, or water until the summer sun

[*Editor's Note: More than 1.2 billion pounds of pesticides were exported from the U.S. in 1995 and 1996, according to a new report by the Foundation for Advancements in Science and Education (FASE). U.S. policy allows the export of banned pesticides, as well as pesticides that have never been evaluated by the U.S. EPA. FASE found that 9.4 million pounds of never-registered pesticides were exported in 1995 and 1996, a 40 percent increase since the period 1992-1994. The U.S. exported more than 28 million pounds of pesticides designated as "extremely hazardous" by the World Health Organization, a 500 percent increase since 1992. Many pesticides shped from U.S. ports are destined for developing countries. The report recommends changing U.S. policy to eliminate double standards of safety. Given the scientifically acknowledged reality of global distillation, it is a fallacy to think that American citizens will escape contamination from pollutants that U.S. companies are permitted to sell abroad. Sadly and dangerously, it is more likely that once again, our chickens will be coming home to roost.}

revaporizes them and air currents blow them further toward the poles. They then drift downward once again.

The rising and falling movements of global distillation explain not only why chemicals used in rice paddies and cotton fields eventually end up in the skin of Arctic trees but also why the bodies of seals in Siberia's Lake Baikal — the world's oldest and deepest lake-contain the same two contaminants as the alpine soils of New Hampshire's Mount Moosilauke. Indeed, the levels of PCB and DDT in the blubber of these animals are high enough to render them illegal for consumption by U.S. standards. Of all the component aspects of the environment, air is the one with which we, who inhale about a pint of atmosphere with every breath, are in most continuous contact. Air is the element most diffuse, most shared, most invisible, least controllable, least understood.

The phenomenon of global distillation, along with more local forms of atmospheric deposition, shows that not all of the dangers from carcinogens in our air supply come from breathing. Some also come from eating. Poisons dumped and plowed into the earth are fed, molecule by molecule, into the air, where they redistribute themselves back to the earth and into our food supply.

In short, because of air, we each consume suspected carcinogens released into the environment by people far removed from us in space and time. Some of the chemical contaminants we carry in our bodies are pesticides sprayed by farmers we have never met, whose language we may not speak, in countries whose agricultural practices may be completely unfamiliar to us. Some of the chemical contaminants we carry with us come from long-defunct products of industry — objects manufactured, used, and discarded by people of a previous generation.

Air is by far the largest receptacle for industrial emissions. Of all of the toxic chemicals released by industry into the nation's environment each year, more than half is released into air. These emissions include about seventy different known or suspected human carcinogens. When vehicle exhaust and emissions from power plants are added to the mix, the number and amount of carcinogens in air rise further. According to the International Agency for Research on Cancer, ambient air in cities and industrial areas typically contains a hundred different chemicals known to cause cancer or genetic mutations in experimental animals. And while air pollution in the United States has markedly improved over the past quarter century, more than a hundred urban areas still fail to meet national air quality standards. In other words, nearly one hundred million Americans breathe air that is officially illegal.

These are facts not in dispute. How much airborne carcinogens actually contribute to human cancer, however, remains an elusive question. As two leading researchers have noted, airborne carcinogens create an "epidemiological dilemma": we know they exist, but we have no good method for linking them directly to disease because air... can evade the rigors of scientific analysis.

There is more to the story of lung cancer than cigarettes. The first comes from doctors' offices. Oncologists who specialize in lung cancer are reporting increasing numbers of nonsmokers among their patients, as well as increases of a specific kind of lung malignancy not strongly associated with tobacco. Called adeno-carcinoma, it is distinguishable from oat cell and squamous cell carcinomas, both strongly linked to smoking. "Although better laboratory methods for classifying tumors have enabled doctors to identify more adenocarcinomas, scientists believe that environmental pollutants and carcinogens are also causing an increase in cases." So concludes a recent review article published by the Harvard Medical School.

Meanwhile, epidemiologists have been focusing on understanding the urban factor in lung cancer. Many ecologic studies from countries around the world have revealed urban lung cancer rates two to three times higher than those of surrounding countrysides.

However, city folk also tend to smoke more than their rural counterparts. When smoking habits are taken into account, the excess of lung cancer incidence in urban areas is much less, but is still higher than in rural areas. Areas with chemical plants, pulp and paper mills, and petroleum industries also show elevated rates of lung cancer. One recent cohort study of more than eight thousand U.S. adults showed that air pollution was positively associated with death from lung cancer after cigarette smoking, age, education, body mass, and occupational exposures were all accounted for. And a cohort study of five thousand chimney sweeps in Sweden found increased mortality from lung cancer and other tumors that was not explainable by smoking habits but that was related to exposure to carcinogenic soot.

Water is regulated much the same way food is. Just as food has tolerances, drinking water has maximum contaminant levels. These represent the highest limits allowable by law of particular toxic substances in public water supplies.

Like an accountant who proficiently measures and records individual values but fails to sum the results, this system of regulating contaminants in water suffers from the same constricted one-chemical- at-a-time vision as the parallel system of regulating pesticide residues in food. It ignores exposures to combinations of chemicals that may act in concert. Radon gas and arsenic, for example, occur naturally in some aquifers tapped for public drinking water. Both are considered human carcinogens. Maximum contaminant levels have been established for each, and each is supposed to be regulated below those levels. However, if water containing these two elements is also laced with traces of herbicides, dry-cleaning fluids, and industrial solvents — even at concentrations well below their respective legal limits — the resulting mixture may well pose hazards not recognized by a laundry list of individual exposure limits. Exposure to one compound may decrease the body's ability to detoxify another.

In 1995, in the first study of its kind, researchers sampled water from faucets in kitchens, offices, and bathrooms every three days from mid-May through the end of June in communities throughout the Corn Belt. Herbicides turned up in the tap water in all but one of twenty-nine towns and cities. Atrazine, the suspected breast carcinogen, exceeded its maximum contaminant level in five cities.

Biologically speaking, we live only in the present. Our bodies do not respond to contaminants on the basis of averages; they must cope the best they can with the load of contaminants already received as well as with those streaming in at any given moment. If, during the period of April through June, a woman living in rural Illinois drinks enough weedkiller to overwhelm her body's ability to detoxify it, and if, as some animal evidence suggests, these chemicals are capable of initiating and/or promoting genetic lesions in her breast tissue, then the damage has been done, regardless of what happens during the months of August, October, or January.

This issue is even more critical for infants and children. Many researchers believe that exposure to even minute amounts of carcinogens at certain points in early development can magnify later cancer risks greatly. What are the implications for the unborn child who happens to reach one of these key points at the same time a bevy of farm chemicals in the local water supply is reaching its peak? What are the implications for the adolescent girl whose breast buds start to form during this particular quarter of the calendar year?

Happily, a recent right-to-know clause added to the Safe Drinking Water Act makes this information more accessible to the public. Under 1996 amendments, water utilities must tell customers, in their water bill and at least once a year, what pollutants have been detected in their drinking water and whether water quality standards have been violated. The law also mandates the creation of a national database of contaminants found in drinking water. Previously, national records did not tally contaminants unless they constituted actual violations.

Exposure to waterborne carcinogens is more commonplace than many people realize. In the same way that intake of airborne pollution involves the food we eat as well as the air we breathe, intake of contaminants carried by tap water involves breathing and skin absorption as well as drinking. These alternative routes are especially important for the class of synthetic contaminants called volatile organics — carbon-based compounds that vaporize more readily than water. The solvent tetrachloroethylene is a common one. Most are suspected carcinogens.

The simple, relaxing act of taking a bath turns out to be a significant route of exposure to volatile organics. In a 1996 study, the exhaled breath of people who had recently showered contained elevated levels of volatile organic compounds. In fact, a ten-minute shower or a thirty-minute bath contributed a greater internal dose of these volatile compounds than drinking half a gallon of tap water. Showering in an enclosed stall appears to contribute the greatest dose, probably because of the inhalation of steam.

The particular route of exposure profoundly affects the biological course of the contaminant within the body. The water that we drink and use in cooking passes through the liver first and is metabolized before entering the bloodstream. A dose received from bathing is dispersed to many different organs before it reaches the liver.

We know that someone, sooner or later, ends up consuming whatever poisons are spread on the earth and above it.

In full possession of our ecological roots, we can begin to survey our present situation. This requires a human rights approach. Such an approach recognizes that the current system of regulating the use, release and disposal of known and suspected carcinogens — rather than preventing their generation in the first place — is intolerable. So is the decision to allow untested chemicals free access to our bodies, until which time they are finally assessed for carcinogenic properties. Both practices show reckless disregard for human life.

The estimate (of cancer deaths due to environmental causes) put forth by those who dismiss environmental carcinogens as negligible, is 2 percent. Two percent means that 10,940 people in the United States die each year from environmentally caused cancers. It is the annual equivalent of wiping out a small city. It is thirty funerals every day. None of these 10,940 Americans will die quick, painless deaths. They will be amputated, irradiated, and dosed with chemotherapy. They will expire privately in hospitals and hospices and be buried quietly. We will not know who most of them are. Their anonymity, however, does not moderate this violence. These deaths are a form of homicide.

Why should we assume that we will not number among them?

RECOMMENDED READING:

G. B. Dermer (1993) *The Immortal Cell: Why Cancer Research Fails*, Avery Publishing, Garden City Park, NY.

D. Pimentel and H. Lehman (eds.) (1992) *The Pesticide Question: Environment, Economics and Ethics*, Chapman & Hall, New York.

T. Heller, et al (eds.) (1992) *Preventing Cancers*, Taylor & Francis, Bristol, PA.

HOW CORPORATIONS COMBAT

ENVIRONMENTALISM

SCIENCE FOR SALE*

Dan Fagin & Marianne Lavelle

> "Junk science can be described as a report from scientists on the stern of
> the Titanic who insist the ship isn't sinking. They would say that thirty minutes
> ago the stern was 200 meters out of the water and now it's 250 meters, which
> proves it's actually rising. They are correct. But rising to a vertical position
> before the whole thing sinks to the bottom of the ocean."
>
> **B. Ishmael**

Buying Legitimacy

The nation's chemical industry, by pouring billions of dollars into
the nation's research universities, exercises an overwhelming influence
on the direction — and even the results — of chemical research in America.
Collaboration between industry and academia is nothing new, nor is it
inherently troublesome. DuPont, for example, worked with university
researchers during World War II — in a successful and beneficial
partnership — to develop nylon.

But in those early years of the chemical revolution, when relatively few
profitable synthetic compounds were on the market, DuPont and other
companies focused their energies on developing new ones and bankrolled
academic researchers who were doing the same. Today, so many immensely
profitable chemicals have established niches in the marketplace that there is
little incentive for companies to take a chance and support researchers who
are looking for innovative products that might turn out to be far less profitable.

Instead, the industry pours its money into research that defends
existing products — particularly those products that are dangerous to
human health and the environment, says Nicholas Ashford, a professor
of environmental health at the Massachusetts Institute of Technology.
Instead of research on safer and cheaper alternatives, the academic
community focuses on saving the market for chemicals in use today
because that is where the money is.[1] The consequences are far-reaching.
The chemical industry's influence is so pervasive that for any particular
chemical there are invariably only a few, if any, independently financed
studies to weigh against the juggernaut of corporate-backed science. As
a result, even when regulators do view industry-sponsored research
skeptically, there is little opportunity to get an alternative view. "The
problem is there is no evidence on the public side," Ashford says.
"Public-interest science has become underfunded, and the number of
really independent, good academics is a small number so industry is

* Reprinted with permission from *Toxic Deception: How the Chemical Industry Manipulates
Science, Bends the Law and Endangers Your Health,* by Dan Fagin, Marianne Lavelle and
the Center for Public Integrity, Carol Publishing Group, New Jersey, 1996. Second
edition to be published in 1999 by Common Courage Press, Maine.

able to overwhelm the science. And industry's way of looking at the science is very unbalanced."

Just how unbalanced becomes clear through a review of recently published industry-sponsored studies of alachlor, atrazine, formaldehyde, and perchloroethylene.

At least 43 studies assessing their safety were financed by corporations or industry organizations (such as the Formaldehyde Institute) and published from 1989 to 1995 in major scientific journals in English, according to the National Library of Medicine's MEDLINE database. Six of them returned results unfavorable to the chemicals involved, and five had ambivalent findings. The other 32 all returned results favorable to the chemicals involved. In short, manufacturers batted .744 when they paid for the research.[2]

Their average would have been even higher except for some circumstances surrounding four of the studies about formaldehyde that reported unfavorable results. Three of them were bankrolled by the Health Effects Institute, which is financed not by formaldehyde manufacturers but by automobile manufacturers and the Environmental Protection Agency. The institute is studying the dangers of reformulated gasoline, which often emits formaldehyde. The fourth was financed by the Center for Indoor Air Research, founded by the tobacco industry, which seeks to show that chemicals are a more serious indoor air pollutant than cigarette smoke.

When nonindustry scientists did the research, the results were quite different. Governments, universities, medical and charitable organizations (the March of Dimes, for example), and other nonindustry groups sponsored 118 studies published on the four chemicals during the same six-year period. Except for the insurance fund of the Amalgamated Clothing and Textile Workers Union (the labor union now known as UNITE!), which sponsored two of the studies, these were organizations without a stake in any specific outcome. They included such federal agencies as the U.S. Department of Agriculture and the U.S. Air Force, state governments, foreign governments, and the United Nations. The results when government agencies and other nonindustry entities were paying for the research: About 60 percent of the studies (71) returned results unfavorable to the chemicals involved; the rest were divided between studies with favorable results (27) and results that were ambivalent or difficult to characterize (20)

How can the underwriters of studies have such an overpowering influence on their results? A close-up look at the arcane discipline of weed science suggests some important reasons.

Weed scientists — a close-knit fraternity of researchers in industry, academia and government — like to call themselves "nozzleheads" or "spray and pray guys." As the nicknames suggest, their focus is actually much narrower than weeds. As many of its leading practitioners admit, weed science almost always means herbicide science, and herbicide science almost always means herbicide-justification science. Using their clout as the most important source of research dollars, chemical companies have skillfully wielded weed scientists to ward off the EPA, organic farmers, and others who want to wean American farmers away from their dependence on atrazine, alachlor, and other chemical weed-killers.

The numbers tell part of the story. About 1,400 weed scientists in the United States work for chemical companies, compared to just 75 for the federal government and 180 for universities around the nation, according to James Parochetti of the U. S. Department of Agriculture. Independent scientists "are a very small part of the picture, because herbicides dominate weed science so totally," says John Radin, who oversees weed science research at the USDA's Agricultural Research Service.

But industry's dominance of weed science is even more overwhelming than those numbers suggest. The reason is money. The USDA usually pays the salaries of researchers who work at the land grant universities that are at the heart of the nation's academic agricultural research. However, a scientist needs much more than a salary. Lab equipment, supplies, technicians, and graduate students are just as essential. Researchers must fend for themselves in securing the financial support for those extra expenses. If they are unable to win federal grants, there is generally only one other place to turn.

Chemical companies have the leverage to dominate research even with relatively small grants of a few thousand dollars per year. "Our universities are like a limousine with a well-trained chauffeur," Kent Crookston, who heads the department of agronomy at the University of Minnesota, says. "We have the limo and the chauffeur, but no gas money. When someone comes along with a little gas, they determine where we drive for a few thousand bucks."

"The large federal grants are difficult to obtain, and there's probably many more weed scientists out there who are relying very heavily on these smaller grants that come from the agricultural chemical industry," Alex G. Ogg Jr., a USDA researcher and past president of the Weed Science Society of America, says. "If you don't have any research other than what's coming from the ag chem companies, you're going to be doing research on agricultural chemicals. That's the hard, cold fact."

Some researchers have soured on the system. Take Orvin Burnside of the University of Minnesota, a past president of the Weed Science Society. A prominent weed scientist since the early 1960s, Burnside learned early that he could do little with the money his school provided: "If I was going to do something beside twiddle my thumbs after two weeks, I had to go out there and find some money." He found it in the chemical industry, which for years sponsored about 80 percent of his research, mostly on the herbicide 2,4-D.

But in 1993, Burnside had an epiphany. He wrote a scathing article for a weed science journal in which he called the discipline a "stepchild" of agricultural research and said that "public weed scientists must redirect their activities after four decades of largely herbicide-focused research."[3]

He also stopped accepting grants from the industry. "They like to direct what you do, and finally I decided I want to direct what I do," he says. "I'm going to work on what I feel is going to be most profitable and productive for the taxpayers, and that's not developing a new herbicide that the farmer's going to have to pay through the nose for." Burnside believes that herbicide use could be reduced by 50 to 75 percent without affecting crop yields. The nation's independent weed scientists, he says, ought to focus their energies on figuring out how to make that happen.

Conventional pesticide-based farming "is the only plate we offer farmers as a research community," Dennis Keeney, the director of Iowa State University's Leopold Center for Sustainable Agriculture, says. "The land grants themselves know that's wrong, and you read piece after piece by the land grant researchers that say they have to change. They talk, but they don't walk the talk. Not yet."

Few weed scientists have followed Burnside's example. Like most of his colleagues, Alex Martin, a professor of agronomy at the University of Nebraska, uses industry grants to pay for both his doctoral students and much of his research. "There is no question that availability of money influences research," he says. "You don't go to Monsanto to fund some kind of ecology problem. They're going to fund something in their area."

"The words 'conflict of interest' come to mind," the USDA's Radin says. "These weed scientists are honest people, but it's a very difficult position that they occupy when they don't have a very steady source of funding that allows them to be independent. It's an endemic problem."

Some forms of industry influence over academia cannot be measured concretely, such as the presence of chemical company executives on the advisory boards of the USDA research stations that are associated with major state universities. Others are theoretically measurable, except that no one is keeping track.

There is no way to know how much money chemical companies are funneling to universities and foundations for research, for example, but it's undoubtedly more than a billion dollars a year. The nation's land grant universities alone get more than $134 million a year in industry funds for agricultural research, with pesticides constituting a major share. In fact, just one of the thousands of potential sources of industry dollars, the corporate-financed Formaldehyde Institute, was spending at least $300,000 a year on research in the late 1970s and early 1980s — for just one chemical.

Even the chemical industry's charitable contributions — which, unlike its research grants, must be publicly reported — appear to be carefully targeted to serve corporate interests. Dow Chemical Company, for example, donated at least $24 million to universities and research foundations from 1979 to 1994, and the charitable arms of its smaller colleagues in the perc business — Occidental Petroleum Corporation, PPG Industries, Inc., and Vulcan Materials Company — gave a total of $7.5 million. During the same period, Monsanto Company's foundation gave at least $18.5 million, including $7.8 million to Washington and St. Louis Universities in the company's hometown. But most other beneficiaries of Monsanto's largesse were far-flung schools that specialize in agriculture and biotechnology, the fields where the company's business is concentrated. Formaldehyde users Hoechst Celanese and Weyerhaueser made research grants totaling more than $13.2 million. Borden's total was $2.8 million, while Georgia-Pacific gave $1.2 million and DuPont $500,000.[4]

The chemical industry also showers money on the nation's cancer research establishment. The American Cancer Society, the Arthur James Cancer Hospital & Research Center, the Ohio Cancer Foundation, the R. David Thomas Cancer Research Foundation, and the Sloan Kettering Cancer Center all receive support from chemical manufacturers.

Industry officials say that their charitable giving is aimed not at influencing scientific research but at improving the quality of life in the communities in which their companies operate, although they acknowledge that there are public relations benefits as well. "Certainly the well-being of the communities where we operate and the company's image as a good citizen and a contributor to that community is very important, "Monsanto's Kenworthy says. "It's very important. It's important to the welfare of our workers and their job satisfaction and how productive they are, and it's important as to how easy it is to operate in that community."

But it is difficult to ignore that, over the years, the nation's cancer research establishment has done much to bolster the industry's position that chemicals should not be blamed for disease. When the EPA was looking at formaldehyde in the early 1980s, Alan C. Davis, a lobbyist for the American Cancer Society, sent a letter to James Ramey, the chairman of the Formaldehyde Institute, in which he said that the society knew of no evidence that formaldehyde caused cancer in humans.[5] The Formaldehyde Institute presented this stamp of approval to the EPA.

In its December 1995 newsletter on research, the society criticized the Delaney Clause of the *Food, Drug, and Cosmetics Act*, which, until it was repealed in 1996, banned additives of carcinogens to processed food.[6] This zero-tolerance level, the newsletter said, ignores "the virtual absence of risk at very low-dose levels and the economic and nutritional benefits which can accompany the use of synthetic chemicals (preservatives, pesticides, sweeteners, etc.) in the production and processing of food. The newsletter article opened with the statement, "Risk is a natural part of life." It echoed the words of the Chemical Industry Institute of Toxicology, which in its 1994 annual report said, "Risk is a fact of life."[7]

Industry's friends in academia are similarly helpful. At the height of the controversy over a possible alachlor ban, in 1985, the Weed Science Society of America submitted a letter to the EPA that used some notably unscientific language to argue against any new restrictions. "Eliminating competition by regulation in the free enterprise system is not the American way," it said. The society's letterhead listed eight board members who worked for universities or the USDA as well as its president at the time, J. D. Riggleman, who was then the president of Dupont's agricultural chemicals department.[43]

A decade later, the society leapt to the defense of atrazine when *Tap Water Blues*, a newly released report from the Environmental Working Group, a Washington-based environmental organization, was drawing nationwide attention to the problem of herbicide contamination of drinking-water supplies. The society issued a "position statement" that took a hard-line stance against stricter regulation of atrazine.* It should not have been a surprise. A "senior research fellow" at Ciba-Geigy, Homer LeBaron, had been the society's president in 1989. LeBaron had written his doctoral thesis on atrazine and spent much of his 30-year career at Ciba-Geigy trying to expand use of the weed-killer until his 1991 retirement.

A friendly academic institution, the University of Missouri, even issued a press release in February 1995 that unblushingly began this way: "Atrazine, one of the Corn Belt's favorite chemicals, could get yanked from farmers' weed control arsenals by the Environmental Protection Agency because tiny amounts are getting into reservoirs and rivers used as public drinking water sources. Agricultural specialists have argued that the amounts found pose no health risks and that losing the chemical would double corn farmers' weed control costs." At a time when Ciba-Geigy was working feverishly to flood the EPA with testimonials for atrazine before the official comment period ended on March 23, the university even took the extraordinary step of including in its press release the EPA address to which public comments could be sent.[9] That same year, another university press release attacked *Tap Water Blues*, calling it "unfounded."[10]

The university's agricultural researchers may not have considered their pro-atrazine efforts particularly unusual. After all, they were used to working closely with agribusiness interests, including Ciba-Geigy. Over the previous five years, the University of Missouri's College of Agriculture, Food, and Natural Resources received more than $2.2 million in research funds from for-profit corporations, including more than $137,000 from Ciba-Geigy, university records show.[11]

The college's dean, Roger Mitchell, says that he worries about whether the school's farming research is too heavily focused on chemical use. "I accept that as appropriate criticism," he says. "I spend a lot of time thinking about it." He added, however, that the aggressively pro-atrazine press releases issued by the college were prompted not by a desire to repay Ciba-Geigy for its research grants, but by the college's view that Missouri farmers are reluctant to adopt non- or low-chemical farming techniques. "Those are very thoughtful choices that we make, and it isn't the chemical company that's driving that choice, it's our trying to evaluate what seems to be working for the farmer." Critics, of course, argue that farmers have been slow to embrace alternative farming techniques precisely because manufacturers and their allies in academia control the flow of information to growers.

[*Editor's Note*: Atrazine, a known endocrine disrupter, is restricted for use in Germany, the Netherlands, and several Nordic countries. However, in the United States, atrazine is used on about two-thirds of all cornfields.]

A closer look at the work of the one of the most prolific researchers on herbicide pollution — David B. Baker of tiny Heidelberg College in Tiffin, Ohio — shows how industry funding can help to shape the scientific debate.

In the early 1980s, before chemical companies became a major source of Baker's funds, Baker's studies of surface-water contamination in northwestern Ohio were among the first to identify atrazine, alachlor, and other weed-killers as major pollutants. The EPA relied heavily on his work when the agency decided to begin its special review of alachlor in 1985. A review of Baker's papers from those early years shows that they tended to emphasize the seriousness of the herbicide problem. A typical sentiment was the first sentence he wrote in his summary for a paper he presented at a 1983 conference: "Relatively high concentrations of many currently used herbicides and insecticides are present in rivers draining agricultural watersheds."[12]

By the 1990s, Baker's tone had changed significantly. The herbicide contamination problem had become a major issue, and Baker was getting almost a third of his research money from Monsanto, Ciba-Geigy, DuPont, and other industry sources. In April 1994, for instance, in an "issue paper" he co-wrote for the industry-financed Council for Agricultural Science and Technology, Baker concluded that the regulatory system "appears to be capable of maintaining overall risk at acceptably low levels" and that "there is no conclusive evidence that allowed concentrations are the source of any human health effects."[13]

Six months later, Baker was a key part of the industry's damage-control strategy after the Environmental Working Group's release of *Tap Water Blues*. The National Agricultural Chemical Association (later renamed the American Crop Protection Association) and other industry groups urged reporters writing about *Tap Water Blues* to call Baker, and many of them did so. A week after the report's release, the association flew Baker to Washington, D.C., where he delivered three speeches attacking *Tap Water Blues* — two to industry groups and the third to journalists at the National Press Club. Within four weeks he had written a 39-page rebuttal, and in the cover letter accompanying the critique he called the environmental group's report "erroneous and slanted" and "designed to frighten the public."[14] The pesticide association, as well as farm bureaus and manufacturers, also touted a paper Baker had not yet published in which he concluded that herbicides do not pose a significant health risk in water supplies. Less than a month after the release of *Tap Water Blues*, the EPA placed atrazine in special review, and Ciba-Geigy was facing an unprecedented threat to its 30-year-old gold mine. It was thrilled with Baker's study, published in early 1995, in which he and colleague R. P. Richards of Heidelberg, using unusually blunt language, wrote that "atrazine exposure through drinking water does not represent a significant human health threat."[15] In 1996, Baker was working on a sharply worded critique of another study by the Environmental Working Group of weed-killers in drinking water.

There is no factual contradiction between Baker's early and late work on herbicide pollution. What changed is his tone. His early government-financed research played up the extent of contamination; the later industry-financed studies played down the health threat posed by that contamination.

"There's certainly been some change in his tenor and approach," says George Hallberg, now the chief of environmental research at the University of Iowa Hygienic Laboratory and, like Baker, a pioneer investigator of herbicide contamination of drinking water. "One contrasts the rhetoric of the articles from early on to today and there certainly seems to be a difference."

Adds Richard Wiles of the Environmental Working Group, a frequent Baker antagonist: "He was one of the guys way back when who identified this as a problem. And the tone of his data, the interpretation of his data, has changed as his funding source has changed."

Baker denies that his findings have been influenced by industry grants to Heidelberg College's water quality laboratory, which he heads. The lab's 1994-95 budget projected the receipt of about $143,000 in industry funds out of a total budget of about $502,000, with the rest coming mostly from government grants.[16] However, while Baker says that he never allows funding considerations to influence the way he interprets scientific data, he acknowledges that funding can sometimes affect the way he presents that data. An industry-funded study, for instance, might stress the relatively low number of cancer cases linked to a chemical exposure, he says. "Do those [funding] concerns color the way I present the data? They may. But I try as hard as I can not to let those kinds of concerns influence the way I interpret the data." He also points out that scientists who rely on activist-oriented foundations for research support are vulnerable to the same criticism. "The funding issues work both ways with both kinds of groups," he says.

J. Alan Roberson, an engineer and director of regulatory affairs for the American Water Works Association, calls Baker an able scientist but says that his recent work grossly understates drinking-water contamination. "Industry spinning is more subtle than faking results," Roberson says. "These guys usually pull the samples and analyze the results straight up. The spin comes in the way they present the data."

An even more fundamental way that manufacturers spin academic research is by carefully selecting the scientists they finance. Like an investor picking a mutual fund, manufacturers tend to choose researchers who have a proven track record of publishing studies that reflect the industry's point of view. Ciba-Geigy, for example, turned to one of the chemical industry's most outspoken defenders in 1996 in an effort to counter new research by independent scientists who found that atrazine changes the way the body metabolizes estrogen in a way that increases cancer risk. Stephen Safe of Texas A&M University, whom Ciba-Geigy hired to conduct its hormonal studies of atrazine, has been one of the most prominent critics of the growing body of evidence that many chemicals dangerously mimic natural hormones or alter the way those hormones function in the body. Safe frequently debates the issue at scientific conferences and gatherings of journalists.

While corporations usually decide in private which scientists to finance, records of Formaldehyde Institute meetings provide a rare window into the process and show how carefully manufacturers shop for friendly researchers.

It is clear that before the institute wrote a $20,000 check to Dr. Leon Goldberg, who had just left his post as the CIIT's president, for example, it knew the gist of his forthcoming paper "Risk Assessment of Formaldehyde: A Scientific Viewpoint."[17] The institute had already made plans to videotape Goldberg and to condense his paper for distribution to the news media.

Members of the Formaldehyde Institute also financed a pet project of Harry Demopoulos, then in the Pathology Department of New York University Medical Center. Demopoulos appeared on PBS's *MacNeil-Lehrer Report* to discount the first tests that linked formaldehyde to cancer. Demopoulos had asked the institute to provide $50,000 of the $220,000 in funds he needed for a symposium on "thresholds"— the theory that low doses of carcinogenic chemicals are safe.[18] On May 11, 1981, soon after the first rat cancer results on formaldehyde, Demopoulos wrote a letter to Ramey, the institute's chairman, in which he expressed outrage over this and other cancer scares. Demopoulos also noted that his fellow NYU researchers were working on a formaldehyde carcinogenicity study but that it was so poorly conducted it was unpublishable.

Arthur C. Upton, the chairman of NYU's Environmental Medicine Institute, was so disturbed by what he termed the "inaccuracies and implications" in Demopoulos's letter that he wrote a letter to the Consumer Product Safety Commission. Upton, the lead author of the paper that would, indeed, be published and back up the original evidence of formaldehyde's carcinogenicity, said that Demopoulos's statements were "not only unprofessional but intolerable in their implication that the work in question was incompetently performed."

When industry-financed studies are released to the public, there is often no disclosure of who paid for the work. For example, when formaldehyde manufacturers persuaded a federal court to throw out the Consumer Product Safety Commission's ban of their insulation products in 1983, the court cited the work of five scientific researchers who had found little evidence of formaldehyde-related health problems. The court pointed out that one of them, John Higginson, a scientist at the Universities Associated for Research and Education in Pathology, Inc. in Bethesda, Maryland, had formerly headed the prestigious International Agency for Research on Cancer in Lyon, France.[19] It did not, however, mention that two of the researchers, Gary Marsh of the University of Pittsburgh and Otto Wong of Tabershaw Occupational Medical Associates, a private firm, had worked for the Formaldehyde Institute and its member companies.[20]

Steering Science

Scientific manipulation by the chemical industry is, in its own way, effective in warding off regulators and keeping dangerous products on the market. They can cheat. They can manipulate research results. They can create front groups, co-opt academic researchers, and attack independent scientists.

Using these techniques, chemical manufacturers set the fundamental direction of research and define the terms of the scientific debate. This is the most far-reaching effect of the chemical industry's misuse of science. Sometimes, the agenda-setting process is as simple as putting money in one place and not another. Pesticide manufacturers, for example, have been far more eager to spend money on developing new chemicals than on finding old ones in water and in soil. In fact, those chemicals had been in drinking water for decades before the scientific techniques to detect them were finally developed in the 1970s. Atrazine's most important metabolites (breakdown products), for example, are toxic and thought by some researchers to be as common in water as atrazine itself. But until the late 1980s, there was no way to know because Ciba-Geigy had not developed a way to test for them. Testing for a third important metabolite, hydroxyatrazine, did not begin until 1992.

The long fight over the carcinogenicity of atrazine is a case study in how the industry exercises its power to set the scientific agenda. Whenever new information has surfaced about atrazine's dangers, Ciba-Geigy has been able to shift the debate in a process that resembles nothing so much as a jittery suspect being questioned by an inept police detective: when one alibi falls flat, the company simply offers a different one, and each time the government spends years laboriously checking each story out.

For many years, the manufacturers of atrazine had a simple position: atrazine did not cause cancer — at any dose, in any kind of animal, under any circumstances. It was a hard position to contest since there was virtually no publicly available data on the issue. The only early rat study of atrazine's cancer-causing potential, conducted in 1961 — two years after atrazine's introduction in the United States — by a consultant to Geigy, was a failure. Almost all the rats died, apparently of unrelated infections, long before the end of the two-year period.[21] The only other test, conducted by the National Cancer Institute in 1969, concluded that atrazine did not cause tumors in mice.

By the mid-1970s, the situation had finally changed, but Ciba-Geigy did not have to change its company line that atrazine does not cause cancer. Under the new Federal Insecticide, Fungicide, and Rodenticide Act, the company had to conduct a cancer study and many other tests and submit the results to the EPA, which would decide whether to allow atrazine to stay on the market. This time, the new cancer study suggested an "oncogenic [tumor-causing] response," as an EPA scientist would later state in a vaguely worded memo. But the company was able to keep arguing to the contrary because, as it turned out, the test had been conducted by scandal-ridden Industrial Bio-Test Laboratories.[22]

That an industry-financed study would find cancer was particularly noteworthy. "If IBT finds a positive carcinogen, you can believe that more than if IBT finds a negative one," Reto Engler, the former senior science adviser in the EPA's pesticides office, says. "It's easier to sweep negative studies under the carpet, but if it's positive it hits you in the face. If an IBT study was positive, you could bet your child's head that the chemical was positive."

When the EPA finally got around to reviewing the atrazine study in 1982, it concluded that Ciba-Geigy would have to redo the study. That was another break for Ciba-Geigy, because it bought the company more time. "If we had had a cancer study of sufficient integrity at that time, we might well have started the risk assessment process sooner," says the EPA's Fenner-Crisp, currently the deputy director of the pesticides office.

Ciba-Geigy finally had to shift its position in 1984, when, as required by law, it told the EPA that the new two-year study was halfway along and that high numbers of mammary tumors were appearing in female rats — the same kind of tumors that appeared in the discarded Industrial Bio-Test study conducted almost ten years earlier. Far from settling the question, however, the new study launched the company on a two-pronged scientific quest that still continues today: to prove that the breast tumors are a fluke confined to a single strain of rat, the Spragiae-Dawley, and that low levels of atrazine pose no cancer risk because a threshold dose is needed to cause tumors. Monsanto has embarked on a similar quest for alachlor and has come up with three elaborate explanations about why each of the three kinds of cancer that alachlor causes in rats — nasal, thyroid, and stomach tumors — would not occur in humans exposed to the chemical. The EPA still has not formally changed its view that alachlor is a probable human carcinogen, but in 1996 it adopted a new way of calculating alachlor's cancer risk that has the effect of playing down that risk.[23]

As with alachlor, formaldehyde, and perc, the EPA has allowed the atrazine debate to take a detour to arcane questions of the mechanisms of carcinogenicity because conventional cancer studies did not go the company's way. Ciba-Geigy has spent millions of dollars on a series of studies aimed at showing that atrazine is safe because, the company argues, it is carcinogenic only above a threshold dose that is never exceeded in lakes, rivers, and aquifers tainted with the herbicide. It has also fed atrazine to mice and a different type of rat, concluding each time that the chemical does not cause tumors. So far, the EPA still says that there is not enough evidence that a threshold exists for atrazine. That isn't surprising, because the notion that toxic chemicals require a certain dose before they "switch on" and become carcinogens is extremely controversial. The only compound that the federal government has ever regulated based on a carcinogenic threshold, in fact, is saccharin. "At this point in time, they [Ciba-Geigy] have not yet convinced us their threshold hypothesis is correct, although quite frankly we have always thought it has merit," the EPA's Fenner-Crisp says. "We've been quite open about that." Even so,

Ciba-Geigy has succeeded in creating enough uncertainty to forestall regulation, despite the long-standing evidence that the herbicide poses a cancer risk. Thirty-six years after atrazine went on the market and began leaching into the nation's water supplies, the EPA still says that the question of whether it causes cancer in humans is unresolved.

"The total number of pesticide poisonings in the United States increased from 67,000 in 1989 to the current level of 110,000 per year."

C.M. Benbrook *et al.*
Pest Management at the Crossroads, 1996

"The average exposure to carcinogens from automobiles in Los Angeles is as much as 5000 times greater than the level considered acceptable by the EPA."

E. Wilkin, "Urbanization Spreading,"
in **L.R. Brown** *et al, Vital Signs:*
The Trends That are Shaping our Future.

"The United States uses over 2700 billion kg of chemicals each year, of which at least 20 billion kg are considered hazardous."

World Resources Institute
World Resources 1994–95

"Environmental factors, including various chemicals, ultraviolet and ionizing radiation, and tobacco smoke, are estimated to cause roughly 80% of all cancers."

C.J.L. Murray and A.D. Lopez
The Global Burden of Disease, 1996

"Estimates are that every 1% decrease in the ozone layer increases cancer-inducing UV-B radiation by 1.4. Exposure to sunlight, including UV-B radiation, accounts for 70% of skin cancers in the United States."

A.J. McMichael
Planetary Overload: Global Environmental Change
and the Health of the Human Species, 1993

The above are cited in: **David Pimentel** *et al*
"Ecology of Increasing Disease:
Population growth and environmental degradation,"
BioScience, **Vol. 48, No. 10, October 1998**

NEWS POLLUTION*

John Stauber & Sheldon Rampton

> *"It's to the point where Brit Hume, the ABC correspondent at the White House, plays tennis with George Bush. Tom Friedman of the New York Times is very close to Jim Baker. You find these relationships are so close that reporters don't challenge the subjects of their stories, they just tell you what the government is saying. In other words, they have become stenographers for power and not journalists."*
>
> **Jeff Cohen, Executive Director,**
> **Fairness & Accuracy In Reporting (FAIR)**

If popular culture is any guide at all to the imagination of the American people, there is something sacred about the news. Journalists, along with private detectives and police, seem to occupy a special place as the ministers of truth and wisdom in our society. The archetypal image for all three professions is the "little guy" with the common touch. Picture Colombo, Lou Grant or Phil Marlowe dressed in cheap clothes, cynical, smoking a cigar, fond of a drink at the local tavern, working odd hours, bothering people, persistent, smart beneath that rumpled exterior, piecing together clues, finding contradictions, relentless and inquisitive, refusing to let go of an investigation until the truth is exposed and the villains receive their just punishment. This image of the journalistic profession has been the backdrop for a number of popular plays, novels, films and television shows, including "Citizen Kane," "His Girl Friday," "Meet John Doe," "The Front Page," "The Paper" and "Murphy Brown." Of course mild-mannered reporter Clark Kent is the alter-ego of Superman, the ultimate comic-book hero, who spends much of his time rescuing fellow reporters Lois Lane and Jimmy Olsen when their journalistic curiosity gets them into trouble. Hollywood turned to real life in *All the President's Men*, in which Robert Redford and Dustin Hoffman portray *Washington Post* reporters Bob Woodward and Carl Bernstein in their investigation of President Nixon's role in the Watergate scandal. The final scene of the movie visually dramatizes the power of the press when the camera zooms in on a clattering newsroom teletype as it prints out a sequence of bulletins from the Watergate affair, culminating in the terse headline: "NIXON RESIGNS."

Today more than 20 years after Watergate, the saga of Woodward and Bernstein is related in high school textbooks as an example of the power of muckraking journalism. To perpetuate the mythology of the crusading press, many newspaper mastheads carry mottos such as Thomas Jefferson's statement that "the only security of all is a free press." Americans

* Excerpted with permission from John Stauber & Sheldon Rampton, *Toxic Sludge Is Good For You: Lies, Damn Lies and the Public Relations Industry*, Common Courage Press, 1995. For more information, see their website, www.prwatch.org.

grow up believing that the free press, so cherished and constitutionally protected, is a fierce watchdog of the public interest and that when societal or political wrongs are splashed across page one of the *New York Times* or aired on *60 Minutes*, our democratic system in some automatic way responds to right the wrongs. But school books often fail to mention that Woodward and Bernstein were virtually alone in their dogged pursuit of the Watergate scandal, which occurred in the midst of a presidential election yet had absolutely no impact on the election outcome. According to Project Censored, a phone call from the Nixon White House was all it took to persuade CBS chair William Paley to scale back Walter Cronkite's attempt to do an extraordinary two-part series about Watergate on the CBS Evening News before the election. Nixon was re-elected by an overwhelming margin, and wasn't forced to resign until two years after the burglary. Even then, the real "heroes" include a still-unknown whistleblower dubbed "Deep Throat" and Nixon's own arrogance in leaving behind expletive-loaded tape recordings of his self-incriminating involvement in a cover-up. Even Woodward and Bernstein were never able to explain important aspects about the scandal, such as White House motivations behind the Watergate break-in, or Deep Throat's reasons for coming forward.

Hard Pressed

The romantic mythology surrounding the journalistic profession attracts many more would-be reporters than there are jobs available. In reality, as most working reporters readily admit, the profession is a far cry from its image. Reporters are notoriously underpaid and overworked. While researching this book, we encountered one reporter with a small-town daily paper who was earning an annual income of $13,000 in 1994 while working 60-hour weeks — less than he could have earned flipping hamburgers. Sitting below a poster of Rush Limbaugh which the newspaper management had mounted on the wall of his break room, the reporter described how the paper had ordered him and his fellow reporters to falsify time cards so it would appear they were only working 40 hours per week thus enabling management to violate minimum-wage laws. We contacted a state labor official who assured us that if the reporter kept good records, he could document the fraud and force the paper to pay him for his uncompensated labor. The reporter, however, was afraid that he would be fired if he filed a complaint, and we were left wondering; Is someone who collaborates this easily with covering up his own exploitation even *capable* of investigating and exposing the larger wrongs in his community? Today's reality, however, is ever more distant from this lofty ideal.

In a democracy, a free and independent press is counted upon to provide the information and opinions that fuel public debate, expose corruption, illuminate major social issues, and enable an informed citizen to make participatory decisions. Journalism is in fact in demise, and its collapse is opening ever more opportunities for PR practitioners to increase their influence in the news room.

To begin with, the media itself is a huge, profitable business, the domain of fewer and fewer giant transnational corporations. "Modem technology and American economics have quietly created a new kind of central authority over information," writes media critic Ben Bagdikian in *The Media Monopoly*, his landmark 1982 exposé. "By the 1980s, the majority of all major American media newspapers, magazines, radio, television, books, and movies were controlled by fifty giant corporations. These corporations were interlocked in common financial interest with other massive industries and with a few dominant international banks." Bagdikian concedes that "There are other media voices outside the control of the dominant fifty corporations," but "most are small and localized . . . their diminutive sounds tend to be drowned by the controlled thunder of half the media power of a great society. "[1]

When Bagdikian updated his book in 1993, he was alarmed to find that during the ensuing decade media concentration had accelerated so that fewer than 20 giant corporations owned over half of all media. When we interviewed him in August 1995, he said the situation is "worsening so quickly and dramatically that it's hard now to even get a number to compare. Suddenly there are big new actors in the media business, super-giant corporations like Disney, Time-Warner, TCI cable TV, and telephone companies. The magnitude of the players is incredibly large. Increasingly corporate giants and super-giants are working together in joint ventures; for instance, Turner is partly owned by Time-Warner and TCI. Journalism, news and public information have been integrated formally into the highest levels of financial and non-journalistic corporate control. Conflicts of interest between the public's need for information and corporate desires for 'positive' information have vastly increased."[2]

"I generally agree with those who bemoan the decline of journalism for sensationalism and titillation," said Buck Donham, a former newspaper editor who has worked for papers in Arkansas and Hawaii. "I blame at least part of this decline on the trend of large corporations to buy up both major and minor news media. Back in the good old days, most medium or small newspapers were owned by families, some of whom lovingly passed along their publications to the next generation. I worked for one such newspaper, the *Arkansas Gazette*, which at the time was the oldest newspaper west of the Mississippi. The pay was atrocious, $85 a week in 1965, but the prestige of working for such a news organization more than made up for the small salary."

When corporations buy up a local paper, Donham says, standards usually decline:

> They practice what I call bottom-line "journalism" which means, to quote the late Don Reynolds, editorial material is the 'gray matter that fills up the space between the ads'. Here's how these corporations work: After buying up a small or medium-sized newspaper with much fanfare, the companies make a lot of noises about local editorial control, promising not to interfere with the editorial content of the publication.

They frequently keep on the old editor Slowly, however, they exert their control, and the editor usually leaves after six months to a year after purchase. They cut the newspaper s staff to the bare minimum it takes to put out the publication on a daily or weekly basis; . . . they stop reinvesting any of the profits in the product and instead ship all profits made back to corporate headquarters; they de-emphasize the news and emphasize advertising and circulation revenue.

After about a year or so, you have an extremely streamlined operation. What remains of the editorial staff is humping so hard just to get the paper out, there is no time to do in-depth or investigative reporting. Needless to say, it makes for superficial journalism and news by press release. If they don't quit, hard-pressed and bitter reporters resign themselves to covering the superficial and the easy, to relying on press releases or on breaking, easy-to-cover stories, such as the O.J. Simpson case. Reporters simply don't have the time, and after a while, the inclination to do anything in depth.[3]

This environment may be demoralizing to journalists, but it offers a veritable hog's heaven to the public relations industry. In their 1985 book, *PR. How the Public Relations Industry Writes the News,* authors Jeff and Marie Blyskal write that "PR people know how the press thinks. Thus they are able to tailor their publicity so that journalists will listen and cover it. As a result much of the news you read in newspapers and magazines or watch on television and hear on radio is heavily influenced and slanted by public relations people. Whole sections of the news are virtually owned by PR.... Newspaper food pages are a PR man's paradise, as are the entertainment, automotive, real estate, home improvement and living sections... Unfortunately 'news' hatched by a PR person and journalist working together looks much like real news dug up by enterprising journalists working independently. The public thus does not know which news stories and journalists are playing servant to PR."[4]

Today the number of PR flacks in the United States outnumbers working journalists, and the gap is widening. A working reporter is deluged daily with dozens if not hundreds of phone calls, letters, faxes and now e-mailed press releases. Pam Berns, the publisher of *Chicago Life* magazine, estimates that her office receives at least 100 PR contacts each and every day. "It's annoying and overwhelming," says Berns.[5]

PR Newswire claims to be "the world's acknowledged leader in the distribution of corporate, association and institutional information to the media and the financial community" for forty years. The firm has 19 offices in the U.S. and distributes some 100,000 news releases a year to some 2,000 newsrooms for more than 15,000 clients. Other PR distribution services specialize in placing stories in news papers through the mass distribution of PR-written feature articles and opinion pieces which are simply picked up as "real" news.[6] The North American Precis Syndicate, for example, sends camera-ready stories on behalf of most of the top PR firms and most Fortune 500 companies to 10,000 newspapers, almost all of whom reprint some of the material. The stories are designed to promote products, or serve clients' political agendas. "Lobbyists love it," claims NAPS' promotional material. "You generate tons of letters to legislators."[7]

A similar business, Radio USA, "supplies broadcast quality news scripts to 5,000 radio stations throughout the country... We'll write, typeset, print and distribute broadcast quality scripts. We give you usage reports based on station verifications." Hard-pressed radio journalists greet these canned scripts with relief rather than suspicion. "When your job is to come up with hundreds of story ideas every month, Radio USA helps," says Suzan Vaughn, news director at KVEC in San Luis Obispo, California. Max Kolbe, the news director at KKIN in Aitkin, MN, describes Radio USA as "a lifesaver on a slow news day."[8]

Even the media itself is getting into the PR distribution act. The Associated Press now makes money distributing electronically digitized PR photos to over 400 newspapers that have agreed to receive them. On June 24, 1994, for example, the *New York Times* ran a prominent article announcing that Federal Express had formally changed its name to "FedEx" — not exactly an earth-shattering exposé. In fact, the story resulted from a PR campaign by FedEx. A photo accompanying the story showed a FedEx jet, and the photo credit simply read "Associated Press." Actually, Federal Express paid AP to electronically distribute the PR photo, which was staged and taken by the Federal Express company rather than an AP photographer.[9]

Is It Real Or Is It Memorex?

The use of radio and video news releases is a little-known practice which took hold during the 1980s, when PR firms discovered that they could film, edit and produce their own news segments — even entire programs — and that broadcasters would play the segments as "news," often with no editing. When Gray & Company began producing a radio program for its clients called "Washington Spotlight," the Mutual Radio Network came to Gray and asked to carry it. "PR firms would not send out packaged radio and television stories if no one was using them," notes author Susan Trento. "Not only technology, but economics made things easier for PR firms in the 1980s."

Video news releases, known as VNRs, typically come packaged with two versions of the story the PR firm is trying to promote. The first version is fully edited, with voiceovers already included or with a script indicating where the station's local news anchor should read his or her lines. The second version is known as "B-roll," and consists of the raw footage that was used to produce the fully-edited version. The receiving station can edit the B-roll footage itself, or combine it with other footage received from other sources. "There are two economics at work here on the television side," explains a Gray & Company executive. "The big stations don't want prepackaged, pretaped. They have the money, the budget, and the manpower to put their own together. But the smaller stations across the country lap up stuff like this."[10]

MediaLink, a PR firm that distributed about half of the 4,000 VNRs made available to newscasters in 1991, conducted a survey of 92 newsrooms and found that all 92 used VNRs supplied free by PR firms and subtly slanted to sell a client's products and ideas while appearing to be "real" TV news. On June 13, 1991, for example, the CBS Evening News ran a segment on the hazards of automatic safety belts. According to David Lieberman, author of a 1992 article titled "Fake News," the safety belt tape "was part of a 'video news release' created by... a lobby group largely supported by lawyers."[11]

"VNRs are as much a public relations fixture as the print news release," stated George Glazer, a senior vice-president of Hill & Knowlton. "In fact, many public relations firms are well into the second generation of VNR technology. We use satellite transmissions from our own facilities almost on a daily basis, and wait eagerly for fiber optics systems to allow us to dial into nationwide networks... With few exceptions, broadcasters as a group have refused to participate in any kinds of standards establishment for VNRs, in part because they rarely will admit to using them on the air... There are truly hundreds of examples of self-denial on the part of broadcasters when it comes to admitting that VNRs are used." Following a beverage-tampering scare on the West Coast, for example, a VNR was mailed out to all three TV stations in the first city to report the problem. All three stations used the VNR in at least one newscast the following day, along with five other stations in the region. When asked later, however, all three stations denied that they had broadcast the material.[12]

In 1985, Trento reports, Gray & Company distributed a VNR featuring a canned interview with one of its clients, the ruthless King Hassan II of Morocco. The segment 5 airing on CNN provoked a scandal with reporters claiming they had been tricked into airing paid propaganda. An executive at Gray & Company scoffed at the media's hypocrisy: "I used to read in Broadcasting the cache of letters from news directors after the story broke about electronic news releases saying, 'How despicable. Never in a thousand years!' And they were people I had talked to who had called me back so that they had the right coordinates on the satellite so that they could take the feed. They knew exactly who we were. They called us all the time. They asked us for stuff. They told us they couldn't get it. They forgot to turn their downlink on, and could we send them a hard copy FedEx overnight because they'd use it tomorrow night."

"I was personally aggrieved at all this sort of self-righteousness of the media when that story broke," said another Gray & Company executive. "They are free to use it. Not use it. Use it for B-roll. Write their own scripts. Most of them take it straight off the air and broadcast it. Rip and read. Rip and read."[13]

Watching the Detectives

In theory, journalism is a "watchdog" profession, which serves the public by finding and reporting on abuses of power. In practice, reporters live under closer scrutiny than the people they are supposed to be monitoring.

Former *Wall Street Journal* reporter Dean Rotbart has carved a niche for himself within the PR industry by compiling dossiers on his former colleagues so that his corporate clients know how to manipulate individual members of the media. Rotbart's firm, called TJFR Products and Service, publishes this information in high-priced newsletters and delivers customized workshops and reports.

Rotbart told a 1993 meeting of the Public Relations Society of America that his workshops and newsletters help PR professionals know "what a journalist is thinking... One of the services we provide is taking biographies of reporters from all over the country — something like 6,000 bios — in our computer system, and if at any point you get a call from a journalist and don't know who it is, call up and we will fax you that bio within an hour."[14]

These bios are a regular feature in a new Rotbart publication, the *TFJR Environmental News Reporter.* Promotional literature boasts that this $395-a-year PR resource is "tailored to serve the needs of communications professionals who deal with environmental issues... Let us be your eyes and ears when the environmental media convene... Gather vital information on key journalists... Who's the boss?... How do you break the ice?... Not only will you find news on journalists, we'll tell you what they want from you and what strategies you can employ with them to generate more positive stories and better manage potentially negative situations."[15]

The premier issue of Rotbart's newsletter includes a long piece on CNN's Environment Unit, with biographies of all its top staff. It explains, for example, that Peter Dykstra worked for Greenpeace for 11 years and attended Boston University's College of Communications. The issue also contains an interview with Emilia Askari of the *Detroit Free Press*. The accompanying bio explains that Askari is president of the Society of Environmental Journalists and "enjoys all kinds of outdoor activities and tutors illiterate adults with Literacy Volunteers of America." In addition to this information, the story tells PR managers whom to contact if they want to complain about something that Askari writes: "Chain of command: Reports to Bob Campbell, assistant city editor." [16]

Some PR firms specialize in tracking specific issues and compiling reports on the journalists writing stories. Rowan & Blewitt, conducted in depth analyses for the dairy industry, analyzing media coverage of the controversial milk hormone rBGH issue to "help answer these questions: Has the coverage been sensationalist... Has the coverage favored the anti-rBGH views?... How does the coverage of [rBGH] compare with the volume of coverage on Alar? Air emissions? ... Alaskan oil spills?" Detailed charts and graphs examined virtually every story on rBGH over an extended period of time.[17] Another media monitoring firm, CARMA International, also worked on the rBGH account and ranked individual reporters based on whether their stories were "favorable" or "unfavorable" to rBGH.[18]

The February 1995 issue of the newsletter *Environment Writer* reports on another PR effort to get inside the heads of journalists — laboratory research using 12 real journalists as paid guinea pigs to help develop a PR strategy for DuPont pesticides. DuPont flacks recruited participants by sending an invitation to "selected members of the media," which promised: "This learning endeavor will be used to help DuPont establish new policies regarding pesticides: their use and information important to consumers, the government, farmers and the press... Your ethics as a journalist (and that of your news organization) will not be violated or jeopardized in any way... the goal is to make better pesticide policies."[19]

One participant who asked not to be named said, "They would give us small pieces of paper which would say something like, 'DuPont makes very wonderful chemicals, and no one needs to worry.' "Journalists were then told to develop a storyline based on the information on the slip of paper, while DuPont researchers observed from behind a mirrored window. When their work was done the reporters were handed envelopes that contained $250 cash." I came out of there and I felt really disgusted that I had to earn money in this kind of way," said one journalist who participated in the study.[20]

PR firms also hire real journalists to participate in training sessions so flacks can hone their skills in handling media situations. In *Sierra Magazine*, reporter Dashka Slater describes her experience working for Robert J. Meyers and Associates, a Houston-based consulting firm that hired her and two other journalists to help ARCO Petroleum practice its PR plan for handling the news media following environmental disasters. In a staged run-through of an oil spill, Slater and the other reporters were assigned to play the part of the "predatory press." Professional actors were brought in to play the part of environmentalists. ARCO employees and government officials played themselves. "The drills give company flacks the opportunity to practice varnishing the truth just in case the mop-up doesn't go as planned," wrote Slater. "Mostly the company and government spokespeople did what they had learned to do in numerous media-training workshops: convey as little information as possible in as many words as possible." In the past 6 years Meyers and Associates have conducted more than 400 such training drills.[21]

A number of firms offer clipping and database services used to monitor the media. For a fee of $1,000 per month, for example, organizations can subscribe to NEXIS/LEXIS, which contains the full text of media stories appearing in a wide range of newspapers, magazines, newsletters and TV and radio programs. A skilled researcher can search for mention of a specific word and obtain articles from hundreds of publications. Clipping services are also available which charge a reading fee to provide virtually instant reports on stories appearing in the media which relate to clients' interest.

To monitor TV coverage, for example, Video Monitoring Services advertises that it "records all news and public affairs programs on local TV stations in more than 130 markets, local radio stations in 14 markets and national broadcast TV, cable TV and radio networks." VMS also monitors stories in more than 20 countries including Australia, Canada, Germany, Israel and Japan. The company notifies its clients immediately which stories it has snagged off the airwaves that correspond to a list of "keywords" the client has provided. The keywords can be "executives, company names, brand names, events, general or specific topics, etc." Such intelligence can assist PR firms in identifying sympathetic or cooperative reporters and editors, pressuring or punishing reporters who file unsympathetic stories, and measuring the impact of coverage upon public opinion.[22]

Careful overnight telephone surveys can also help a corporation decide how, or even whether, to respond to a breaking story on a-news program like *60 Minutes*. Before such instant surveys were available, a corporation exposed on national TV for crimes or corruption might feel compelled to immediately hold a news conference. Today, if overnight polling shows that the *60 Minutes* expose had little impact on viewers, PR consultants advise their clients to simply ignore the report rather than risk drawing attention to it. Surveys can also reveal strategic approaches to crafting a response — perhaps the polling shows some sympathy for the corporation or other opinions that can be exploited in a carefully designed response the next day.

Experts Agree

The advertising industry learned years ago that one of the best ways to influence an audience is to put its message in the mouth of a publicly-trusted expert such as a scientist, doctor or university professor. A whole genre of TV commercials has evolved featuring actors dressed in white laboratory coats who announce that "research proves" their brand is the best product on the market. The PR industry has also mastered the art of using "third party" experts, a ruse which almost never fails to hoodwink supposedly cynical reporters.

Via the internet, for example, public relations representatives "assist" the news media through an on-line service called Profnet, owned by the PR Newswire company. Journalists in search of information are invited to simply e-mail their request to Profnet, which distributes it to over 1,000 flacks at corporations and in large PR firms who pay to receive the journalists' queries sent to Profnet. The flacks then find their own "experts" to answer the questions. Needless to say, this "free" information helps shape the spin of the story in a direction the PR representative is trying to promote.[23]

Corporations also fund "nonprofit research institutes" which provide "third party experts" to advocate on their behalf. The American Council on Science and Health (ACSH), for example, is a commonly-used industry front group that produces PR ammunition for the food processing and chemical industries. Headed by Elizabeth Whelan, ACSH routinely represents itself as an "independent," "objective" science institute. This claim was dissected by Howard Kurtz of *The Washington Post* in the March 1990 *Columbia Journalism Review*, which studied the special interests that fund ACSH. Kurtz reported that Whelan praises the nutritional virtues of fast food and receives money from Burger King. She downplays the link between a high fat diet and heart disease, while receiving funding from Oscar Mayer, Frito Lay and Land O'Lakes. She defends saccharin and receives money from Coca-Cola, Pepsi, NutraSweet and the National Soft-Drink Association. Whelan attacks a Nebraska businessman's crusade against fatty tropical oils — the unhealthy oils in movie popcorn — while she is in the pay of palm oil special interests. "There has never been a case of ill health linked to the regulated, approved use of pesticides in this country," she claims, while taking money from a host of pesticide makers. And Whelan speaks harshly of mainstream environmentalists such as the Natural Resources Defense Council. Speaking to the *Bangor Daily News*, Whelan described the NRDC as an "ideologically fueled project" whose "target is the free-enterprise, corporate America system. I think they hate the word 'profit' and they'll do anything that will involve corporate confrontation."[24]

Whelan defends her "scientific" views by saying that her findings have undergone "peer review" by experts among the scientists affiliated with her group. But Michael Jacobson of the Center for Science in the Public Interest dismisses the bona fides of such "peer review" scientists: "They don't exactly publish in leading scientific journals. They publish pamphlets that are reviewed by their professional cronies of the regulated industries. It's science that's forced through a sieve of conservative philosophy."

Journalists rarely check the background of sources, so Whelan and the American Council on Science and Health are often quoted in the news as "scientific experts." For example, in a show hosted by Walter Cronkite titled "Big Fears, Little Risks," Cronkite introduced Whelan as one of "a growing number of scientists who fear that overstating the risk of environmental chemicals is actually threatening the health of Americans." In *Fortune* magazine, Whelan appeared as the source in a story by Ann Reilly Dowd which stated, "A big part of the problem is that America's environmental policy making has increasingly been driven more by media hype and partisan politics than by sensible science... Despite the waves of panic that roll over America each year, some 500 scientists surveyed by the American Council on Science and Health have concluded that the threat to life from environmental hazards is negligible." Neither Cronkite nor Dowd explained that the ACSH is an industry front group.[25]

Rhys Roth of the Northwest Atmosphere Protection Coalition continually goes up against industry "science" as he tries to raise public concern about the greenhouse effect and ozone depletion. Roth says scientists like Whelan — who speaks of the "allegedly depleting ozone layer" — are "atmosphere confusionists" who achieve their goal by "simply sowing enough confusion in the minds of Americans about the science behind the greenhouse effect to defuse our collective concern and outrage, rendering us politically mute."[26]

The Truth About Tumors

Whelan and other "experts" use creative manipulation of statistics to obscure the rising rate of cancer in industrialized nations. "We also know there is no cancer epidemic," she says. "Most cancer rates have been constant for decades... What a marvelous time to live, and to be born! We are giving ourselves and our children the gift of better and longer lives."

This claim was repeated by David Shaw of the *Los Angeles Times* in a series of articles examining environmental health risks. Shaw took his information from Resources for the Future (RFF), a pro-industry group that he described as a "think tank that specializes in environmental issues." Shaw quoted RFF vice-president Paul Portney: "If everything is as harmful as we're told, how come we're healthier and living longer... than ever before?" Shaw also turned to the National Cancer Institute, a government agency with close ties to the chemical and pharmaceutical industry. According to the Institute," the age-adjusted mortality rate for all cancers combined except lung cancer has been declining since 1950, except for those 85 and over. "[27]

These statistics, however, paint a misleading picture. Research by Samuel Epstein at the University of Illinois School of Public Health, published in the *American Journal of Industrial Medicine*, shows that the incidence of all types of cancer, excluding lung cancer, rose by 29.1 percent during the period from 1950 to 1988. Contrary to the National Cancer Institute's claims, the British medical journal *Lancet* reported in 1990 that the death rate from brain and other central nervous system cancers, breast cancer, multiple myeloma, kidney cancer, non- Hodgkin's lymphoma, and melanoma has been increasing over the past 20 years in persons age 55 and older in the US and five other industrialized nations. [28]

Improved medical care, not a decline in cancer rates, has kept cancer mortality rates from jumping dramatically. "The extent to which mortality rates can obscure trends in the incidence of cancer is clearly and tragically demonstrated by childhood cancer statistics," states author David Steinman. According to the National Cancer Institute, deaths from childhood cancers decreased between 1973 and 1987. Yet between 1950 and 1988, the incidence of childhood cancers among whites increased 21.3 percent.[29]

Notwithstanding the comforting assurances from industry front groups, most independent scientists recognize that cancer rates are increasing and that industrial chemicals play a critical role in this increase. The National Cancer Institute itself recognizes asbestos, benzene, arsenic, aromatic amines, coal tars, vinyl chloride, chromium and wood dust as carcinogens.[30] A growing body of scientific evidence links pesticides to escalating rates of certain kinds of cancer in farmers.[31] "We are just beginning to understand the full range of health effects resulting from the exposure to occupational and environmental agents and factors," admits a recent NCI report. "Lack of appreciation of the potential hazards of environmental and food source contaminants, and laws, policies and regulations protecting and promoting tobacco use, worsen the cancer problem and drive up health care costs."[32]

Revolving Doors

Media critics note that the media habitually fails to report on itself; it also fails to report on the PR industry. To do so would reveal the extent of its dependency on PR for access, sources, quotes, stories and ideas. According to authors Jeff and Marie Blyskal, "the press has grown frighteningly dependent on public relations people. Outsiders — the reading and viewing public — would have a hard time discovering this on their own because the dependence on PR is part of behind-the-scenes press functioning... Meanwhile, like an alcoholic who can't believe he has a drinking problem, members of the press are too close to their own addiction to PR to realize there is anything wrong. In fact, the press, which has a seemingly inborn cynical, arrogant, down-the-nose view of public relations, seems sadly self-deceptive about the true press/PR relationship."[33]

Canned news and industry-supplied "experts" are effective because they appeal to budget-conscious news organizations. When a TV news show airs a video news release, the PR firm that produced the segment pays for all the costs of scripting, filming and editing. Likewise, PR-supplied experts enable reporters to produce authentic-sounding stories with a minimum of time and effort. The public rarely notices the self-serving bias that creeps into the news along with these subtle subsidies.

Sometimes the financial pressures that influence the news are more direct. In Canada, PR giant Burson-Marsteller's work for the British Columbia timber industry became the subject of investigation by Ben Parfitt, forestry reporter for the daily *Vancouver Sun*. In 1991, however, Burson-Marsteller picked up the *Sun* as a client, and editorial policy shifted. Before Burson-Marsteller went to work for the *Sun*, the paper employed five full-time reporters to cover forestry, fisheries, native affairs, energy and mines, and environment. Today only the environment position remains, and the reporter on that beat has been instructed to cover environmental issues in Greater Vancouver and the lower mainland, an area which is conveniently distant from the Clayoquot Sound, where Burson-Marsteller is helping fell one of the last large areas of intact coastal temperate rainforest in the world.[34]

Parfitt sold an article to another publication, *The Georgia Straight*, that discussed Burson-Marsteller's past history, such as the company's PR work to clean up Argentina's international image at a time when the Argentinian military was murdering thousands of political dissidents. Parfitt also reported that Ken Rietz, a senior Burson- Marsteller employee and timber industry consultant, was a key Watergate conspirator. Following the publication of Parfitt's article, the *Sun* pulled him from the forest beat. "My personal experience in trying to cut through Burson-Marsteller was not greeted with favor by the paper," he says.[35]

Corporate advertisers have enormous power to influence news coverage, despite editors' statements to the contrary. Large corporations pump $100 billion per year in advertising dollars into the coffers of the US media alone. Ben Bagdikian points out that "selecting news in order to make advertising more effective is becoming so common that it has achieved the status of scientific precision and publishing wisdom." PR executive Robert Dilenschneider admits that "the notion that business and editorial decisions in the press and media are totally separate is largely a myth."[36] Mergers, buyouts and new electronic technologies are all hastening the crumbling of walls that supposedly separate news reporting, advertising, and PR. Two of the biggest global PR firms, Burson-Marsteller and Hill & Knowlton, are owned by two of the biggest advertising conglomerates, respectively Young & Rubicam and the WPP Group. These two PR/advertising giants purchase billions of dollars of media print space, TV and radio time. Their clients include Philip Morris, McDonald's, Ford Motor Company, Johnson & Johnson, AT&T, Pepsi, Coca-Cola, NutraSweet, Revlon, Reebok, and hundreds of other major advertisers.

In 1992 the nonprofit Center for the Study of Commercialism invited some 200 journalists to a Washington, DC, news conference where the Center released a report titled *Dictating Content: How Advertising Pressure Can Corrupt a Free Press*. The scholarly report documented dozens of instances of media self-censorship "imposed by advertisers and advertising-related pressures." Almost none of the invited journalists came to the news conference, and the report was virtually ignored in the news, prompting "Project Censored," a media watchdog project of Sonoma State College, to name *Dictating Content* as one of the ten "best censored" stories of 1992.[37]

Corporations have found that one good way to curry favors with the media is to court individual journalists who have become media celebrities, offering them large sums of money for a brief appearance and talk. During the 1993-94 debate over health care reform, the *National Journal* reported that drug companies and trade associations were "practically throwing money at journalists to get them to speak at their events." Media figures including Fred Barnes of the *New Republic*, Eleanor Clift and Jane Bryant Quinn of *Newsweek*, Dr. Bob Arnot of CBS and Dr. Art Ulene of ABC collected speaking fees ranging up to $25,000.[38] *The Political Finance & Lobby Reporter* noted in June 1995 that "*ABC News'* Cokie Roberts accepted a $35,000 fee for a speech last May to the Junior League of Greater Fort Lauderdale that was subsidized by JM Family Enterprises, a privately-held $4.2 billion company that distributes Toyotas.

Roberts refused to discuss her speaking fee. "She feels strongly that it's not something that in any way, shape or form should be discussed in public," ABC spokeswoman Eileen Murphy said when *American Journalism Review* reporter Alicia Shephard requested an interview.[39]

Most reporters, of course, never achieve the celebrity status that enables them to cash in on the speaking circuit. Corporate conglomeration and "downsizing" has brought hard times to the newsroom. Many journalists find themselves forced out of their chosen profession when they enter their thirties and find it difficult to support a family, save for retirement and fund college for the kids on a reporter's limited salary. They see former schoolmates and colleagues leaving journalism to earn more money in public relations, and the original dream of becoming another Woodward or Bernstein begins to look naïve and ridiculous. "The revolving door also contributes to the blurred reality projected by the powerhouse PR firms," writes Vermont newspaper reporter John Dillon. "This door not only spins between the government and lobbies but between the press corps and the PR firms. Like Capitol Hill aides who trade in their access and expertise for a lobbyist's salary, burned-out or broke reporters can be tempted by the greener and more lucrative pastures offered by PR companies."[40]

According to author Susan Trento, this revolving door — and the collaboration it fosters among elite groups in Washington — account for much of the gridlock in America's political process. "Nothing seems to change. Nothing seems to get done. Nothing seems to get cleaned up. From Watergate to Koreagate to Debategate to the HUD scandals to BCCI, it seems that the same people are doing the same things over and over, and never getting punished — and no one seems to care. The triangle — the media, the government, and the lobbying and PR firms — protect each other."[41]

The Information Superhypeway

The news media is presently undergoing a technological transformation as the "information superhighway" enters the mainstream of American culture. Its backbone, the internet, began as a military communications system and evolved into an inexpensive, government-subsidized melange of arcane computerese and loosely-organized data on obscure academic topics ranging from bee migration patterns in Brazil to verb valence structures in Old Saxon. As computer technology brings a user-friendlier version of the internet to a wider spectrum of users, it has become an object of intense corporate interest.

Hyped as the ultimate in "electronic democracy," the information superhighway will supposedly offer "a global cornucopia of programming" offering instant, inexpensive access to nearly infinite libraries of data, educational material and entertainment. In some circles, the hype surrounding the information age has reached evangelical proportions, as its enthusiasts predict a revolutionary new utopian era in which "the technologies of communication will serve to enlarge human freedom everywhere, to create inevitably a counsel of the people."[42]

Other observers see darker possibilities on the horizon, and point out that similar hype surrounded the introduction of older media technologies such as the telephone, radio and television. Given that a handful of corporations now control most media, media historian Robert McChesney writes that it is "no surprise that the private sector, with its immense resources, has seized the initiative and is commercializing cyberspace at a spectacular rate, effectively transforming it into a giant shopping mall." He predicts "a flurry of competition followed by the establishment of a stable oligopoly dominated by a handful of enormous firms... a world of information-haves and have-nots, thereby exacerbating our society's already considerable social and economic inequality. "[43]

PR firms are jumping on the online bandwagon, establishing "world wide web" sites and using surveys and games to gather marketing and opinion information about the users of cyberspace, and developing new techniques to target and reach reporters and other online users.

The information superhighway is only one of the technologies enabling PR firms to "reach audiences more directly and efficiently than ever before," writes Kirk Hallahan in the Summer 1994 *Public Relations Quarterly*. "Today, with many more options available, PR professionals are much less dependent upon mass media for publicity... In the decade ahead, the largest American corporations could underwrite entire, sponsored channels. Organizations such as Procter & Gamble might circumvent public media altogether and subsidize programming that combines promotional and otherwise conducive message news, talk shows, infomercials, or sponsored entertainment or sports... Shows such as 'Entertainment Tonight' stand to become the prototype for programming of tomorrow, in which the source doubles as the deliverer of the message... Channel sponsors will be able to reach the coveted user with a highly tailored message over which they exert control."

Ironically, Hallahan worries that the growing interpenetration of news and advertising is "troublesome" because it weakens the credibility of the traditional news media. "Every time that a newspaper produces an advertorial section that offers free puff pieces to advertisers," he writes, "and every time that a television station presents an infomercial in the guise of programming... media organizations cheapen the value of their product... When a news medium covered a story in the past, the information sponsor gained more than mere exposure. The client, product or cause gained salience, stature and legitimacy." That legitimacy will be lost, he warns, if the public ceases to see a difference between news and paid propaganda." While PR people might circumvent the press occasionally, we aren't going to want to do so all the time," Hallahan writes. "We can't kill the goose that laid the golden egg. A loss of public reliance upon and confidence in the mass media could be devastating."[44]

RECOMMENDED READING:

Cutlip, Scott (1994) The *Unseen Power: Public Relations: A History.* L. Erlbaum Associates.

Dilenschneider, Robert L. (1990) *Power and Influence: Mastering the Art of Persuasion.* New York: Prentice-Hall.

Reece, Ray (1979)*The Sun Betrayed: A Study of the Corporate Seizure of Solar Energy Development.* Boston: South End Press.

LEGALIZED EXTORTION
The "Takings" Threat to People, Property & the Environment

Glenn P. Sugameli[1]

"For greed, all nature is not enough." **Seneca**

What do the following have in common: (1) a county's denial of a permit to operate a hazardous waste facility; (2) federal enforcement to prevent a coal company from continuing to undermine and collapse occupied homes; (3) state enforcement of hunting licenses and bag limits; and (4) a judicial order to remove a fence that caused the starvation of pronghorn? They all have been unsuccessfully challenged in lawsuits alleging that the actions constituted a Constitutional "taking" of private property.[2] Moreover, powerful special interests are pushing state and federal takings bills and litigation that could reverse the results of the court decisions rejecting these claims by drastically altering the nature of property rights.

The Fifth Amendment Takings Clause requires that just compensation be available through the courts when private property is taken for public use; it states, "nor shall private property be taken for public use, without just compensation." However, as John McCoy wrote in the April 1995 *Field & Stream*, "today's `property-rights' advocates want `takings' to be defined much more broadly, especially in regard to government-imposed regulations -- in essence, they want to have the right to sue any agency that forces them to do anything they don't want to do."

The value of property is affected by the enactment, amendment and enforcement of virtually any law: from Antitrust, Bankruptcy, Copyright, Drug, Environmental, Food safety, through the alphabet all the way to Zoning. This fact explains why, in deciding what is a taking, the Supreme Court has consistently used a balanced approach that builds on Justice Oliver Wendell Holmes' warning that "[g]overnment hardly could go on if to some extent values incident to property could not be diminished without paying for every such change in the general law."[3]

The Constitution requires payment of just compensation for "physical takings," where the government builds a road or a dam that physically appropriates or invades private property. Justice Holmes issued his warning in the 1922 Supreme Court majority opinion that extended this constitutional requirement to "regulatory takings," rare cases where government regulation essentially eliminates the use and value of private property so that it can be considered the equivalent of a physical taking. Even in those unusual cases, the Supreme Court has stressed the need to look at other factors, such as what the property owner reasonably expected and whether the government action prevents a public or private nuisance that would harm the rights of other people.

Environmental, zoning, natural resource and other laws that protect against threats to health, safety, and private property are too popular to repeal on the merits. As a result, special interests have launched major Trojan Horse, back-door efforts to undermine or repeal indirectly these safeguards through lawsuits and proposed legislation that seek to alter drastically the very nature of "property rights."

Under the guise of protecting property rights, the takings assault actually would undermine the private property and other rights of the vast majority of Americans. Developers and corporations would for the first time be given a new right to profit at the expense of neighboring people and property.

If successful, such efforts could also radically redefine and enshrine special, unjustifiable private privileges in public resources. By tranforming these privileges into new private property rights, takings bills could insulate from reform such practices as below-cost timber sales, public land grazing fees, and agribusiness irrigation water; subsidized federal flood and barrier island insurance; and *Mining Law of 1872* claims that allow companies to acquire billions of dollars of publicly owned gold and silver for $2.50 an acre. As Professor Richard J. Lazarus testified:

> [T]he property rights bills ignore the significant rights that the public has to many natural resources, such as air, oceans, and wildlife, that have long been considered not susceptible to private ownership... government must be able to protect the public's rights to its natural resources. The government need not allow individual property owners unilaterally to convert public rights into private property.[4]

For example, takings bills could essentially reverse decisions by courts that reject lawsuits arguing a taking of alleged private property rights in public land grazing permits,[5] and claims under the *Mining Law of 1872*.[6] Senator Russ Feingold (D-WI) detailed how, "As the author of legislation to reduce Federal spending on water subsidies, I am especially concerned with provision[s] of S. 605 [an unsuccessful *Contract With America* takings bill] that could expand the rights of agricultural water users at considerable cost to the taxpayer."[7]

The takings movement is likely to continue despite the fact that grassroots opposition has managed to deflect or defeat most of the major "takings" assaults. Special interests recognize the difficulty of promoting the alternative—outright repeal of anti-pollution and other laws. Companies can more easily make the erroneous claim that pollution and other laws result in a taking of their property rights that requires payment or special procedures. For example, as discussed below, coal mining companies that want to open up National Parks to strip mining and which claim the right to undermine (and collapse) occupied homes have repeatedly turned to takings litigation and legislation rather than directly seeking to repeal the *Surface Mining Control and Reclamation Act of 1977* (SMCRA) provisions that protect sensitive public and private resources.

Imagine living in a world where a sweeping form of "piracy," a kind of extortion, has been legalized. Corporations are free to ignore environmental, natural resource, public health, worker safety, zoning, civil rights, and other laws— unless they are paid to comply. Your rights to drinkable water, to the peaceable enjoyment of your home, rights that you take for granted, no longer exist. Long-established local, state and national safeguards all suddenly do not apply, unless taxpayers agree to write blank checks to companies and developers. A kind of legalized blackmail creates perverse incentives for developers to propose extremely destructive land uses solely to force neighboring citizens and communities to pay to protect public and private downstream and downwind property and resources.[8]

Recently, special interest groups have launched massive lobbying and litigation campaigns, fueled by misinformation, that would create just such a world. State and federal "takings" bills and lawsuits would radically redefine property rights to allow companies to profit at the expense of their neighbors' property, health, safety and civil rights. Taxpayers or individual ordinary citizens would be forced to pay to prevent corporations from polluting the local water supply, fouling the air, collapsing homes from underground mining, and ruining the property value and moral values of communities.

Under a radical "takings" agenda being pushed by the misnamed "property rights" movement, the rights of neighboring property owners and entire local communities could be held hostage by big development interests, and subjected to a hefty "ransom." Directly or indirectly, many major takings bills would force U.S. taxpayers to pay companies to comply with anti-pollution and other laws. These radical proposals threaten to undermine actions at all levels of government that protect health, property, and the environment.

As Senator Patrick Leahy (D-VT) stated in successfully opposing a *Contract With America* takings bill:

> The clean air laws say that a polluter cannot use his property to cause a child to get asthma. The occupational health statutes say that an employer does not have a right to use his property in a way that injures or kills his employees. The labor laws say that an employer does not have the right to use his property to exploit children. (Parenthetically, the opponents of child labor laws claimed they interfered with the private property of the mill owners.) Wetland laws say that you cannot use your land to flood my land or lower the water table and dry your neighbor's well. Many of the so-called property rights bills disagree with this premise of our legal heritage. Their premise is that a citizen must be *paid* not to use his property in a way that injures his neighbor.[9]

Takings Lawsuits Attempt to "Rewrite" the Constitution

The takings assault includes well-financed litigation brought by industry that seeks to convince federal and state courts to reinterpret the Takings Clause of the U.S. Constitution and similar provisions in every state constitution. To the extent these industry claims succeed, legislation cannot reverse adverse decisions; only later court decisions or constitutional amendments can change judicial interpretations of a Constitutional provision.

At least 12 non-profit legal foundations, with a combined annual budget of more than $15 million, litigate takings cases on behalf of corporate and other property owners. The Pacific Legal Foundation alone, with offices in 5 states and a budget of over $4 million, is currently involved in dozens of state and federal takings cases. Scores of law firms bolster these efforts by donating "pro bono" legal services to these foundations, while the National Association of Home Builders and other major industry trade associations file briefs supporting many takings claims.[10]

Takings lawsuits have sought radical revisions that are more extreme than those in even the most outlandish major takings bills. For example, takings bills typically include a modest exception that does not require payments to prevent uses of land that would cause nuisances. Judge-made "common law" defines nuisances as activities that unreasonably interfere with public rights or that invade the property rights of neighboring landowners. In the 1992 United States Supreme Court case *Lucas v. South Carolina Coastal Council*,[11] the plaintiff, David H. Lucas, and "friend of the court" amicus briefs by the mining and timber industries and the American Farm Bureau Federation, argued that there is no "nuisance exception" to the Fifth Amendment requirement that private property not be taken without just compensation.[12] This new right to cause a nuisance would have drastically expanded the profits of developers and companies at the expense of the property rights of everyone else. Neighboring homeowners and communities would either have had to allow a wide variety of nuisances or pay to prevent them. The Supreme Court was badly divided on other issues in the *Lucas* case: Justice Scalia wrote an opinion for a bare, five Justice majority of the Court; Justice Kennedy wrote an opinion concurring in the judgment; Justices Blackmun and Stevens wrote dissenting opinions; and Justice Souter filed a separate statement. The Court unanimously rejected the nuisance argument, however, with Justice Scalia's majority opinion holding that it can never be a taking to prevent a nuisance, because laws cannot take away property rights that never existed, such as the non-existent "right" to cause a nuisance.

In other cases, takings advocates have achieved at least initial success in convincing trial or intermediate appellate courts to issue aberrational decisions. These lawsuits have required an organized response by those who support the Supreme Court's consistent interpretation of the Fifth Amendment Takings Clause as requiring an approach that balances everyone's rights. For example, the National Wildlife Federation (NWF) filed several friend of the court briefs that supported successful appeals of very dangerous lower state court takings decisions that threatened protections for critical resources.[13]

Origin of Takings Bills

The origin and history of the takings movement indicate that it was designed to undermine environmental and other protections. Former Solicitor General Charles Fried (1985-89) has described the Reagan Administration Justice Department's

> specific, aggressive, and, it seemed to me, quite radical project...to use the Takings Clause of the Fifth Amendment as a severe brake upon federal and state regulation of business and property. The grand plan was to make government pay compensation as for a taking every time its regulations impinged too severely on a property right... there would be, to say the least, much less regulation.[14]

Government compensation for actual takings of property is required by the Fifth Amendment. The "radical" agenda of the Reagan Justice Department, however, was to require compensation or to force repeal of regulations, not only for actual takings, but in thousands of other instances "as for a taking." President Reagan's Executive Order 12,630 embodied this agenda. The Order requires that all federal regulations be approved under a takings test which, as the Congressional Research Service and others have demonstrated, severely misrepresents Supreme Court rulings.[15]

Types of Takings Bills

Until recently, takings bills have generally assumed two forms: so-called "compensation" or "payment" bills, and "assessment" bills. As described below, payment bills would use taxpayer funds to pay those who claim a partial loss in property value due to government regulation, but who have not lost any property rights according to every Justice of the Supreme Court.

"Assessment" Bills

Takings assessment bills are typically based on an impossible premise. They require that each regulating agency perform costly studies to make a takings determination and a dollar takings estimate when any regulation is proposed. Such a requirement is inconsistent with Supreme Court rulings and would generate costly and obstructive red tape that can block or delay needed protections for people and property.[16]

Even Roger Marzulla, who authored President Reagan's Executive Order on Takings which is the source of the concept of takings assessment or "liability planning" bills, admits that these bills do not yield meaningful information about potential liability. (Roger Marzulla was Assistant

Attorney General, Land and Natural Resources Division from 1987-1989, and is now Chairman of Defenders of Property Rights). He has recognized that:

> Planning bills do have a serious weakness, however. As Maryland [deputy] Attorney General Ralph S. Tyler points out, "no meaningful analysis can be done" of the liability at stake in a taking when so much depends not just "upon the particular circumstances" of the case, but on the philosophy of the particular judge hearing the case... When judges take this *ad hoc* approach to takings law, liability planning becomes a shot in the dark.[17]

While such efforts cannot yield any useful estimates of potential government liability or the number or cost of takings of private property, they are not harmless. These efforts can incur high costs in time, effort and expense and can function, intentionally or not, to delay or block implementation of laws that protect people, property and communities.

Payment Bills

Proposed takings payment bills would create an unlimited new entitlement for those who have not lost any property rights according to every Supreme Court Justice. Thus, payment bills would pay developers, factory owners and others to obey laws that protect the rights of all Americans. In this sense, these bills are "payoff" bills that would create a system of legalized extortion (legalized privateering as opposed to illegal piracy).

The most prominent takings payment bill originated in 1995 as a relatively obscure part of one of the *Contract With America* bills promoted by the new Republican majority in the House of Representatives. In the early, heady days of the 104th Congress, the Republican leadership rushed it through the House, which passed it on March 3, 1995, over bi-partisan opposition.[18] As approved, it would have created a new entitlement requiring payments whenever federal wetlands, endangered species and certain water laws diminished the value of any portion of land or water rights by 20 percent or more.

On March 23, 1995, Senate Majority Leader and Republican Presidential candidate Bob Dole (R-KS) introduced S. 605, an even more sweeping Senate takings payment and assessment bill. It covered all federal laws, all types of property, actions or inactions by federal agencies, and actions by state agencies carrying out regulatory programs. The bill would have required federal taxpayers to pay for any covered regulatory action that reduced by one-third or more the speculative value of any affected portion of a real, personal or intangible piece of property. President Clinton promised to veto S. 605 or any similar legislation, in a December 13, 1995 letter to the Senate Judiciary Committee: "S. 605 does not protect legitimate private property rights. The bill instead creates a system of rewards for the least responsible and potentially most dangerous uses of property. It would effectively block implementation and enforcement of existing laws protecting public health, safety, and the environment."[19]

The high point for payment bills occurred on December 21, 1995, when the Senate Judiciary Committee, under heavy pressure from Senator Dole, voted S. 605 out of Committee. As described below, payment bills were largely discredited by widespread bi-partisan opposition to the Senate bill. As a result, Majority Leader Dole repeatedly failed to bring his bill to the Senate Floor. After Dole's resignation from the Senate, the new Majority Leader, Sen. Trent Lott (R-MS), who was a cosponsor, also failed to bring it up.

As more than 370 law professors wrote Congress regarding the takings test in the House and Senate bills: "Not only has the Court never adopted that radical view of the Fifth Amendment; no single past or present Justice on the Court has."[20] In 1993, the Supreme Court's *Concrete Pipe* decision unanimously reaffirmed the Court's long-standing rejection of three premises at the heart of these and other takings bills. The Court ruled that because regulatory takings decisions must consider many factors, including impacts on neighboring homeowners and the public, "our cases have long established that mere diminution in the value of property, however serious, is insufficient to demonstrate a taking."[21] Second, the Court reaffirmed that takings analysis must focus on the overall property, not just the affected portion.[22] Third, the Court reiterated the importance of looking at specific facts, including what the property owner reasonably expected.[23]

In contrast, takings bills like S. 605 require payments when there is: (1) a specific diminution in the value of (2) any affected portion of property, (3) regardless of reasonable expectations or other factors. This radical redefinition of takings fails to consider impacts on other people and property. S. 605 would have extended this new entitlement to impacts of a broad sweep of federal actions on any property, including contract rights and other intangible property.

Enactment of S. 605 would have caused a flood of costly litigation. Looking at only the "affected portion" would trigger claims whenever erosion or flood prevention laws affect one acre of streamside buffer or floodplain out of a 10,000 acre development or whenever worker safety rules affect one machine out of a thousand in a factory. As Joseph L. Sax, Counselor to Interior Secretary Bruce Babbitt, testified during the Senate Judiciary Committee hearings on S. 605:

> [A]nybody who thinks when you pass a law that says you can be compensated by the Federal taxpayers when your property is reduced, any affected portion of your property, is reduced by 33 percent, thinks that isn't going to create a great burgeoning of lawsuits must be smoking something pretty strong.[24]

Voters have consistently and overwhelmingly rejected legislatively approved takings bills in statewide referenda. By the same 60-40% margin, voters repealed a Washington State takings payment bill in November, 1995, and an Arizona takings impact assessment bill in November, 1994.

(In each state, takings bill supporters outspent opponents by 2-to-1).[25] Takings bills prompted bi-partisan vetoes in 1996 from western governors (Gov. Philip E. Batt (R-ID) vetoed a takings bill that was limited to billboards and Gov. Roy Romer (D-CO) vetoed a takings assessment bill).

Supporters of takings recognize that the American people oppose these bills: the *Seattle Times* reported that "R.J. Smith of the conservative Competitive Enterprise Institute, a Washington, D.C. think tank, said the defeats in Washington and Arizona may have taught another lesson—that property rights leaders shouldn't take the issue directly to voters through initiative or referendum."[26]

Grassroots opposition to takings bills reflects the fact that these bills would force taxpayers either to give up needed protections or to pay billions of dollars to maintain health, safety and other measures that do not take any property.

Takings Bills Threaten Communities and Private Property Rights

Takings advocates would have you believe that environmental protection, zoning, and a wide variety of other laws endanger property rights. They base their "solution" to this falsely perceived problem on two unfounded claims. First, they claim that proposed, radical new "takings" laws would protect private property. Second, they claim that such laws are needed to codify and implement the Constitution's Fifth Amendment Takings Clause: "nor shall private property be taken for public use, without just compensation."[27] Neither claim is true.

In fact, takings bills would harm property and other rights of average Americans *because* they would radically redefine property rights by using standards that are contrary to the Fifth Amendment's balanced approach.[28] Congress could constitutionally require payments to companies and developers who have not lost any property rights under established Supreme Court decisions. The result, however, should be recognized for what it would be— a new entitlement that would impose massive costs on taxpayers, trigger a litigation explosion, generate more bureaucracy and cause an inability to enforce protections for people, private property and public resources.[29]

Takings bills would create a new right in developers, factories and others to be bad neighbors: extracting profits at the expense of nearby people, property, and the environment. As Senator Patrick Leahy (D-VT) said in opposing a 1997 takings bill: "All of us live downstream, downwind, or next door to property where pollution or unsuitable activities can harm our health, safety or property values."[30]

Since 1990, the author[31] and others at the National Wildlife Federation (NWF),[32] the nation's largest membership-supported conservation organization, have been in the forefront of the broad coalition opposed to state and federal takings bills. NWF and others who oppose takings bills are the genuine private property protection movement, not the self-styled "property rights" advocates. NWF strongly supports the Fifth Amendment's balanced protection of private property. If a court determines that a government limit on the use and value of private property goes so far as to be a taking, just compensation must be paid.

NWF opposes takings bills because they threaten a wide range of protections of private property, people and public resources which do *not* take private property rights. As discussed below, takings bills would delay, block, or be so prohibitively expensive as to force repeal of, these protections.

Takings bills are a reaction to environmental, zoning, health, safety and other laws that restrain inappropriate uses of private property.[33] Legislators enacted environmental and zoning laws in part to protect against threats to public health, safety, and private property.[34] Examples abound. Soil conservation laws enacted in response to the massive loss of farms during the Midwest Dust Bowl of the 1930s prevent disastrous impacts on land, water, and surrounding development. Hazardous waste laws protect homes from future Love Canal toxic disasters that can force evacuation of entire neighborhoods. Coastal zone laws protect sensitive shoreline areas from erosion and prevent hazardous coastal development; watershed protection laws preserve fresh water resources. Anti-pollution laws protect the value of homes, businesses and entire communities by safeguarding the air we breathe, the water we drink, and the land on which we live, work and play. Zoning laws enable us to ensure that neighboring land uses are compatible, prohibiting inappropriate developments like a factory or slaughterhouse in a residential neighborhood or an adult business next to an elementary school.

Takings bills pose a real threat to all of these safeguards. Republican State Senator Richard Russman of New Hampshire detailed these concerns in his testimony on behalf of the National Conference of State Legislatures before a U.S. House Judiciary subcommittee.

> The potential impact of takings bills is dramatically illustrated by looking at how they could reverse the results of actual takings claims that have been rejected by the courts under the Constitution's balanced approach. For example, proposed state bills could require local governments and states to abandon needed protections or to pay unsuccessful takings plaintiffs in the following cases:
>
> • proprietors who challenged city restrictions in Tampa, Florida, and Mobile, Alabama, on topless and exotic dancing;[35]
>
> • a chemical company that challenged Guildford County, North Carolina's denial of a permit to operate a hazardous waste facility;[36]
>
> • a landfill operator who contested a county's health and safety ordinance prohibiting the construction of additional landfills;[37]
>
> • a tavern owner in Arkansas who claimed state sobriety highway checkpoints for drunk driving reduced his business;[38]

- an outdoor advertising company that challenged a Durham, North Carolina ordinance that limited the number of billboards in order to preserve the character of the city;[39]

- an existing sand/gravel mine that challenged a Town of Hempstead, New York ordinance prohibiting excavation within two feet of the groundwater table that supplied water for the town;[40]

- a Minneapolis sauna, in which four prostitution arrests had occurred, that contested the city's order to shut down for one year;[41]

- a nightclub company that challenged Philadelphia's denial of a license to operate an all-night dance hall;[42]

- companies contesting denials of state mining permits for small portions of farms that contained historic graves and archaeological artifacts (there was no limit on continued farming) in Iowa and Indiana;[43] and

- operators of "to-go beer windows" who challenged Austin, Texas restrictions on consumption of alcohol "in or on any public street, sidewalk, or pedestrian way" in designated areas of the city.[44]

Some of the main targets of the takings movement have been mining and wetland laws that protect homes and other private property. Passage of the *Surface Mining Control and Reclamation Act of 1977* (SMCRA) was prompted by the 1972 collapse of a massive coal company waste refuse pile which had dammed a stream.[45] The resulting 20-to-30-foot tidal wave killed over 125 people, devastated sixteen communities and destroyed 1,000 homes.[46]

Subsequently, coal companies have filed repeated, unsuccessful lawsuits claiming that they have a constitutionally protected property right to destroy homes through strip mining and underground mining.[47] These included:

- a sweeping takings claim against the SMCRA provision that protects homes and other sensitive areas by prohibiting coal mining in backyard, schoolyard, churchyard and graveyard buffer zones. The Supreme Court rejected the claim that the law had taken the coal that could not be mined. The Court subsequently rejected another takings claim against a state law requiring that 50% of the coal be left in place under protected homes to prevent collapse and "to protect the public interest in health, the environment, and the fiscal integrity of the area;"[48]

- even more recently, the M & J Coal Company in West Virginia removed so much coal from an underground mine that huge cracks opened on the surface of the land, rupturing gas lines, collapsing a stretch of highway, and destroying homes. When the federal Interior Department required M & J Coal to reduce the amount of coal it was mining to protect property and public safety, the company sued. The court rejected M & J Coal's claim that, despite the company's 34.5 % annual profit, mining regulations had "taken" its property.[49]

Takings lawsuits and bills attempt to reverse the common-sense, balanced, Constitutional results of these cases in favor of a development-at-any-cost agenda. These bills would delay, block, or be prohibitively expensive so as to force repeal of mining, wetland and other laws that protect both these valuable resources and private property.

Wetlands protect private property from pollution and flood damage. Wetland ecosystems filter contaminants from the water moving through them and slow the flow of water to allow sediments to settle out. They are a form of natural pollution control, providing an economic benefit to downstream property owners. By absorbing excess water, wetlands reduce flood damage and protect coastal properties from storm damage.[50]

Mythical Takings -- No Need for Takings Bills

Much of the pro-takings bill rhetoric consists of argument by mythical anecdote. Proponents of takings bills have repeatedly failed to substantiate their assertions of widespread takings.[51] This is especially true of assertions that widespread takings of property rights result from conservation and other laws that protect private property, public health and safety by limiting destruction of wetlands, floodplains and other natural resources.

For example, in the 1996 Vice Presidential candidate debate, Republican nominee Jack Kemp spoke vaguely of an unnamed Oregon farmer who allegedly had voluntarily declared 25 percent of his property to be wetlands and then found that:

> [T]he bald eagle began to use it as a habitat. The Corps of Engineers, the Bureau of Wildlife and Fisheries [sic], all of the federal agencies came onto his property, declared it a federal wetland and said he couldn't drive. They took away the road. He couldn't mend his fences. And they wouldn't pay the value of the loss...[52]

Attempts to locate the source of this story were unsuccessful.[53] Anecdotes such as the one related by Jack Kemp appear to be versions of what a court described in a different context as:

> the "urban legends" that Jan Harold Brunvand writes about in "The Vanishing Hitchhiker," and "Curses! Broiled Again!" (1989). According to Brunvand, a distinguishing feature of an urban legend is that no one is ever able to produce an eyewitness to the actual event -- only a "friend of a friend." Another is that the stories crop up over and over again, in many different settings.[54]

In fact, wetlands, species protection and other conservation laws are extremely unlikely to take private property. For example, in the over

20 year history of the *Endangered Species Act* (ESA), courts have only decided four Fifth Amendment takings cases on the merits, all of which have found that the ESA did not take private property.[55] In Fiscal Year 1995, the U.S. Army Corps of Engineers only denied 274 (or 0.5 percent) of the over 62,000 *Clean Water Act* section 404 applications it received for permits to dredge and fill wetlands.[56] Even in the rare cases where permits are denied, there is no limitation on use of upland portions of property or on grazing, agricultural or other uses of wetland areas for which a permit is not required.

Takings Bills Would Favor Irresponsible Developers at the Expense of Homeowners and Established Businesses

Supporters of takings bills focus heavily on fanciful, unsubstantiated tales of small property owners who allegedly suffer from takings. However, As Rep. Sam Farr (D-CA), stated in debate on the *Contract With America* takings bill, H.R. 925:

> Who are the special interests supporting this? The National Mining Association, the Chemical Manufacturers Association, the National Association of Manufacturers, the American Petroleum Institute, the American Independent Refiners Association, American Forest and Paper Association, and International Council of Shopping Centers. Those do not sound like small landowners to me.[57]

Indeed, I have reviewed virtually every federal and state takings bill that has been introduced, not one of which was in any way limited to small property owners or even to real people, as opposed to corporations. The Fifth Amendment Takings Clause, of course, applies to all property owners. Takings bills like H.R. 925, however, would establish a new entitlement for those who have not lost any property rights. Thus, the fact that the bills omit any limitation offers an insight into the intention of the sponsors.

If any such bills were to pass, the vast majority of payments would be to big corporations and developers who are the subject of most legal requirements and who have the lawyers, appraisers and experts necessary to demonstrate a right to payment under the vague standards of these bills.[58] This fact was dramatically illustrated in May 1996, when an Exxon subsidiary filed a lawsuit claiming that the $125 million Exxon Valdez tanker had been taken.[59] The claim challenged a provision of the *Oil Pollution Act of 1990*[60] that allowed the ship to operate anywhere in the world except Prince William Sound, where the Exxon Valdez had spilled 10.6 million gallons of crude oil.[61] Maritrans, Inc. subsequently filed a takings claim for more than $200 million to cover the loss of 37 single-hull tank barges that would be forced from service in 2003 by the double-hull requirements of the same Act.[62]

Payments regarding land would also reflect the highly concentrated nature of land ownership. "[I]f one combines the land holdings of the large farm operators and timber operators, 2.1 million land owners own 1,035 million acres of land. That means that 2.65 percent of all private land owners own 78 percent of all private land. Their size also implies a likely sophistication in dealing with government programs."[63] In contrast, the roughly 60 million owners of residential property own 3 percent of all private land.[64]

Takings bills would benefit developers, and the small minority who own large tracts of land, at the expense of homeowners who depend upon clean air, safe drinking water, zoning and other laws. In fact, most home property values benefit from land use regulations.[65]

In an editorial opposing a Florida takings bill, the *Tampa Tribune* pointed out that:

> Under this measure, a factory could be built beside a retirement village, a massage parlor next to a church, or a night club on a quiet residential street. One irresponsible developer could spoil a neighborhood and ruin property values of its residents without fear of governmental interference. This, apparently, is some legislators' idea of property rights.[66]

Takings Bills Are Budget Busters

Proposed takings legislation would force repeal, or block implementation, of basic protections for people, property, and natural resources by making them too expensive to enforce. The Office of Management and Budget (OMB) concluded that the costs of Senator Dole's takings bill, S. 605, would be "several times the $28 billion [over seven years] of the House-passed legislation."[67]

Taxpayers for Common $ense, a budget watchdog group, issued a May, 1996 report stating that the cost of Senator Dole's takings bill could be $100 billion over seven years, or, more likely, a virtual blank check.[68] Former Senator Paul Tsongas, a strong advocate of a balanced budget, presented very powerful testimony on this issue before the Senate Environment & Public Works Committee.[69] The Senate Judiciary Committee Minority Report on S. 605 cited a study by the University of Washington Institute for Public Policy Management which revealed that Washington State's defeated takings legislation (Referendum 48) could have cost local governments up to $1 billion annually for takings studies alone and exposed them to payments of as much as $11 billion.[70] As I stated at the time, "The only limit to how much this bill [S.605] would cost taxpayers is any limitation on corporate greed and ingenuity."[71]

Because of their impact on the budget, payment bills would thus block enforcement of environmental and conservation laws as well as a wide variety of public health, safety, financial,[72] and civil rights laws.[73] These bills would create a new entitlement requiring blank check payments to corporate and individual property owners. As a result, these bills would compel avoidance of these costs through repeal or non-enforcement of needed protections for people, neighboring property, and public resources. The cost of takings studies mandated by many assessment bills would have a like effect.

National Association of Home Builders Takings Bill Targets Zoning

The power of special interest lobbying was demonstrated by a different kind of takings bill, *The Private Property Rights Implementation Act of 1997*, H.R. 1534. The leading Senate opponent, Senator Patrick Leahy (D-VT) "said, 'In my 23 years [in the Senate] I've rarely seen anything so arrogantly special interest as this,' adding that the bill 'wouldn't pass the smell test in any town in America.'"[74]

There is only one reason this developers' dream bill existed and enjoyed such a remarkable initial success: an all-out campaign by the National Association of Home Builders (NAHB).[75] The *Ventura County Star* reported that "Rep. Elton Gallegly said his bill to give landowners expedited access to federal courts was written for the benefit of 'ordinary landowners,' but, in fact, its author was an attorney for the National Association of Home Builders."[76] *The Baltimore Sun* observed that "the bill served as an object lesson in how special interests can direct legislation from its inception to its conclusion. The bill's principles were drafted by the home builders' lobby, which welcomed its introduction last fall with $173,000 in campaign donations. Grass-roots lobbying pushed it through the House, and a phalanx of lobbyists greeted senators last evening as they rushed to cast their votes."[77]

The NAHB is a special interest group that strongly opposes a range of legal requirements, from federal protections for endangered species and wetlands,[78] to local requirements for fire sprinklers in new homes. For example, "nearly every fire department and firefighters organization in the United States has endorsed this requirement" for sprinklers in new homes, the latest data show that over 80% of all U.S. fire deaths occur in house fires, and the cost of sprinklers is about 1 percent of the total sales price. An NAHB construction codes official, however, reiterated that NAHB "doesn't support any mandatory requirements for sprinklers because they don't believe the amount of money required is a good use of money and it can keep first-time buyers out of houses."[79]

Congressional Quarterly described NAHB's campaign for H.R. 1534 as "a classic tale of Washington influence and how a single association responsible for $295,250 in campaign contributions in the first six months of 1997... mobilized support with a small army of lobbyists..."[80] On September 23, 1997, Senator Paul Coverdell (R-GA) introduced S. 1204, the *Property Owners Access to Justice Act of 1997*, the Senate version of the NAHB bill. Two days later, in the House, a Judiciary subcommittee held the first hearing on H.R. 1534. "The next day the home builders' political action committee distributed $173,000 to federal candidates, the most money the PAC had given out in a single day this election cycle, according to the non-partisan Center for

Responsive Politics."[81] *National Journal* characterized the state and local government opposition to the NAHB bill as a "David v. Goliath" fight in which "local officials are vastly outgunned... The Home Builders political action committee gave $594,250 in campaign contributions last year [1997] alone. It also doled out $90,000 in 'soft money' to the political parties."[82]

The Baltimore Sun related that "[Jerry M.] Howard, the home builders chief lobbyist, denied any connection between campaign contributions and the legislation, noting that any donation predicated on the recipient's position on an issue would be illegal. 'I want to get this bill passed very badly, but not so badly that I'd do something illegal,' he said."[83]

Enactment of the NAHB takings bill would have undercut increasing efforts by state and local governments to manage growth and sprawl. This would have supported political and legislative efforts by developers at the state level. For example, *The Washington Post* reported that "[m]any developers are contributing generously to local county candidates throughout Maryland in an effort to protect their right to build amid growing public concerns over the pace of suburban sprawl... Susan S. Davies, co-director of government affairs for the Home Builders Association of Maryland... said builders are seeking to support those who `understand the role growth plays in terms of the vitality of a county.'"[84]

Two September 17, 1998 press reports indicate the increasing recognition of the need for communities and states to control growth and the key special interest role of home builders in opposing these efforts. *The Philadelphia Inquirer* reported that "Suburban sprawl is Pennsylvania's most pressing environmental problem, a state commission has told Gov. Ridge. The panel urged that local governments be given more power over developers to reduce environmental, social and economic impacts... The home builders' representative on the commission, John M. DiSanto of Harrisburg, was the only one of the 40 members to vote against approval of the final report... The Pennsylvania Builders Association yesterday issued an 11-page critique of the report, arguing that suburban development has been a lifestyle choice, not an environmental hazard." Governor Thomas Ridge (R-PA) created "The 21st Century Environment Commission... to set the state's environmental priorities..."[85] *The Washington Post* reported that "Loudoun and a growing chorus of other fast-growing counties in Virginia are pressing the General Assembly to allow them to shut off residential development in areas where there are not enough schools or roads.... Michael L. Toalson, executive vice president of the Home Builders Association of Virginia, said the industry is mobilizing to oppose the measure... He also said requiring adequate facilities is a bad idea because schools and other facilities follow development -- they don't precede it."[86]

The federal NAHB bill would undermine local zoning and federal environmental protection by radically changing the procedures by which private property takings claims are decided. Contrary to existing law, the bill would allow developers and companies to file premature suits in federal court without first fully using opportunities to resolve land use conflicts at the local level. Federal cases could be filed once a single land use proposal is denied, even if local or federal officials would approve other equally valuable uses that would not violate local zoning and even if the proposal would harm neighboring homeowners. In effect, the bill would reverse established Supreme Court decisions by allowing developers to sue local governments directly in federal court, bypassing local zoning procedures and state courts. The bill would also force cases into federal court without all of the information that the Supreme Court has identified as necessary to decide whether a local, state or federal regulation results in a taking of private property.

The NAHB bill is the first federal takings bill to be aimed at local governments. Because the bill would allow federal court intervention much earlier in the decision making process, it could undermine local control of local development issues.

In a statement reiterating a presidential veto threat, Vice President Al Gore said, "Under H.R. 1534, developers would have little incentive to cooperate with local governments because they could take their case to federal court. As a result, this bill would give developers a new club to wield in their negotiations with local officials and citizens groups --the threat of premature and costly litigation in federal court..." Threatened federal lawsuits could force localities to abandon fundamental safeguards and allow inappropriate development, such as toxic waste dumps, nightclubs, and strip malls in residential neighborhoods.

The Senate Judiciary Committee Minority report on H.R. 1534 stated that:

> As we pointed out in the hearing, the top four residential developers in the country have annual revenues in excess of $1 billion per year. Most of our small towns generate less than $10 million a year in tax revenues. As Mayor Curtis of Ames, IA, testified in the hearing, 90 percent of cities and towns in America have less tha[n] 10,000 people. These towns cannot support even one municipal lawyer, much less the number that would be required to battle billion-dollar developers. Under the threat of battling large corporate developers with deep pockets, more local governments would opt to settle the case at inflated compensation standards or let the development go ahead.[87]

The NAHB Bill Would Put Neighboring Citizens at a Disadvantage

The NAHB bill would allow developers to bypass local administrative appeals, where neighbors can participate without hiring a lawyer, and state courts. As the Senate Minority Report concluded (at 44), this would eliminate the most convenient and inexpensive forums for neighbors who are concerned about a proposal's impacts on their property, health, safety, community and environment.

From a religious perspective, the U.S. Catholic Conference, the National Council of Churches, Evangelical, and Jewish groups have opposed the NAHB bill, in part because it would deny average citizens the equal opportunity to participate in local land-use decisions. In a letter opposing the companion Senate bill, these groups explained that for many people in poor, rural areas, "the nearest federal courts are often a considerable distance away and thus participation in such courts can be a great financial burden.," while "removing the opportunity to participate in local or state administrative procedures... effectively eliminates the easiest point of local citizen access to land-use decisions. Consequently, property owners with sufficient financial resources will be heard— but residents affected by the use of that property who do not have similar financial resources will not."[88]

A Stealth Campaign Temporarily Obscured the Radical Nature of the NAHB Bill

NAHB ran a stealth campaign, avoiding publicity and claiming that the bill was a non-controversial, procedural change. NAHB hired a law firm to help draft the bill, which was introduced without fanfare by Representative Elton Gallegly (R-CA) as the only original sponsor. Intensive office-to-office contacts by NAHB lobbyists and members quietly lined up House sponsors,[89] eventually convincing 239 Members (44 Democrats and 195 Republicans), a majority of the House of Representatives, to co-sponsor the bill.

As the National Wildlife Federation worked with a broad coalition to expose the true nature of the bill, the House Republican leadership rammed the measure through. A September 25, 1997 subcommittee hearing, and votes in the subcommittee, in committee, and on the House Floor, were all squeezed into less than a month. Rep. William Delahunt (D-MA), a strong opponent of the bill, "said that he had 'no doubt' that the bill would have failed had it not been rushed through the House."[90]

The dangerous implications and special interest nature of the NAHB bill were illustrated by statements of those who tried unsuccessfully to amend the bill. First, Rep. Jerrold Nadler (D-NY) described how, in the House Judiciary Committee, he had "offered an amendment to ensure that in cases where public health and safety are involved, the plaintiff cannot circumvent State and local courts to get the Federal courts. And the bill's sponsor rejected it. It appears then that supporters of this bill would deliberately seek to undermine the health and safety of our Nation's communities."[91] Second, Rep. Bruce Vento (D-MN) detailed how the House Rules Committee rejected his request to be allowed to offer a Floor amendment that sought "to balance this bill with adequate protection for the 65 million Americans that own their own homes... It would have prevented this bill from going into effect in States that have not passed laws that protect homeowners' property rights. These laws will have to provide families with adequate notice when adverse development is moving in to affect their property. The intent was to provide homeowners with guaranteed access to the courts when their property is devalued by harmful developments nearby."[92]

NAHB's bill passed the House by a vote of 248-178 on October 22, 1997.[93] Bi-partisan opposition, however, ensured that there were insufficient votes for the two-thirds margin needed to override a threatened Presidential veto.

There was a vast difference between how the bill was portrayed and what it would do. This was reflected in the extraordinary actions of 9 Republican and 4 Democratic House cosponsors who had the integrity to examine the bill in detail and then vote against it on the Floor. For example, after the vote, Rep. John Porter (R-IL) "said the bill was described to him differently when he was asked to co-sponsor it last month. After his staff had a chance to analyze it, Porter said he decided 'it really is not good federal policy and it probably is unconstitutional.' `I think a lot of members were simply told this was a procedural change that made it fairer and easier to have these matters litigated, but they didn't understand the mechanism by which the sponsor was trying to do that,' he said."[94]

In the House of Representatives, the heart of the support for the NAHB bill came from Members who professed to be the strongest advocates of: (1) devolution of power from the Federal government to local governments; (2) Federalism and states' rights; (3) and reining in the power of unelected Federal judges. The bill would undermine every one of these principles, indicating its special interest character. As Rep. Barney Frank (D-MA) argued: "I do not think in recorded parliamentary history there has ever been a greater gap between people's professed principles and what they voted for than here in this bill."[95]

The radical nature of the NAHB bill has generated extensive opposition from all three branches of the federal and state governments.[96] Administrative branch opposition includes President Clinton, Vice President Gore and numerous federal officials, as well as the National Governors Association, and 41 State Attorneys General. Legislative branch opposition includes Members of Congress, and the National Conference of State Legislatures. Judicial branch opposition includes both the Conference of Chief Justices on behalf of the state courts that the bill would bypass, and the Judicial Conference of the United States, chaired by Chief Justice Rehnquist, on behalf of the federal courts that would be inundated with premature takings claims.

Bipartisan opposition includes virtually every organization representing state and local governments, including the National Association of Counties (which had never before opposed a takings bill), the National League of Cities, the U.S. Conference of Mayors, the National Association of Towns and Townships, and the International Municipal Lawyers Association.

The National Wildlife Federation and a broad range of other national, state and local conservation and environmental groups strongly oppose the bill, which would undermine environmental and other safeguards. The coalition in opposition includes: the American Planning Association, National Trust for Historic Preservation, League of Women Voters, and labor organizations, including the United Steelworkers of America and the American Federation of State, County and Municipal Employees (AFSCME).

Major newspaper editorial boards denounced the bill, including *The Washington Post, The Dallas Morning News, The Arizona Daily Star, The News and Observer* (Raleigh, NC), and *The* [Manchester, N.H.] *Union Leader*.[97] *The New York Times* editorial stated that "[The bill] is a dangerous piece of work that would threaten local zoning laws, reshape time-honored principles of federalism and make Federal judges the arbiters of land-use decisions everywhere. It would be a dream come true for developers but a nightmare for rational community planning... [T]he threat of an expensive Federal lawsuit brought by a well-heeled developer could pressure local officials into decisions that help the developer but harm the larger interests of the community."[98]

The near-success of this legislation in the face of such widespread opposition is largely the result of a massive lobbying effort from large development interests, primarily the NAHB.

Related Takings Bill Targets Federal Safeguards

On February 26, 1998, the Senate Judiciary Committee approved a broader bill that combined the NAHB bill with H.R. 992, another "procedural" takings bill. H.R. 992, the *Tucker Act Shuffle Relief Act*, sponsored by Rep. Lamar Smith (R-TX), subsequently passed the House on March 12, 1998, by a vote of 230-180, far less than the 2-1 margin needed to override a threatened Presidential veto.

H.R. 992 would unjustifiably change the rules to promote challenges to federal safeguards for people, property, communities and the environment. It would do this by granting the U.S. Court of Federal Claims (CFC) the power to invalidate federal statutes, regulations and enforcement actions. The highly specialized CFC decides individual takings and other money claims against the United States, but lacks the experience and legal precedents in determining whether federal government regulations are lawful or should be struck down. Both the CFC and (on appeal) the Federal Circuit would be free to ignore prior rulings by other courts that upheld vital site-specific or national safeguards for neighboring homeowners, communities and the environment.

Under H.R. 992, by simply including a takings claim for money in their lawsuit, companies could judge-shop and revive in the CFC a variety of failed legal claims. Even if the CFC rejected the takings claim, it could invalidate federal environmental protections on separate claims that other courts have previously rejected. As a result, H.R. 992 would invite companies to bypass long-settled decisions in their local federal district and appellate court and to re-open a wide range of failed claims by asking the CFC to invalidate agency rules and actions.

Now, neighbors and the community can rest assured if they have successfully helped to defend a federal anti-pollution regulation or decision to deny a permit to destroy wetlands that would increase flooding and harm groundwater. They know that a decision to reject an identical proposal by another company will be upheld by the same court. Under this bill, however, the neighbors and community would be forced to start all over again in the CFC in Washington, D.C. with no precedent on point.

Rep. Melvin Watt (D-NC) and Rep. Sherwood Boehlert (R-NY) led the bi-partisan House Floor opposition to H.R. 992. The Judicial Conference, chaired by Chief Justice Rehnquist, joined NWF and other conservation groups in the ultimately successful opposition to H.R. 992 and similar Senate provisions.

Senate Showdown

In July, 1998, Senate Majority Leader Trent Lott (R-MS) decided to bring to the Senate Floor S. 2271, the *Property Rights Implementation Act of 1998*, sponsored by Judiciary Chairman Orrin Hatch (R-UT). S. 2271 was a slightly revised version of the bill that Sen. Hatch's Committee had approved: both versions combined the two House-passed takings / "property rights" bills. Like H.R. 992, S. 2271 would have allowed polluters to re-open long-settled issues by giving the U.S. Court of Federal Claims the authority to invalidate environmental protections that had been upheld by other federal courts. Like H.R. 1534 (the NAHB bill), S. 2271 would have enabled developers to evade local remedies and state courts and to intimidate towns, cities and counties into approving inappropriate projects such as corporate hog farms in floodplains or residential areas. As the Clinton Administration stated in promising a veto, "The bill would subject local communities to the threat of premature, expensive Federal court litigation that would favor wealthy developers over neighboring property owners and the community at large."[99]

NAHB pushed the bill with last-minute radio ads in ten states, "blast faxes" to 17,000 builders, and "a phalanx of lobbyists [who] greeted senators... as they rushed to cast their votes."[100] These efforts, however, were not enough to overcome grassroots messages from constituents who urged their Senators to oppose this dangerous bill that would have allowed developers to undermine local and federal safeguards for people, property and the environment.

Fortunately, on July 13, 1998, the Senate's 52-42 vote fell well short of the 60 votes needed to limit debate on the issue, stopping the bill for the 105th Congress. Senators Patrick Leahy (D-VT) and John Chafee (R-RI) led the successful bi-partisan Floor opposition.

NAHB has vowed to return with an even stronger effort in 1999 and beyond, so it is important that Representatives and Senators continue to hear from their constituents who oppose this bill.[101]

Real Protections for Property Rights

Individuals and legislators who are truly concerned with protecting property rights should oppose takings bills and support initiatives to protect and empower individual private property owners against threats from pollution and inappropriate nearby development.

Protection of Homeowners

One federal legislative effort to focus on the property rights of average Americans was introduced in 1996 as S. 2070, *The Homeowners Protection and Empowerment Act*, by Sen. Ron Wyden (D-OR) and Sen. John Warner (R-VA). This bill would have established a "homeowner right of access to information" about activities with the potential to reduce the value of citizens' homes. Giving homeowners access to information about developments with potential impact on their property values *before those developments occur* would allow homeowners to take legal action to stop undesirable and environmentally damaging projects. The bill would also have allowed homeowners to sue the holders of federal permits for compensation for lost property value. Defendants would have been strictly liable, if the owners of one or more private homes could demonstrate that the federally authorized activity caused a reduction in the value of their home(s) of $10,000 or more.[102]

Although this bill was not passed, it highlights a basic fact about private property ownership in the United States. Homeowners represent 75% of all landowners in this country. While the total amount of land owned by homeowners is only roughly 3% of private land, these homeowners hold the majority of all private real estate values. Most land use laws benefit this large class of property owners by protecting the value of their homes.[103] Empowering homeowners to protect their property from harmful neighboring development would represent a true victory for private property rights, in contrast to the approach embodied in takings legislation being pushed by so-called property rights advocates.

Preserving Communities

Many communities are recognizing the need for increased safeguards to protect health and safety, the environment, and private property. A host of local and regional planning problems arising from rapid economic and population growth have inspired concern over the impact of continued development. New protections, which generally fall within the zoning and taxing powers of local governments, address questions ranging from open space preservation to pollution prevention to protection of property values.

Preserving livable communities in the midst of urban sprawl is a key issue for areas throughout the United States that are experiencing rapid economic growth and development. In many areas, open space and historic character are important commodities that help attract development in the first place.

Increasingly, communities across the country are recognizing the need for some form of growth management to protect individual property values and the public interest.

Pollution Harms Private Property

Pollution can be a severe threat to private property. A December, 1997 report compiled by the minority staff of the United States Senate

Committee on Agriculture, Nutrition, and Forestry for Senator Tom Harkin (D-IA) illustrates the potential impacts of animal waste pollution on individual health, safety and property.[104] The immense scope and concentration of corporate livestock production in this country has made it increasingly difficult to manage animal waste responsibly. Serious spills of animal waste into waterways have occurred frequently in recent years, resulting in fish kills, nutrient pollution, and in some cases serious threats to human health. Factory farm operations and their associated animal waste can cause problems with odor and ground water infiltration that impact nearby and downstream property owners. This growing problem has prompted action at the local, state and federal level. In many cases, those advocating local control are seeking more stringent protections than are provided by state law, and are working under the legal authority of local zoning and health ordinances.

There has been a veritable explosion of activity in State legislatures to address impacts of animal waste pollution on downstream private property and the environment. Proposals include new regulatory standards, improved inspection, moratoria on the building of new large-scale farms, and mandates to create new zoning plans for such operations. In September, 1998, the Clinton Administration announced a crack-down "on a major source of pollution in rivers and streams by requiring the nation's largest livestock farms to develop plans to store animal waste as a condition of remaining in business... The long-awaited 'national strategy' for managing livestock waste will call for tougher oversight of the nation's increasing number of factory-like animal feedlots, from Eastern Maryland's pervasive poultry farms to warehouse-size hog barns in North Carolina."[105]

Conclusion

Those who truly care about protecting private property rights, as well as our health, safety, communities and environment, will join the National Wildlife Federation and a broad coalition of others in opposing radical takings bills and lawsuits and in supporting measures that continue our nation's tradition of balanced protection for the rights of all Americans.

RECOMMENDED READING:

Echeverria, John (1998)"The Politics of Property Rights." 50 *Oklahoma Law Review* p. 351.

Sugameli, Glenn P. (1996)"Environmentalism: The Real Movement to Protect Property Rights," in Philip D. Brick and R. McGregor Cawley, eds., *A Wolf in the Garden: The Land Rights Movement and the New Environmental Debate*. Lanham, MD: Rowman & Littlefield Publishers, Inc.

MOBILIZING AGAINST ENVIRONMENTALISM
The 'Wise Use' Movement & the Insurgent Right Wing[*]

Tarso Luís Ramos

"We have to struggle with the old enemies of peace: business and financial monopoly, speculation, reckless banking... They have begun to consider the government of the United States as a mere appendage to their own affairs, and we know now that government by organized money is just as dangerous as government by organized mob."

Franklin D. Roosevelt, 1936

"Jobs versus the environment," "people before spotted owls," "protect property rights" and "Clinton's war on the West" are all familiar slogans of the so-called "wise use" movement. This movement, comprised of numerous groups acting individually and in concert, is waging campaigns across the West and the nation to defend the rapacious practices of natural resource industries and to protect the property interests of developers, realtors and other powerful business interests. The movement has successfully tapped corporate coffers to fund and otherwise support pro-industry, anti-environmental campaigns that utilize community organizing and advocacy techniques to advance a conservative, pro-corporate agenda. In its efforts to build a grassroots base, the movement relies on fear, deception and manipulation. By characterizing environmental regulations as a principal cause of job loss and economic dislocation and promoting activism as the only defense against economic ruin and government tyranny, wise use organizers have spawned anti-environmental citizens groups in communities throughout the West and the nation, dominating local politics with an industry agenda and building power at the state and national levels.

Much like the extractive practices of some of their corporate sponsors, wiseuse campaigns leave deep scars on communities. Directing widespread economic insecurity and political powerlessness into anger at an identifiable, if contrived, enemy — the environmental movement and its alleged lackeys in the federal government — wise use has deeply polarized communities, setting them at war with themselves and with governmental institutions. The battles, fought over such perceived

---

[*] First published as "An Environmental Wedge: The 'Wise Use' Movement and the Insurgent Right Wing," in Eric Ward, ed., *Conspiracies: Real Grievances, Paranoia, and Mass Movements*, Peanut Butter Press, Seattle, 1996.

dichotomies as environmental protection versus jobs, distract individuals and communities from the true threats to economic and environmental health: irresponsible corporate practices and the public policies that sanction them. The wise use movement represents a rancid populism, mobilizing communities not to take control of their own destinies, but rather to eliminate any remaining community controls on big business — often enlisting workers and their communities as agents of their own demise.

Drawing strength from some of the same leaders, sentiment, and strategy as the Sagebrush Rebellion,[1] the wise use movement involves a much broader coalition of ideological and economic interests that stand to profit from the deregulation of industry and the weakening of environmental laws than did its predecessor. Wise use has launched campaigns to defend the interests of the mining, grazing and timber industries, open public lands to the private exploitation of resources, defeat controls on real estate development, elect and recall public officials, and eliminate a broad array of regulatory restrictions on business practices, among others. Today wise use leaders boast of some 1500 groups and make the clearly exaggerated claim of three million participants. (This latter figure may bear some relation to the number of individuals receiving the Movement's direct mail.) As the movement grows in strength and breadth, even more is at stake than environmental protections, economic stability, and the public health. Wise use is building a base of support for a range of reactionary policies, and the infusion of its activists into the political process is contributing to a realignment of elected governmental bodies throughout the West, with dire implications.

In the years since its official launch at the 1988 Multiple Use Strategy Conference in Reno, the wise use movement has successfully worked to make environmental protections a highly charged issue around which to mobilize a reactionary political force. Indeed, wise use is a pointedly political movement. The principal financial backers and beneficiaries of wise use campaigns are natural resource and development industries who, as a group, are also the largest political campaign contributors in the West. Western politicians beholden to these industries have long proven to be their faithful representatives in state government and in Congress, and since these industries began backing wise use campaigns, such politicians often have become mouthpieces for or even leaders within the wise use movement. To cite but one example, Senator Larry Craig (R-Idaho) uses wise use rhetoric, advances movement legislation, and even assigns his staff to organize wise use events.[2] As a result of such relationships, the movement is often an inside player in national politics even as it works to foster an underdog image. In the electoral arena, wise use has served as a kind of "Southern strategy" for the West, mobilizing traditionally Democratic blue collar workers and resource communities behind anti-environmental politicians, who, it turns out, are nearly always right-wing Republicans. Wise use was one of the converging forces that delivered the November 1994 congressional putsch, sending such ultraconservatives as Idaho's Helen Chenoweth to Congress for the first time, and has contributed to the political realignment of western state legislative bodies over the last three election cycles.

The dramatic rightward shift of domestic politics, exemplified by the '94 election, has created space for increasingly militant groups on the rightmost edge of the political mainstream. (Witness the Senate platform provided to militia groups by Sen. Arlen Specter.) In the West, significant factions within the wise use movement have become increasingly militant, and militant factions increasingly influential. The results include an escalation of violent rhetoric and action, and ties between wise use and far right groups.

While the wise use movement remains distinct from white supremacist and paramilitary groups like the militia, they are linked by crossover leaders, an increasingly overlapping constituency, and some common ideological views — most notably belief in the illegitimacy of the federal government and assertion of state and county "rights" over federal authority. Largely untarnished by its associations with various factions of the hard right, the wise use movement effectively launders some of their views to its steadily growing and comparatively mainstream base. In turn, hard right groups have adopted wise use themes and recruited from wise use ranks. In this manner, and by its involvement in the electoral arena and its association with (and even sponsorship by) powerful political figures, wise use has served as a vehicle for mainstreaming the right. And yet the movement is ill understood by civil rights and other groups concerned about the national and international insurgency of the right wing. The wise use movement is indeed one component of this international phenomenon. This paper will explore the wise use movement and examine its position and function in relation to other right-wing movements in the U.S.

Beginnings

To better understand the wise use movement, it is useful to explore something of its origins. What amount to the blueprints for the wise use movement were drafted by Ron Arnold, a Seattle-area public relations consultant to the timber industry, in a series of articles published by *Logging Management* magazine in 1979 and 1980.[3] In these articles, Arnold urged the timber industry to adopt the cornerstone of what became the wise use strategy: "combine our traditional approaches with the same activist techniques that have been so devastating in environmentalist hands." Advising industry to take a lesson from community organizer and author Saul Alinsky and go outside the experience and expectations of its adversaries, Arnold insisted that "the forest industry can ultimately win [the battle against environmentalism] only by expanding its experience and by striking back with its own brand of activism."[4]

Arnold identified pro-industry citizen activism as the key to defeating the environmental movement. "Citizen activist groups, allied to the forest industry," he wrote, "are vital to our future survival. They can speak for us in the public interest where we ourselves cannot. They are not limited by liability, contract law or ethical codes. They can provide something for people to join, to be part of, to fight for." For "our" part, he argued, "[i]ndustry must come to support citizen activist groups, providing funds, materials, transportation, and most of all, hard facts."[5] Speaking to representatives of the Canadian timber giant MacMillan Bloedel some years later, Arnold would make his point more bluntly: "Give them [the pro-industry 'citizen' groups or coalitions] the money. You stop defending yourselves, let them do it, and you get the hell out of the way. Because citizen's groups have credibility and industries don't."[6]

In his *LM* articles, Arnold proposed that careful investment in and cultivation of existing and future industry-identified (even if not yet activist) citizen groups such as Women in Timber chapters could yield not only a grassroots constituency (lending new legitimacy to perennial industry claims of advancing the best interests of society) but a pro-industry counterculture to compete with and overtake the environmentalist counterculture. It was Arnold's goal to create an actual pro-industry citizens movement and not just a movement facade. With this objective in mind, Arnold set about devising a strategy for industry activism that involved electing public officials, promoting citizen activism, and pursuing an aggressive public relations campaign.[7]

A prolific writer as well as an energetic activist, Arnold's fervent anti-environmentalism landed him a contract with Paul Weyrich's Free Congress Research and Education Foundation to bolster the image of besieged Interior Secretary and Sagebrush Rebellion icon, James Watt, with a flattering biography, published in 1982.[8] In 1984, Arnold became executive vice-president of the Center for the Defense of Free Enterprise, joining CDFE president Alan Gottlieb, a well-known direct-mail fund-raiser for a variety of right-wing causes, especially gun rights.[9] Arnold used the Center to build relationships with right-wing think tanks, legal centers, and activist groups that today participate in the wise use movement.[10] However, despite Arnold's years of work for the forest products industry, timber and other resource companies did not immediately line up to support his "industry activism" strategy. In his articles for *Logging Management*, Arnold remarked that the organizations that to him represented "the first stirrings of a genuine pro-industry citizen activist movement" discovered that "any such constituency first has to fight industry almost as hard as it fights environmentalists." "Trade associations in particular," he noted, "jealously guarded their power turf from the new interlopers."[11] It would take more than a compelling argument to lure industry to the wise use agenda and Arnold's posse of right-wing activists. With Reagan's election to the White House, there was little perceived need for the outsider strategy Arnold advocated. Not until resource industries faced major internal crises and/or critical legislative challenges were they moved to take action on Arnold's unorthodox ideas.

The Oregon Project

The first major wise use campaign got under way in late 1988, soon after the Movement's founding conference. Called the "Oregon Project," the effort was organized by the Colorado-based Western States Public Lands Coalition (now the National Coalition for Public Lands and Natural Resources), headed by former Oregon legislator and Sagebrush Rebel, Bill Grannell.[12] Grannell and his wife, Barbara Grannell, were invited to Oregon by Republican Senator Mark 0. Hatfield,[13] and quickly won the backing of other prominent politicians beholden to the timber industry. Their mission was to create broad-based support for unrestricted industry access to natural resources (and particularly timber) on public lands. The Grannells established their project headquarters in the offices of Associated Oregon Industries and solicited timber industry contributions to fund the campaign, even selling seats on the WSPLC board of directors for $15,000 each to a number of Oregon-based timber companies.[14]

An Oregon Project brochure circulated in April 1989 boasts, "Across western Oregon a rare phenomenon is happening, which if it takes root in other states in the west [sic] could bring about a revolution in the management of federal public lands. The phenonmenon [sic] is grassroots, community organizing."[15]The "revolution" the Grannells had in mind was deregulation — opening all public lands to commercial exploitation — and a fundamental strategy of their Oregon Project was to stress the dependence of county budgets, schools, families and communities on federal timberland harvests.[16] With this message in hand, the Grannells approached timber-related union locals, school associations, local governments, and other institutions for endorsement of their pro-timber industry campaign. The Oregon Project also served as a catalyst for other wise use projects. Timber mills initiated their own internal organizing drives, creating company activist groups that participated in the Grannells' project[17] and Ron Arnold scheduled an Oregon wise use convention for the same weekend as a major Oregon Project rally in Salem.[18] The result of all this activity would be an explosion of pro-timber industry organizing, and the first massive wise use organizing campaign.

Events of the previous eight or so years set the stage for the Grannells' Oregon Project. In the 1980s, the Northwest timber industry pursued short-term profit over long-term stability, liquidating private timber holdings,[19] restricting reinvestment to the automation of a few mills, busting unions, and otherwise reducing costs (especially labor) wherever possible. Over that decade both employment and real wages in Oregon's timber industry, the largest in the country, dropped by around 20 percent.

Throughout the Northwest these shifts produced a large pool of unemployed and underemployed timber workers and struggling small business owners who over generations had come to depend on the timber industry for their livelihoods, but who now feared for their economic futures. At the same time, environmental groups were becoming increasingly successful in garnering public support for their campaigns to save the region's remaining ancient forests, and activists like Ron Arnold were beginning to organize a backlash movement. Faced with timber supply problems, a gloomy economic forecast, and the likelihood of massive popular resentment from workers as well as environmental groups, some timber firms signed on to the Grannells' Oregon Project.

In 1989 a decline in the demand for forest products sent shock waves through timber towns, as some statewide firms closed down and national conglomerates like Georgia-Pacific moved to exploit a lower-paid, non-union work force in the South. This crisis strengthened industry support for the Oregon Project, now underway. However, the Grannells got their most important boost in the spring of 1989, when the northern spotted owl was listed as an endangered species. Injunctions against logging in spotted owl habitat were a windfall for the Grannells. The owl provided the scapegoat for a crisis precipitated in large measure by irresponsible industry practices such as overcutting, automation, and raw log exportation. It also became the national symbol for a conflict painted by wise use as between jobs and owls, humans and birds, working people and environmental elites.

By the spring of 1989 the Oregon Project claimed the endorsement of 11 county governments, four members of Oregon's congressional delegation (Senators Hatfield and Packwood, and Representatives AuCoin and Smith), a number of business associations representing timber and banking, two timber-related labor unions (the International Woodworkers of America and the International Brotherhood of Lumber and Sawmill Workers) and the Confederation of Oregon School Administrators.[20] By the summer of 1989, the Grannells boasted chapters in half the counties in the state, as well as an invitation to send "official observers" to Governor Neil Goldschmidt's Forest Summit.[21] Boosted by the cooperation of timber industry managers and the chilling experience of mill closures (15 in the first months of 1989 alone), the Oregon Project added a growing number of pro-industry "citizens" groups to their coalition of industry and local governments.[22] One of these, the Yellow Ribbon Coalition, popularized its namesake as the symbol of solidarity with timber communities allegedly 'held hostage' by an extremist environmental agenda. In the spring and summer of 1989, yellow ribbons could be seen on lapels and store windows in timber-dependent communities, or flying from the antennas of logging trucks in caravans across the state and circling the state capitol building with their horns blaring.[23]

In April 1990, the Yellow Ribbon Coalition sponsored a rally in Portland that drew 10,000 timber industry supporters. Lending his political authority to the rally, onetime wilderness advocate Senator Bob Packwood, now firmly in the wise use camp, told the crowd that he had "reached his limit of patience" with advocates of more wilderness.[24] More than 300 companies around the state gave employees the day off to attend the gathering, and transportation to Portland was provided by many employers and pro-industry groups.[25] Oregon State AFL-CIO president

Irv Fletcher, who had been invited to speak to the crowd, at the last minute declined to do so when rally organizers prohibited him from speaking in favor of a raw log export ban, insisting that he limit his remarks to the spotted owl.[26] The goal of an export ban would be to save jobs by requiring domestic milling of Northwest timber. Large exporters like Weyerhaeuser oppose a ban because it would cut into their very profitable trade in raw logs to Pacific rim countries such as Japan. Wise use coalitions, even those involving mills reliant on local timber, have generally remained silent on this issue, or rejected it as an environmental red herring.

If this weren't a clear enough indication that the Oregon Project — and wise use generally — represented the interests of industry over and against those of timber workers, that same year, during a strike at Roseburg Forest Products, scabs brought in by the company to break the strike flew the emblematic yellow ribbon from their car antennas.[27] The incident was especially significant because Roseburg Forest Products was a sponsor of the Oregon Project, as well as of the group Timber Resources Equal Economic Stability (TREES), organized by one its employees.[28] Although this union busting episode had an initial chilling effect on grassroots (and particularly union) support for the Oregon Project and other pro-industry efforts, many timber workers found themselves with few options; timber industry activism seemed to be the only hope for freeing up enough timber supply to keep the mills running a few more years. For others, joining a company sponsored activist group and turning out for pro-industry rallies was a way to curry favor with management during a time of seemingly perpetual layoffs.

The network of pro-industry groups that the Oregon Project helped to develop evolved into the Oregon Lands Coalition (OLC). OLC leaders split with the Western States Public Lands Coalition when it became clear that the Grannells wanted to extend the Oregon Project to other states while maintaining tight control over Oregon's wise use movement.[29] Following the split, the Grannells launched a new campaign, People for the West!, based in the mining industry and focussed on stemming reform of the 1872 Mining Law. The Oregon Lands Coalition carried the work of the Oregon Project forward with campaigns to exempt timber sales from the injunctions against logging in spotted owl habitat, and, when she opposed these efforts, to recall Oregon Governor Barbara Roberts.[30] The OLC has had a decided affect on the complexion of the Oregon legislature, and has flexed its muscle at the national level as well, sponsoring annual "Fly-In-For-Freedom" lobby days in Washington, D.C. since 1990.[31] When late in his reelection campaign Bush abandoned his "environmental president" façade and openly embraced the wise use movement, OLC leaders coordinated a Bush campaign stop in rural Oregon[32] and Ron Arnold claims that Dan Quayle's office requested his wise use mailing list for last-minute campaign support.[33] Following the 1992 election, the OLC played a significant role in rallying timber forces during President Clinton's 1993 Forest Conference, held in Portland, Oregon, and remains highly active. The OLC has maintained closer ties with Ron Arnold of the Center for the Defense of Free Enterprise than did the Grannells.[34] Notable OLC corporate sponsors include Boise Cascade and Weyerhaeuser.[35] Oregon Lands Coalition leaders have expanded the influence of the coalition regionally and nationally through a wise use umbrella group, the Alliance for America. The Alliance was created by OLC leaders, Northeast property developers, Louisiana shrimpers (upset about endangered turtle exclusion devices required on their nets), and other wise use activists in 1991 to give a national presence to the movement and forge alliances between various wise use constituencies.[36] OLC co-founder and Communities for a Great Oregon director Tom Hirons became the founding treasurer of the Alliance for America. The Alliance has assumed responsibility for the OLC's "Fly-In-for-Freedom"[37] events and has spread the influence of the OLC as far afield as New England, where the property rights wing of the wise use movement has been highly active.[38] In the West, the Alliance for America has endeavored to replicate the Oregon Lands Coalition in other states such as Washington, where it helped to form The Umbrella Group. Among its other activities, in 1993 the Alliance co-sponsored the Center for the Defense of Free Enterprise Wise Use Leadership Conference in Reno, Nevada.[39]

Over the years, the movement has expanded in size and depth, building national associations and cultivating a locally based network of groups and individuals ideologically committed to wise use goals. The Oregon Project is an example of what may be considered the "industrial model" of wise use organizing, where workers of a particular industry are recruited as the foot soldiers in industry driven campaigns. This model has been replicated in the mining industry and has also been applied, with some significant variations, to public lands grazing. A second major wing of the wise use movement is built around the doctrine of regulatory takings and the rhetoric of protecting private property rights. Yet a third wing is focussed on county and states "rights," advancing, for instance, a campaign in which counties assert control over federal lands within their boundaries.

The Hard Right and Wise Use:
Birds of a Feather

Although corporate financial support has underwritten many wise use organizing campaigns, some critics have portrayed wise use as simply a network of industry front groups and discounted the vital role of right wing activists and organizers. But the wise use movement is driven by two main motors: corporations and right-wing ideologues. Natural resource, property development and other business interests expect to profit from new laws promoting private exploitation of public resources, and weakening or eliminating environmental and other government regulations. For their part, right-wing activists of various stripes have used anti-environmental and "property rights" messages as an organizing handle, a means by which to convert widespread economic insecurity and political disenfranchisement into a broad reactionary political force. Although often uncomfortable, the alliance of these two forces has proved remarkably effective, yielding impressive results in the electoral and policy arenas, as well as in mass mobilization and institution building.

The right-wing ecumenical character of wise use leadership is apparent at the movement's conferences, where speakers have declared their opposition to D.C. statehood and compared the struggle against "radical preservationists" to the cold war fight against communism, labeling environmentalists as "watermelons — green on the outside and red on the inside."[40] One speaker at the 1993 Wise Use Leadership Conference censured his colleagues for openly expressing their GOP sympathies, remarking that Republicans had done as much as Democrats to lead the country down the road towards socialism.[41] Other speakers have implicated environmentalists in a global conspiracy to drastically reduce world population by indirectly attacking food supplies, through the banning of pesticides (e.g. DDT) and coolants used in refrigeration (e.g. CFCs). Bigoted remarks about Native Americans are not uncommonly made from the podium. At conference literature tables, in addition to *The Wise Use Agenda* and the late Dixie Lee Ray's *Trashing the Planet*, one can find such titles as *From My Cold Dead Fingers: Why America Needs Guns!*, *Surviving the Second Civil War: The Land Rights Battle... and How To Win It* , and the works of W. Cleon Skousen.

Multiple factions of the right wing have participated in the wise use movement since its inception, and through its various organizing campaigns, publications, leaders and groups the movement has encompassed a broad spectrum of the right. Among the right-wing groups involved in the Movement's founding was the National Center for Constitutional Studies. Formerly the Freeman Institute, the NCCS seeks, among other things, to institute biblical law in the United States. The organization and its director, W. Cleon Skousen, have been supported by the Unification Church through its CAUSA group, which has paid for hundreds of state lawmakers to attend Skousen's constitutional seminars. Skousen is an apologist for slavery and also advances international conspiracy theories akin to those of the John Birch Society. These theories implicate international bankers, presidents Nixon and Eisenhower and others in a plot to take control of the world using "Communist revolution" and other objectionable tactics.[42] Closely aligned with the NCCS, and also involved in the early days of the wise use movement, is the National Federal Lands Conference. The organization principally responsible for advancing the Movement's "county" strategy, the NFLC, also promotes the Militia of Montana and openly endorses the formation of armed citizens militia groups. NFLC literature identifies Ron Arnold and several of his close associates as advisors to the group.

Also listed in Arnold's 1989 "Index of the Wise Use Movement," a list of what he calls movement "founders," is the American Freedom Coalition. Arnold and his employer, right-wing direct mail wiz Alan

Gottlieb, once served as directors of the Washington Chapter of the AFC, a Moon-funded group started in collaboration with Robert Grant of Christian Voice. Arnold and other wise use leaders participated in the AFC's Environmental Task Force, the objectives of which included oil development in Alaska's Arctic National Wildlife Refuge and accelerated harvests on Alaska's Tongass National Forest. Following exposure of wise use ties to Moon, Arnold initially denied, and later admitted his involvement with the Moon group. As the "Moon connection" became a political liability for both Arnold and the broader wise use movement, Arnold came to portray the Washington AFC, of which he was president, as autonomous from the national organization, and his own involvement with the group as brief and inconsequential. The Washington AFC, once housed in Alan Gottlieb's commercial property, moved its address essentially next door.[43] One lasting legacy of this controversy has been the reluctance of some movement activists and groups to be associated with Arnold or even with the term "wise use" (which Arnold coined for the movement), now compromised by its association with Arnold and, by extension, Moon. Some groups instead use "multiple use," while others identify as "property rights" organizations.

Other groups in Arnold's index of movement founders include: Accuracy in Academia, which in the 1980s promised to use students to monitor Marxist influences in U.S. classrooms; California Farm Bureau Federation, which, like the national Farm Bureau Federation, has a broad right-wing political agenda that includes anti-labor, anti-women and anti-environmental positions; Center for the Defense of Free Enterprise, Arnold and Gottlieb's group and the sponsor of the 1988 Multiple Use Strategy Conference; Mountain States Legal Foundation, formerly headed by James Watt and currently under the leadership of New Right activist Perry Pendley; National Inholders Association, headed by Charles Cushman, who, among other things, has been an active participant in organized efforts involving white supremacists and others to undermine the sovereignty and treaty rights of Native American nations;[44] National Rifle Association; and the Northwest Legal Foundation and the Pacific Legal Foundation, which have provided legal service to right-wing activists on a variety of issues.[45]

The growing influence of the wise use movement has attracted other right-wing factions, including the John Birch Society and followers of neo-fascist Lyndon LaRouche. In fact, some wise use themes were developed earlier by LaRouche and his organizations, and today LaRouchians play an active role in wise use domestically and in Europe. In the early 1980s, the LaRouchians attacked the anti-nuclear Clamshell Alliance as terrorist front.[46] This tactic resurfaced later in the wise use movement. Rogeho Maduro, associate editor of *21st Century, Science and Technology* magazine and a contributor to *Executive Intelligence Review*, both LaRouchian publications, has teamed up with private investigator and wise use activist Barry Clausen to publish *Ecoterrorism Watch*, a newsletter devoted to exposing alleged environmental terrorism.[47] (Ironically, the support of some Wise Users for "state's rights" has led LaRouchians to denounce the wise use and environmental movements as twin elements of a Royal British conspiracy to undermine the United States by weakening its strong central government.[48]) Maduro won wise use acclaim for having authored a factually erroneous article about the Convention on Biological Diversity that helped to kill the legislation in Congress.[49]

As Arnold's American Freedom Coalition experience indicates, wise use ties to the hard right are viewed within the movement as little more than public relations problems. For most of these individuals and organizations, the fight against environmental protections is part of a larger struggle against what they view as an unduly liberal or even leftist, tyrannical government, and includes anti-labor, anti-feminist, and anti-civil rights fronts. Environmentalism, in the words of Alan Gottlieb, has simply become "the perfect bogeyman" for society's ills.[50]

Wise Use and the Far Right: Cross-Pollination at the Grassroots

As has been shown, various factions of the U.S. Right have been involved with the wise use movement since its inception. Evidence of cross-pollination between wise use and the hard right is most abundant in two wise use sub-movements: county rule (also known as county supremacy), and private property rights.

The County Rule Movement

The county rule wing of the wise use movement is largely orchestrated by the Utah-based National Federal Lands Conference. One of the founding organizations of the wise use movement, the NFLC promotes its county rule campaign through seminars conducted around the West and the nation. Titled "The Power and Authority of County Government," the seminars frequently are held in conjunction with larger wise use conferences, or are sponsored by local Farm Bureau chapters or timber companies seeking to increase their access to public resources.[51] The cornerstone of the county rule effort is model ordinances sold by the NFLC that promise to confer upon counties authority over federal lands within their boundaries. More than two dozen counties have wholly or partly adopted the ordinances, which the NFLC continues to sell, even though they've been ruled unconstitutional by an Idaho district court.[52]

The NFLC ordinances were originally drafted by Cheyenne, Wyoming attorney and wise use leader Karen Budd-Falen for Catron County, New Mexico ranchers seeking to circumvent federal environmental laws. The "power and authority" of these ordinances are derived from Budd's dubious legal interpretations of national environmental laws. For instance, the *National Environmental Policy Act* requires that federal managers preserve "important historic, cultural, and natural aspects of our national heritage" wherever possible.[53] Budd expands greatly on this NEPA provision, asserting that:

> [l]ocal governments have the opportunity to protect their local tax bases through the protection of custom and culture. Consider those rural communities that, for several decades, have depended upon the "customs" of livestock grazing, mining or timber production for their economic livelihoods. Those customs and cultures, passed through the generations should be "preserved" as required by NEPA.[54]

From NEPA's supposed requirement for the preservation of cultural aspects of the U.S. national heritage, Budd derives "custom and culture" as an inviolable but undefined category — an empty vessel to be filled with the wishes of her clients. The federal government, she argues, is required to preserve whatever a county government determines to be its local custom and culture. The county rule strategy has counties create a series of committees charged with establishing the "custom and culture" of the county as they pertain to the exploitation of natural resources, and to specify the practices of the various resource interests (mining, timber, ranching, etc.) to be protected by county ordinance.[55]

In addition to the NEPA "culture" clause, Budd claims to find support for county rule in a variety of federal regulations that call for agency consultation and coordination with existing local land use plans. However, she makes novel interpretations of such provisions. In a handout that Budd regularly distributes at NFLC events, she offers legal citations that suggest that federal agencies such as the Forest Service are required to "coordinate" procedures with county government, and adds the following notation:

> Coordinate is defined as "equal, of the same rank, order, degree or importance; not subordinate." *Blacks Law Dictionary* 303 (5th ed. 1979).[56]

With this note, Budd implies that county and federal governments have equal authority over the management of public lands. Elsewhere, Budd argues that "agency regulations require that Bureau of Land Management land use plans be 'consistent [sic] with' county land use plans and policies," suggesting, yet again, that county plans will drive federal planning.[57] In May 1992, Budd told a Montana audience that, "According to their own rules, if there is any way the federal agencies can comply with the county plan, they must."[58] That Budd's county rule strategy claims to deliver to counties nothing less than primary authority over public lands within their boundaries also is indicated in the following statement made by Budd in a 1991 speech to the Idaho Falls Soil Conservation Service: "I have a lot of clients that are just waiting for the Forest Service to come in, so they can see them in jail under criminal sanctions."[59]

The assertion of county "rights" over federal powers provides an ideological bridge between wise use and the far right, and by the summer of 1993, the overlap between wise use and far right organizing had become apparent. For instance, a July NFLC seminar held in Jordan, Montana featured Martin "Red" Beckman, a tax protester, notorious anti-Semite, and activist with the Fully Informed Jury Association.[60] Observers of the right wing in Montana report that wise use and militia meetings frequently attract much the same audience.[61] In October of the following year, the National Federal Lands Conference proclaimed its support for militia groups with an article in its monthly newsletter, *Federal Lands Update* (the masthead motto reads, "bringing to the federal land user, helpful information for protecting private rights"). Written by NFLC staffer Jim Faulkner and titled, "Why There Is A Need for the Militia in America,"[62] the article rhetorically asks:

> Do we really need a militia, and why? Because we have scalawags and rascals and mischievous persons and people open to temptation and flat out liars and thieves in places of power in our federal government.

Some sections of the article address the relationship between environmentalism and the need for a militia. In one place, Faulkner asserts, "There is a greater risk of being struck by an errant lavender blue asteroid than there is of dying form [sic] EPA regulated pollutants! Which brings us back to the militia..."[63] The article concludes, "Long live the Militia! Long live freedom! Long live government that fear [sic] the people!," and provides address and telephone information for the Militia of Montana and the *Deseret Political Journal*, published by Samuel Sherwood of the Blackfoot-Idaho-based US Militia Association.[64]

Endorsement of militia groups by the National Federal Lands Conference is a significant development. The NFLC is a staple organization of the wise use movement. In addition to its close association with James Watt protégé, Karen Budd, the group's advisory board includes several important movement leaders, such as property rights guru Mark Pollot and Nevada public lands rancher Wayne Hage, both of whom work for Ron Arnold's Center for the Defense of Free Enterprise, and Arnold himself.[65] Queried by reporters after the Oklahoma City Federal Building bombing about his relationship with the National Federal Lands Conference, Arnold suddenly claimed that his involvement with the group had long since ended and had always been limited.[66] However, Arnold has not publicly repudiated the NFLC, which continues to distribute literature bearing his name. Five days after the Oklahoma City blast, *USA Today* carried an op-ed by Arnold's employer and head of the Second Amendment Foundation, Alan Gottlieb. Gottlieb's opener was, "Anti-gun zealots in the media, White House and Congress are most to blame for the creation and growth of the militia movement in the USA today."[67] Rather than distancing himself from the National Federal Lands Conference, Mark Pollot declared his support for militias, which, he asserted, are sanctioned by the Constitution.[68] Although the NFLC was conspicuously absent from Arnold's summer 1995 Wise Use Leadership Conference, the group continues to be a regular feature at other wise use gatherings, where they conduct "power and authority of county government" seminars and sell literature, including the published works of far right Mormon "constitutionalist" W. Cleon Skousen.

The wise use and militia movements developed separately and mostly function independently of one another. However, in rural northwestern communities, militia organizing has often come on the heels of wise use county rule campaigns to take over local government and declare authority over federally managed public lands. Many militia groups believe county boards of commissioners to be the highest legitimate governmental bodies (and the local sheriff to be the highest law enforcement officer in the land) and call for expanding local power at the expense of the federal government. These views derive from the Posse Comitatus (literally, "power of the county"), forerunner to Christian Patriot and Northwest militia groups. Although the legal theories underlying wise use and Christian Patriot assertions of county rights differ markedly, wise use county rule campaigns have softened the ground for Christian Patriot and militia organizing in the Northwest. For instance, US Militia Association leader Samuel Sherwood has recruited eastern Idaho miners and timber workers to his group, urging them to resist the "green gestapo," and warning of an imminent civil war.[69]

Another county supremacy strategy associated with wise use is led by Nye County, Nevada commissioner Dick Carver. In the last two years, Carver has traveled the West declaring federal ownership of Western lands to be illegal and claiming states' rights to such lands. Under Carver's leadership, the majority of Nevada counties have passed resolutions to this effect. Dramatically acting out his beliefs before an armed band of supporters, on Independence Day 1994, Carver bulldozed open a closed road in the Toiyabe National Forest and threatened the Forest Service ranger who tried to stop him.[70] Carver said later that, "All it would have taken was for one of those rangers to have drawn a weapon... Fifty people with sidearms would have drilled him."[71] Such militancy has made Carver a popular wise use leader.

Carver also is admired by some on the far right and has been a recurrent speaker at events sponsored by the *Jubilee* newspaper, the premier Christian Identity publication in the country.[72] Christian Identity is a pseudo-theology which holds that Jews are the spawn of Satan, people of color are subhuman "mud people," and whites are the true Israelites. Identity doctrine is espoused by the Aryan Nations and by Militia of Montana leader John Trochmann, among others. Although he has spoken at *Jubilee* conferences featuring such notorious racists and anti-semites as Louis Beam and Martin "Red" Beckman, Carver claims never to have heard hate rhetoric at these events.[73]

In the spring of 1995, Carver conducted a 10 day tour of Washington state. A Stevens County wise use organization promoted Carver's visit by reprinting a *Jubilee* article celebrating the Nye County commissioner in its newsletter.[74] On another stop, Carver used rhetoric reminiscent of the Posse Comitatus when at the start of his presentation he secured the commitment of the local sheriff to stop any federal law enforcement agent who might attempt to arrest him.[75] Although Carver and the National Federal Lands Conference promote different strategies for gaining control of federal lands, they participate in many of the same conferences and attract a similar following. Advertisements announcing Carver's appearance in Snohomish, Washington even identified his topic as, "The Power and Authority of County Government," the same title as used by the NFLC. [76]

The Property Rights Movement

As with the movement's county rule strategy, wise use leaders have endeavored to build a network of local property rights organizations across the country. The ideological framework undergirding property rights rhetoric is regulatory takings doctrine. This doctrine holds that government regulatory action that negatively affects the value — actual or potential — of private property constitutes a "taking" of property and, as such, is prohibited under the takings clause of the Fifth Amendment of the US Constitution, unless fairly compensated. The relevant portion of the Fifth Amendment reads, "nor shall private property be taken for public use, without just compensation." Historically, the courts have interpreted the takings clause as pertaining to cases of condemnation under eminent domain — that is, the government cannot confiscate your land or other property without paying you a fair price for it. Only in instances where government regulations have been found to eliminate virtually all economic value of property have some Supreme Court justices supported financial compensation for a "regulatory takings."[77] By contrast, regulatory takings doctrine deems a vast array of public interest and regulatory laws to be illegal.[78] Corporate and right-wing libertarian proponents of regulatory takings doctrine hope to advance their cause by framing takings laws as an antidote to allegedly burdensome and costly environmental regulations.

In the spring of 1992, Ron Arnold declared that "The future of the property rights issue for the next decade will probably be centered in Washington state."[79] Opponents of Washington's recently enacted *Growth Management Act* had just managed to weaken the law, and their earlier, successful ballot fight against a more stringent version of the GMA proposed by environmental groups had served to galvanize corporate support for the burgeoning property rights movement.[80] Working to realize Arnold's prediction, and responding to a GMA requirement that counties develop local land use plans to confine suburban sprawl and protect sensitive natural areas, wise use organizers, in cooperation with the state's building industry, constructed a network of county-level "property rights" groups across the state. Various of these local groups have stymied implementation of the GMA, taken over local governments, launched campaigns to form new counties under their command, and campaigned for the adoption of regulatory takings laws.[81] Washington state property rights groups have embraced crossover figures like Dick Carver, and some groups distribute militia literature. These developments may reflect, in part, the origins of Washington's property rights activism in organized efforts to subvert the sovereignty and treaty rights of Native American nations. Groups such as United Property Owners of Washington and wise use leaders like Charles Cushman and Alan Gottlieb were active in the Anti-Indian Movement prior to the emergence of wise use.[82] However, despite some negative news coverage, Washington's property rights groups have largely been legitimated in the press as mainstream citizens groups concerned with overregulation.

Other examples of cross-fertilization between wise use and the far right include *The Oregon Observer*, a wise use publication that advertises for the Oregon Militia.[83] In another, more significant case, a recent campaign opposing a proposal to join Washington's North Cascades National Park with a park across the Canadian border, that featured a barnstorming tour by national wise use leader Charles Cushman, promoted the idea that the Park was a pretext for the New World Order to subvert U.S. sovereignty. In February of 1995, a principal organizer of that campaign, Don Kehoe, appeared with Militia of Montana leaders John Trochmann and Bob Fletcher at a community meeting in Washington state.[84] In some areas there is little meaningful distinction to be made between wise use and militia activists and groups.

Electoral Politics

Wise use is a profoundly political movement. Backed by natural resource and development corporations that, combined, are the largest political campaign contributors in the West, the wise use movement is a powerful, and usually partisan, political player. This is particularly evident in the timber industry, which, if measured in campaign contributions, is the most pro-Republican industry in the country in both national and statewide races. In 1990 timber gave 74% of its national campaign dollars to Republicans, and was even more partisan at the state level, where it was also a potent political force. In Oregon in 1990, 85% of timber contributions went to Republicans, and 68% of timber-supported candidates won their races. In Washington state 79% of timber contributions went to Republicans, in Montana 86%, and in Idaho an incredible 93.6% went to GOP candidates, with 60% of timber-backed candidates winning election. In Idaho and Oregon, timber was the third largest contributor to political campaigns.[85]

The 1992 electoral season saw heavy wise use involvement in statewide, congressional, and presidential races. At the congressional level, the most unseemly wise use involvement in the elections may have been in Montana, where reapportionment left Republican Rep. Ron Marlenee and Democratic Rep. Pat Williams contesting a single congressional seat. Marlenee, a wise use ally, received the strong-armed support of the timber industry. In one case, workers at Plum Creek Timber's Columbia Falls, Montana plant found their pay envelopes stuffed with flyers urging them to attend an anti-wilderness rally sponsored by the Western Environmental Trade Association's Ad Hoc Coalition, a wise use group. The flyer contained two pages of suggested slogans for signs to be carried at the rally, including "No more Williams, wilderness or wolves," and "You'll need a job, Pat." A county attorney ruled that the flyers violated Montana election laws and imposed the maximum penalty against Plum Creek, which pleaded guilty to the crime.[86] Williams won the election. However, in legislative races Montana progressives suffered a serious setback in the House, which turned from a 61-39 Democratic majority to a 47-53 minority. Political observers assert that wise use was a significant factor in about half of the fourteen seats lost. The practice of stuffing pay envelopes with wise use propaganda is not unique to Plum Creek or the Ad Hoc Coalition. The Oregon Lands Coalition, for instance, has used its newsletter to publish flyers labeled "PAYROLL STUFFER" in bold typeface.[87] In the 1994 election Montana Democrats lost the Senate. Timber financing of elections and wise use pro-timber industry campaign shave worked hand in glove to erode legislative support for environmentally sustainable forest practices.

The 1992 presidential race also was influenced by the wise use movement, particularly around timber-related issues. As the Clinton/Bush/Perot race wore into its final months, Bush sought to undercut Clinton's strong support in the Northwest by adopting the rhetoric and seeking the support of the wise use movement. In September, as wise use activists were rallying in the streets of Washington, D.C. and meeting with Bush aides at the White House,[88] the President made a quick swing through the Northwest, visiting the small timber towns of White City, Oregon and Colville, Washington.[89] In White City, he told timber workers in a local lumberyard, "The balance has been lost. It is time to make people more important than owls. It is time to put the mills back to work."[90] Bush's relationship with wise use involved more than rhetoric. Earlier that same month, Bush ordered federal agencies to expedite the process of harvesting dead timber on public lands, circumventing the normal comment and appeals processes.[91] Following this change of posture, wise use timber groups came out strongly in support of the President. Bush's White City stop was coordinated by Oregon Lands Coalition leaders[92] and Ron Arnold claims that Vice President Dan Quayle's office requested his wise use mailing list for last-minute campaign support.[93]

The wise use movement has found especially strong political support from within the western Congressional delegation. Even before the election of high profile wise use leaders like Helen Chenoweth to public office, western politicians beholden to the natural resource agencies lent their support to wise use campaigns. For instance, in December of 1992, an internal report of the House Subcommittee on the Civil Service revealed collusion between Bush administration officials, wise use and natural resource industry leaders, and members of Congress to destroy a proposed management plan for the greater Yellowstone Park area.[94] National Park and Forest Service officials began work on what became known as the

Yellowstone Vision Document as early as 1985. It was to have been a model for ecosystem management, a relatively new idea calling for public resource managers to view the resources within their areas holistically, assess their role in the ecosystem — and not just within a park's boundaries — and devise a management plan that would result in a healthy environment for plants, animals and people. Independent of that effort, however, top officials in the Bush Administration were lending support to the emerging wise use movement. In June of 1989, American Freedom Coalition leaders Merrill Sikorski, Richard Ichord and Bob Wilson, and Blue Ribbon Coalition Director Clark Collins met with Interior Secretary Manuel Lujan to discuss natural resource issues and the wise use agenda. Afterwards, Sikorski reported, "Our meeting with Secretary Lujan was very positive and under his experienced leadership, prospects for continuing dialogue between 'wise use' advocates and the Department of the Interior appear bright."[95]

A draft of the Vision Document was released in July 1990. Three months later, its opponents held a secret strategy meeting in the office of Wyoming Senator Alan Simpson. Those present included Bureau of Land Management Director Cy Jamison, Park Service Director James Ridenour, Forest Service Chief Dale Robertson, Interior Department Deputy Assistant S. Scoff Sewell, Wyoming Wool Growers Association director Carolyn Paseneaux, Wyoming Farm Bureau associate director Dave Flinter, and Warren Morton of MKM Oil, Wyoming Heritage Association and Wyoming Taxpayers Association.[96] The House Subcommittee report credits a 1990 letter to Interior Secretary Manuel Lujan from Crown Butte Mines, which at the time was planning a major gold mine just east of Yellowstone National Park, with beginning the anti-Vision Document campaign.[97] The report finds further that, at the instruction of former White House Chief of Staff John Sununu, Sewell improperly wrested control of the planning process and gutted the draft plan.[98] Senators Alan Simpson and Malcolm Wallop (both R-WY) and Representative Ron Marlenee and Senator Conrad Burns (both R-MT) were among the politicians leading the attack on the Vision plan.[99] This back-room political effort was reinforced with a regional organizing assault by wise use activists. In November 1990, Charles Cushman launched the Yellowstone Regional Citizens Coalition, representing more than 40 industry, wise use and local government groups from Idaho, Wyoming and Montana.[100] Cushman's National Inholders Association mailed out 20,000 pieces of literature to area residents, attacking the plan as a government land grab that would restrict access to hunting, fishing and recreation on public lands, and even threaten private property.[101] Other groups that actively campaigned against the plan included: Montana Mining Association, Western Environmental Trade Association (WETA), Wyoming Heritage Society, the Blue Ribbon Coalition, and the People for the West! campaign of the Western States Public Lands Coalition.[102]

This coordinated campaign generated a wellspring of opposition to the Yellowstone Vision Document. Public hearings on the plan held between November 1990 and January 1991 were heated. The final hearing, in Bozeman Montana, drew the largest crowd. Yellowstone National Park Superintendent Robert Barbee recalls, "There were 700 people there. You can't imagine the virulence of the outcry. I was Saddam Hussein, a Communist, a Fascist, everything you could think of. One lady got up there, jaw quivering, used her time to say the Pledge of Allegiance, then looked at me and called me a Nazi. They loaded the hall. It represented the very worst of the public participation process. It was grim... revolting... a rout."[103] The Vision Document was effectively killed and National Park Regional Director Lorraine Mintzmyer, who oversaw its development, was transferred from Colorado to Pennsylvania.[104] Charles Cushman, would later offer the following analysis: "We didn't compromise; they got annihilated."[105]

Evidence is building that the wise use movement may provide greater access for militia and other far-right groups to the electoral process and the political mainstream. This is clearly the case at the local level, where Dick Carver is only the most visible example of a wise use county commissioner with ties to the far right. Increasingly common are local politicians such as Chelan County, Washington commissioner Earl Marcellus, who warns Forest Service leaders to cooperate with wise use efforts, which he describes as "ballot box" organizing, or face the militia's alternative of the "cartridge box."[106] The growing influence of wise use and militia groups at the local government level may produce a pool of candidates for higher office. Already, wise use has elected leaders who are also sympathetic to the militia movement to state and federal office.

In Idaho, militia leader Samuel Sherwood claims partial responsibility for the election of U.S. Representative Chenoweth as well as state Secretary of Education Anne Fox. Chenoweth's 1994 campaign against incumbent Democrat Larry LaRocco was rooted in the wise use movement (in one instance, national wise use leader Charles Cushman hosted "endangered salmon bake" fund raisers for her)[107] and also drew strong support from the Christian Right, the John Birch Society and, evidently, militia groups.[108] While Sherwood likely exaggerates his electoral influence, since taking office Chenoweth has injected militia themes into the public debate with her unfounded suggestion that natural resource agencies are using mysterious black helicopters (symbolic to militia groups of "New World Order" forces) to harass property owners, and by sponsoring a bill that would require federal law enforcement agents to gain the approval of local sheriffs before making arrests or performing other duties.[109] Chenoweth has also supported efforts to pass a local ordinance that essentially requires a gun in every household, yet she backs proposed national legislation proposed by Sen. Larry Craig to disarm federal resource agents, an agenda which, if successful, would have everyone in Idaho armed, except for the resource agency workers who increasingly are the targets of wise use and militia aggression.[110] After the Oklahoma City attack, Chenoweth refused to denounce militia groups and said bombing suspect Timothy McVeigh's allegations that the explosion was orchestrated by the federal government as part of a plot to discredit militia groups should not be dismissed outright.[111] Such actions and statements have earned Chenoweth the support of militia groups and others on the far right. For instance, the Militia of Montana features a Chenoweth video in its mail order library, and the white supremacist tabloid, *The Truth at Last*, compares Chenoweth favorably to "whiny liberals like Patricia Schroeder or Jews like Senator Dianne Feinstein."[112] Another Idaho elected official, Secretary of State Pete Cenerussa, has addressed militia events and even discussed the possibility of conferring legal status on the Idaho militia.[113] In 1995 Sherwood's US Militia Association filed a ballot initiative to achieve just that.

The wise use movement's access to the political arena has served as a vehicle for militia groups and others on the far right to move their issues and rhetoric into the political mainstream. The degree to which this has occurred is unclear, and is a matter in need of additional research.

Violence

Wise use activists may be adopting some of the tactics of the far right. Nineteen days before the Oklahoma City bombing incident, the U.S. Forest Service office in Carson City was bombed. Wise use leader Chuck Cushman tried to pin the attack on environmentalists,[114] while Dick Carver suggested it might be the work of the Forest Service.[115] A Bureau of Land Management office in Reno was bombed in October of 1993, and several federal land management agency installations and a property of agency personnel have been bombed since then.[116] In one incident, a bomb planted in the vehicle belonging to National Forest Service District Ranger Guy Pence was detonated outside his home in Carson City, Nevada, destroying the vehicle and shattering his front windows.[117] None of these crimes have been solved. However, federal resource agency employees are experiencing a surge in harassment, mostly directed from wise use supporters. In some areas, federal workers find that they are refused service in local stores, and have taken the precaution of removing identifying labels from government vehicles.[118] Anticipating further harassment, the Forest Service has instructed its employees not to resist arrest by county law enforcement agencies for violations of wise use county rule ordinances that claim authority over federally managed public lands.[119]

Dick Carver's suggestion that the Forest Service might be behind the bombing of one of its own offices foreshadowed similar assertions, now widespread within the militia movement, that the federal government itself arranged the bombing of the Oklahoma City Federal Building. In the summer of 1995 Ron Arnold claimed that, "A faction of the environmental movement is trying to use the horror of the Oklahoma City bombing that killed 168 people as a public relations ploy to smear the wise use movement... It would be all too easy for us to fall into the same trap of frustration... and to accuse every environmentalist of being 'linked' to the recent rise in Earth First! attacks and the bombing death of [California Forestry Association director] Gil Murray [by the 'Unabomber']."[120] In fact, accusations of environmentalist terrorism have proliferated in wise use publications since the Oklahoma City bombing. Charles Cushman's accusation that Earth First! was likely to blame for the Carson City Forest Service office attack reflects the broad effort among wise use groups and leaders to deflect attention away from the Movement's association with far right militants by reviving the specter of "ecoterrorism" raised earlier by the LaRouchians. Blue Ribbon Coalition director Clark Collins even refers to environmental organizations as "hate groups."

Barry Clausen co-edits the newsletter *Ecoterrorism Watch* with LaRouchian Rogeho Maduro. Clausen is also co-author of *Walking on the Edge*, which recounts his infiltration of Earth First! on behalf of timber and ranching interests. [121] He has been promoted at events organized by Arnold and Gottlieb's Center for the Defense of Free Enterprise, Clausen's book is distributed by a company owned by Gottlieb, and Arnold designed the cover for the volume.[122] Since 1993, Clausen has given presentations in struggling natural resource communities, where he brands Earth First! as a terrorist organization and promotes vigilantism against environmental activists by inferring that the entire environmental movement has been infiltrated by Earth First! agents. In response to environmental efforts to halt logging in the Cove Mallard area of Idaho, in April of 1994, a wise use timber company front group called the Gold Hill Resources Coalition announced a meeting featuring Clausen and three men with expertise in surveillance, counter-terrorism and military operations.[123] Environmentalists report a rise in incidents of harassment in the wake of Clausen's visits.

Counter to the image of a green menace conjured by Clausen and others, terror is increasingly the experience of environmentalists and other wise use adversaries. In November 1994, Washington Audubon Society activist Ellen Gray received threats at a Snohomish County Commission hearing on an environmental ordinance. According to Gray, during testimony on the ordinance, several property rights activists referred to environmentalists as "fascists," "eco-nazis" and "ecoterrorists." After her own testimony, a man unknown to Gray "reached under his seat, pulled out a hangman's noose made out of rope, turned in his seat to face me, shook the noose in the air and said, 'this is a message for you!' Immediately afterwards," she recounts, "another man I did not recognize approached me. He leaned toward me and said, 'We have a militia of 10,000 and if we can't beat you at the ballot box we'll beat you with a bullet.'" The man with the noose was Darryl Lord, who a few months later was elected president of the Snohomish County Property Rights Alliance.[124] That same fall, Ron Arnold was among the featured speakers at a wise use conference in Joseph, Oregon, where two local environmentalists were hung in effigy.[125] The preceding April in north central Washington, environmental activist Jerry Payton looked out the window of her home in the small town of Oroville to discover a man dressed in camouflage fatigues marching up and down her street with a rifle.[126] The incident came close on the heels of a visit by Militia of Montana leader John Trochmann to nearby Wauconda.

While wise use-inspired harassment and violence appear to be on the rise, the problem is barely a new one. Charles Cushman, who once delighted in his "rent-a-riot" nickname and compared himself to Nazi tank commander Erwin Rommel, has long been notorious for inflammatory rhetoric and the destruction left in the wake of his organizing campaigns.[127] Cushman even distributes an article detailing the violence directed against the targets of his organizing campaigns as promotional literature.[128] Arnold has urged crowds to "destroy the environmental movement," and use the "sword of political power" to "kill the bastards."[129] Yet, as a public relations ploy following a spate of negative publicity about wise use-related violence, Cushman and Arnold issued a wise use 'Declaration of Non-Violence' at the 1993 Wise Use Leadership Conference.

Conclusion

Having established itself as a mainstream political and social force, seeming to represent the economic concerns of working- and middle-class people, the wise use movement serves an effective vehicle for advancing the agendas of various right-wing political and social movements. Just as right-wing factions within the wise use movement have used anti-environmentalism as a means to further a broader agenda of corporate deregulation and right-wing populism, factions outside wise use, ranging from militia groups to the religious right, have adopted wise use themes as a means of mainstreaming themselves. At the same time, encouraged by the rightward shift of the domestic political "center," and its own organizing successes, elements of the wise use movement have become bolder with scapegoating and promotion of violence. The result is an increased sense of threat and polarization, as well as actual danger. However, the wise use movement's association with the far right may be its greatest vulnerability. Confronted with the bright light of exposure, decent people lured to wise use with the false promise of economic prosperity can be compelled to reconsider their involvement, and politicians susceptible to public scrutiny can be compelled to withdraw their support. But exposure alone will not suffice. The success of wise use to date has demonstrated that, in the absence of progressive economic alternatives, as well as vigilant efforts to defend democracy and protect human rights, the movement will continue to expand its base. We must offer real hope by redoubling our organizing for economic justice, civil rights and environmental protection.

RECOMMENDED READING:

Helvarg, David (1994) *The War Against the Greens*. San Francisco: Sierra Club Books.

Reed, Scott (1993-1994) "The County Supremacy Movement: Mendacious Myth Marketing." *Idaho Law Review*, Vol. 30, No. 3.

Ryser, Rudolph (1992) *The Anti-Indian Movement on the Tribal Frontier*. Kenmore, WA: Center for World Indigenous Studies.

HOW INDUSTRY COMBATS EFFORTS TO PROTECT OUR CLIMATE

John Passacantando

In a recent study, 2,500 of the world's preeminent atmospheric scientists examined current data and concluded that global warming has begun. (IPCC, 1995a). Some of this evidence is now familiar to many Americans: concentrations of carbon dioxide (CO_2), the primary greenhouse gas, have risen nearly 30% in the last 100 years. The average global temperature has risen 1 degree Fahrenheit over the same period. The ten warmest years in the past 100 have occurred since 1980. Glacial ice is retreating on five continents due to rising temperatures. Other evidence includes "increased evidence of drought, above-normal temperatures, winter-time precipitation and heavy rainstorms in many areas of the United States" since 1980. (Stevens). The midwest heat wave during the summer of 1995, which killed 669 people (Star-Ledger Wire Services) came during one of the hottest summers on record.

While an average temperature change of only a few degrees Fahrenheit may not seem like much, consider by comparison that in the depths of the last ice age, when mile-high sheets of ice reached as far south as the Great Lakes, the Earth was only 5 to 9 degrees Fahrenheit cooler than it is today. (Stevens).

While precisely when and where the effects of global warming will occur is uncertain, each of us will face the impacts in one way or another. Some of the impacts are:

Changes in climate due to global warming are expected to have a major impact on human health. More extreme temperatures and precipitation and greater frequency and severity of storms, floods, and droughts will likely lead to increase deaths, illnesses and injuries. Increasing illness and death are predicted from diseases such as malaria, cholera, and dengue fever, whose range will spread as mosquitoes and other disease vectors migrate.

Rising sea levels resulting from warming oceans and melting glaciers, causing massive flooding in coastal areas, where over half of the U.S. population lives and which provide significant revenues and jobs.

Greater extremes in temperatures and precipitation, which will create greater variability in agricultural production. More moderate temperatures or increased precipitation may lead to a marginal gain in agricultural productivity in some regions. But increased heat stress, decreased soil moisture, greater frequency and severity of drought and floods, and the proliferation of harmful insects and disease will likely devastate agricultural yields in many others. These swings will disrupt markets for food and other agricultural commodities with potentially devastating consequences.

Warming waters and changing water flows, which will place numerous fish species ar risk, affecting both commercial and recreational fishermen, the availability of food on the market, and the ecosystems in which the fish play and important role.

Scientists predict that global warming will have a significant effect on the function and composition of forests in many regions. Temperature extremes, changes in precipitation patterns, the increased intensity and frequency of wildlife and storms, pests and diseases, and even increases in air pollution will affect both forest survival and growth rates (Peters and Lovejoy, 1992, p. 245.)

Some analysts have estimated that global warming could cost as little as $59 billion or as much as $438 billion annually. However, the methodology used to determine these figures does not take into account several costly results including the effects on public heath.

Patricia Glick
"The High Costs of Inaction"
Sierra Club website www.toowarm.org/resources/inaction.htm

One proud chapter of American history is the recurring theme of spirited innovation, of overcoming long odds to achieve victory. Examples include throwing the awesomely powerful British army off the continent in the 1770s with a rag tag group of patriots to the rapid rebuilding of our naval fleet, after near total loss at Pearl Harbor, to save the world from Axis aggression. In these, among many other examples, great threats are overcome by an educated and empowered citizenry. However, as the ying must be accompanied by the yang, there is a competing chapter, driven by the narrow special interests of a handful of corporations, their public relations machines, lobbyists and lawyers. One such example is the stunningly effective effort by auto, oil and coal companies to thwart U.S. leadership in addressing the threat of global climate change.

For context, consider the overwhelming consensus that has emerged regarding the threat of global warming. The above excerpt by Patricia Glick sheds succinct light on the problem. Thousands of scientists, including hundreds of Nobel Laureates, and members of the National Academy of Sciences, have lent their names to statements urging the U.S. to lead the world in averting climate change. In the peer-reviewed science journals, new research emerges weekly showing ever-stronger evidence of human-induced climate change and the wide range of negative impacts to our environment and the health of our children.

As a species, we are creating a global problem by our excessive burning of fossil fuels that experts tell us will throw a wrench into the stable climate upon which our civilization must rest for centuries, if not millennia. Impacts include the spread of infectious diseases, a rising sea level, disrupted agriculture and increasingly severe weather. More immediate are the direct, but externalized, costs from the burning of fossil fuels.

Expressed in terms of present day dollars, these direct costs of burning fossil fuels may comprise 2 to 5% of our Gross Domestic Product.[1]

According to the Centers for Disease Control and Prevention:

> The number of Americans who suffer from asthma has risen 75 percent since 1980, to more than 15 million, in part due to pollution and other environmental factors.[2]

Worse still, an estimated 15,000 premature deaths occur each year due to particulate soot pollution.[3] Consider also the unnaturally fast rise of the oceans, projected to rise one to three feet during the next 100 years from global warming. Sea level rise accelerates the pace of beach erosion, which threatens the lifeblood of entire coastal communities. New York, California, Massachusetts, Texas, Florida, Maryland, Delaware, Virginia and every other coastal state already spend millions of dollars, supplemented with taxpayer dollars, pumping sand back on shore, just to give us a day at the beach.[4]

To see just how weak the U.S. response has been to the threat of climate change, we must look deeper than the swell sounding speeches given by President Clinton, Vice President Gore and their cabinet officials. The rubber meets the road at the esoteric international meetings that are part of the Framework Convention on Climate Change, meetings which reached a crescendo in Kyoto, Japan in December 1997 with an international agreement on climate change. The agreement, called the *Kyoto Protocol*, was considered a victory by some, simply because the collective forces of the fossil fuel industry had conspired to insure there would be no agreement. Scientists have told us that fossil fuel emissions must be reduced 50 to 80 percent immediately, simply to keep the greenhouse effect from further disrupting our climate. The Kyoto Protocol's baby step was to secure support from the industrialized countries to reduce greenhouse gas emissions by 5.2 percent (below 1990 levels) by 2012. Add to this the powerful list of loopholes and it is questionable if the Kyoto Protocol will do much at all to avert climate change. The question then becomes why was the U.S. so weak in these negotiations?

We have had a few windows into the efforts by the fossil fuel industry to confuse the public on climate change. Several of these were highlighted in a series of reports and investigations by the environmental organization Ozone Action. The report series was entitled, *Ties That Blind*, and it was an attempt to unmask some of the more egregious attempts by the fossil fuel industry to distort the public debate on climate change. The first report documented the funding sources of two prominent climate change skeptics (Patrick Michaels and Robert Balling) which included U.S. coal, British coal, German coal and the Government of Kuwait. These scientists are well known around the world for showing up at the most important climate meetings, using their credentials and university affiliations (University of Virginia for Michaels and Arizona State University for Balling) to undermine the peer reviewed research showing that climate change is a real threat. According to testimony offered before the Minnesota Public

Utilities Commission on March 15, 1995, Dr. Michaels has received funding from Western Fuels Association (a consortium of coal interests), German and American coal interests and Cyprus Minerals Company. It is difficult to ascertain the amount of money Dr. Michaels has received to publish *World Climate Review* now called *World Climate Report*, one of his primary venues to attack measures designed to avert global warming

Nonetheless, Dr. Michaels' own web page tells us the following about *World Climate Report*. *World Climate Report* is:

> a research review edited by Dr. Patrick J. Michaels. *World Climate Report* provides policy makers, journalists, and the interested public with an ongoing and accurate portrayal of the science of global climate change which will function as an antidote to the vision of apocalypse promoted by the professional environmental community and by the United Nations. Funding for this publication is provided by Western Fuels Association, Inc. with additional funding by associated companies.[5]

This description is corroborated by Western Fuels 1995 *Annual Report*:

> Our publication and distribution of *World Climate Review* has clearly had an impact on the climate change debate. After 11 quarterly editions, we are discontinuing the magazine and replacing it with *World Climate Report*. This bi-monthly newsletter will provide a rapid response to the spurious reports that try to create virtual climate reality, a phony picture of increasing weather catastrophes caused by carbon dioxide emissions."[6]

In addition to support from the Western Fuels Association and other coal and energy interests to fund *World Climate Review* and *World Climate Report*, Dr. Michaels received a $63,000 grant from Western Fuels for research on global climate change. From the German Coal Mining Association, Dr. Michaels has received $49,000 and from Edison Electric Institute he has received $15,000. Dr. Michaels has received $40,000 from Cyprus Minerals Company.

According to Ross Gelbspan's *The Heat is On*, Dr. Michaels has received more than $115,000 from coal and oil interests over the past four years.[7] Ozone Action has confirmed that Dr. Michaels has received $167,000 from undisclosed industry sources. This total does not include $50,000 he received from an anonymous donor or funding for either of the publications discussed above.

According to testimony offered before the Minnesota Public Utilities Commission on March 15, 1995, Dr. Robert Balling has received significant levels of funding since 1989 from the Kuwait government, foreign coal/mining corporations and Cyprus Minerals Company.

The Kuwait government has actively participated in negotiations under the Framework Convention on Climate Change. Due to the large oil reserves in that country, the Kuwait delegation along with other Middle Eastern delegates, have opposed findings made by the Intergovernmental Panel on Climate Change and the effects climate change may have on living systems.

In his spoken testimony, Dr. Balling was cross-examined about funding he has received from the Kuwait government. The following is an excerpt:

> Q: And the last one on this page is from the Kuwait Foundation for the Advancement of Sciences, can you suggest why Kuwait, or rather the Kuwait Foundation might have any interest in global warming?
>
> A: No, I can't. I know a Kuwaiti who did a Ph.D. with our group and theKuwaiti said that he is well connected in Kuwait to funding sources. You have to understand, I'm the director of a research laboratory and there'snever ending pressure to find research money. And when you discover that the Kuwaiti government has a research pool that is being allocated to scientists in my field, you would be crazy not to run out and make some attempt to put a proposal in to obtain funding from that source.[8] [Transcript page 111]

Dr. Balling was further questioned as to his links to the Kuwait government:

> Q: With respect to your book, *The Heated Debate*, isn't it true, Dr. Balling that the Pacific Research Institute for Public Policy was in fact foundedto oppose environmental regulations?
>
> A: I know nothing of their history. I'm aware that they have been a conservative public policy group. But I did not investigate who these people were that asked me to prepare a book for them.
>
> Q: And they're also publishing a Middle Eastern version of your book, is that correct?
>
> A: It has been published.
>
> Q: It has been published. Is it published by this organization?
>
> A: Yes
>
> Q: Is it funded by the Kuwaiti government?
>
> A: The Kuwait Foundation for the Advancement in Science gave Dr. Nasrallah a grant and money from that grant went toward the publication of this book.[9]

The Kuwait Foundation for the Advancement of Science is a "government establishment," according to *Moneyclips*, August 21, 1994, "Private University Gets Initial OK; 'Practical Steps Taken to Implement Project'" and the *Arab Times*.

Aside from funding received by the Kuwait government to reprint *The Heated Debate*, Dr. Balling also has received grants from the Kuwait

Foundation for the Advancement of Science ($48,993) and the Kuwait Institute for Scientific Research (amount undisclosed).

Dr. Balling has received three grants from the British Coal Corporation for a total of $103,544 and two grants from the German Coal Mining Association for a total of $81,780. In addition, Dr. Balling has received five grants from Cyprus Minerals Company (totaling $72,554) and one grant from Cyprus Minerals Chief Executive Officer ($4,904). Cumulatively, Cyprus Minerals funding to Dr. Balling totals $77,458.

According to *The Heat is On*, Dr. Balling has received nearly $300,000 from coal and oil interests over the past six years.[10] Dr. Balling was confronted with Gelbspan's allegations by a reporter from *The Arizona Republic*; the following is an excerpt of that article:

> One of the nation's leading skeptics is Robert Balling, director of the Office of Climatology at Arizona State University. He's treated none too gently in an article in the December issue of *Harper's* magazine. The essay, written by Pulitzer Prize-winning journalist Ross Gelbspan, makes the case that global warming poses a far greater threat than most people realize. The problem, he says, isn't that the media are conspiring to downplay the threat but that we are being misused by people like Balling, whom he disparages as a kind of rent-a-scientist in the pocket of the energy companies.
>
> News reports on the subject, Gelbspan writes, usually "come qualified with the observation that the question of global warming can never be conclusively resolved. The confusion is intentional, expensively gift-wrapped by energy industries." He wrote that oil and coal associations spend millions to spread contrary opinions, calling on Balling and a few others who are "interchangeable ornaments on the hood of a high-powered engine of disinformation."
>
> Gelbspan says Balling "has received more than $200,000 from coal and oil interests." Balling, author of a 1992 book, *The Heated Debate: Greenhouse Predictions Versus Climate Reality*, says that number is way off. "Actually, I've received more like $700,000 over the past five years," he corrected.[11]

At present, Ozone Action is only able to confirm that Dr. Balling has received $311,775 from coal and oil interests.

Another window into the workings of the fossil fuel industry came in 1991, when the strategy document for the Information Council for the Environment (ICE) was leaked out of a public relations firm. The documents stated that the goal of ICE was to "reposition global warming as theory (not fact)."

The ICE President, Gale Klappa of The Southern Company (an electric utility) and ICE Vice President, Frederick Palmer, Western Fuels Association worked with the Edison Electric Company to target key Congressional districts with information about climate change.

ICE also created a Science Advisory Panel which included Dr. Robert Balling, Dr. Patrick Michaels and Dr. Sherwood Idso. According to documents obtained by Ozone Action, several ICE strategies were laid out including: the repositioning of global warming as theory, not fact; achieving broad participation across the entire electric utility industry, and using a spokesman from the scientific community. Media strategies included:

• targeting "older, less-educated males from larger households who are not typically active information seekers" and "younger, lower-income women;"

• using "technical source" (i.e. scientists) because they receive a higher credibility rating from the public;

• and proposed advertisements such as: "The most serious problem with catastrophic global warming is— it may not be true," "If the earth is getting warmer, why is Minneapolis getting colder?" "Who told you the earth was warming... Chicken Little?" and "Some say the earth is warming. Some also said the earth was flat."[12]

Sporting a budget of $500,000, ICE worked with Simmons Advertising, Inc. in Grand Forks, ND to place a sixty second spot on the Rush Limbaugh Show. The text is as follows:

> Global warming. I know you've been seeing more and more stories about the global warming theory. Stories that paint a horrible picture. Stories that say the polar ice caps will melt. Stories that [illegible] for catastrophe. Well get real! Stop panicking! I'm here to tell you that the facts simply don't jibe with the theory that catastrophic global warming is taking place.
>
> Try this fact on for size. Minneapolis has actually gotten colder. So has Albany, New York. And the Department of Agriculture says that on both coasts of this country, winter temperatures are five to ten degrees cooler than previously reported. So folks, grab hold of yourselves and get the whole story before you make up your mind. Right now, you can get a free packet of easy-to-understand material about global warming. Just call this number: 1-800-346-6269 Extension 505. That's the Information Council for the Environment. After you read the free materials they send you, you'll have a better picture of what the facts are all about. That's 1-800-346-6269 Extension 505. Call Today. Because the best environmental policy is based on fact.[13]

The ICE public relations campaign targeted several congressional districts including Champaign, IL (Terry Bruce, House Energy and Commerce Committee) and Fargo, ND (Byron Dorgan, House Ways & Means Committee).

Another *Ties That Blind* report looked into the financing by the fossil fuel industry of a high profile economic model examining the costs of reducing CO_2 emissions. This is a logical next step for the fossil fuel industry to take as it gets increasingly difficult to convince the public that climate change is not a problem. "Industry supports a reduction in greenhouse emissions but — in view of the uncertain science — wants to see better economic justification for energy curbs and wants to see those burdens shared worldwide, says John Schlaes, director of the Global Climate Coalition."[14]

In light of the significant media coverage recently given to the costs involved in reducing global CO_2 emissions, Ozone Action's *Ties That Blind* reports a closer look at one of the models, frequently cited in the popular press, that is reflected in present U.S. climate models. Charles River Associates (CRA) is a Boston based consulting firm, commissioned by the American Petroleum Institute (a trade association comprised of oil companies including Ashland Oil, BP Oil, Chevron, Exxon, Phillips Petroleum and Shell) to develop a computer model to look at the economic costs of climate change. Dr. W. David Montgomery, an economist with CRA, developed the International Impact Assessment Model (IIAM) in collaboration with Paul M. Bernstein, also of CRA and Prof. Tom Rutherford, of the University of Colorado. The model was designed to determine the ways that climate change policies will affect various countries. We will not attempt a detailed rebuttal of this model. That will require another format. We will, however, point out the inherent weaknesses of relying too much on such a model.

Any good economist will tell you that economic models never tell us anything new, they just provide an analytical framework for articulating certain beliefs, assumptions and projections. So when the American Petroleum Institute finances David Montgomery to develop his model, one must look at the underlying assumptions. Only by making certain assumptions about the economy can Montgomery have come to the following conclusions:

> Different studies of the economic cost of returning emissions in the year 2000 to 1990 levels concluded that those costs could range from 0.2% to 4.0% of GDP in 2010. This huge range, a factor of 20, underscores the economic risks of committing to a legally-binding target without having thought through what it will take to fulfill that commitment.[15]

> ...it is always less costly to allow emissions to rise for a while, then to make sufficiently larger reductions in later years to achieve the climate objective, than it is to begin with emission limits... [16]

> The costs of legally-binding emissions targets in the near- to mid-term are likely to be large.[17]

Montgomery's conclusions rest on many assumptions, including the following:
- ecological or economic costs of inaction or delay are not represented;
- the discount rate [it's always cheaper to hang on to your money, invest it, and spend money on the emissions reductions years later];
- technologies to make use of carbon-free energy sources at a reasonable cost are not stimulated by well-structured policies. Rather, they appear many years later simply as a function of time. In other words, policies to reduce CO_2 emissions don't create new markets or incentives and the resulting lower costs.[18]

Montgomery has had considerable exposure with his climate model and yet is rarely identified with the American Petroleum Institute, his financial backer. Ozone Action found numerous citations for Montgomery and his model without any reference to the American Petroleum Institute.

The way Montgomery has set up his model is appropriate for the fossil fuel industry. The longer policymakers wait before CO_2 emissions are undertaken, the better it is for fossil fuel companies. Shareholders in these companies are well-served by Montgomery. Downplaying the enormous potential benefits of averting the worst climate change impacts and ignoring the importance of the rate at which climate change takes place are major flaws. It is not "cheaper" to wait to reduce CO_2 emissions until several small island states have gone under from rising sea-level. It is not "cheaper" to wait until the spread of infectious diseases from climate change is at a fever pitch, so to speak. It is not "cheaper" to wait as property damage along our coasts escalates even further and it is not "cheaper" to have a declining freshwater supply.

Additionally, it is understandable that the fossil fuel industry would not want to incorporate into the model the benefits to society of supercharging the alternative energy industry. Or to make more generous assumptions about the benefits of running an economy in a more energy efficient manner. Ironically, Montgomery's assumption that cost-effective alternative energy sources appear magically in the future goes against recent history.

David Montgomery has considerable international influence as well. For example, at the December 11, 1996 Geneva round of negotiations he spoke about the economic impacts of climate change mitigation. In the advertisement for this briefing, he is not identified with the American Petroleum Institute, but rather, is identified as vice president of Charles River Associates and "a lead author in the IPCC Working Group III Assessment." This briefing was sponsored by the U.S. Council for International Business and chaired by Clement B. Malin who is the head of the International Chamber of Commerce which represents Shell International, Dow Europe, Texaco, Elf Atochem, Korea Electric Power Corporation, and DuPont among others. In this briefing, Montgomery stated that "What happens to the rates of emissions is irrelevant." Ironically, the IPCC couldn't be more clear on this point.

> Decisions taken during the next few years may limit the range of possible policy options in the future because high near-term emissions would require deeper reductions in the future to meet any given target concentration. Delaying action might reduce the overall costs of mitigation because of potential technological advances but could increase both the rate and the eventual magnitude of climate change, hence the adaptation and damage costs.[19]

Montgomery and his fossil fuel backers have also contributed to policy gridlock, which has become more obvious at the international negotiations. While on the one hand, Montgomery uses his model to convince the U.S. to do nothing unless significant commitments are secured from the developing nations, he also uses his model to convince developing countries that any CO_2 reductions will hinder their economic development. This produces an impasse from which the only policy that can emerge is inaction.

Examples abound of industries about to be regulated, crying that the American economy will be dealt a deathblow. In numerous cases, new regulations, when implemented, result in new industrial innovations and significant environmental benefits. Preregulatory cost estimates always seem to exceed actual costs. When it was first learned that CFCs could damage the stratospheric ozone layer and a ban on the chemicals was considered, DuPont and other CFC producers, along with most aerosol-makers, fought any change claiming that the theory was unproven, more research needed to be done and that the jobs of hundreds of thousands of people would be at risk.

> DuPont, the world's leading CFC maker— argued that 'there is no concrete evidence to show that the ozone-depleting reaction with chlorine takes place' and that government regulation would trigger 'tremendous dislocation' in the CFC industry, which he estimated contributed $8 billion to the U.S. economy and employed 200,000 people.[20]

However,

> Despite predictions to the contrary, the 1978 CFC ban has produced significant economic benefits, and net cost savings to the U.S.economy. One mid-1980s study concluded that the switch to hydrocarbon propellants — which in 1986 cost one-third less than CFCs — saved American businesses and consumers more than $1.25 billion (in current dollars) from 1974 to 1983.[21]

What we have learned from history is that our actions and policies help create our future and strong environmental guidelines lead to the creation of previously unseen markets and technologies. If one sets real environmental goals, well structured policies will stimulate innovations and quickly bring down the cost of compliance. Despite the inappropriateness of models funded by the fossil fuel industry, their spokespeople continue to make the rounds at the international negotiations. A similar model by the WEFA Group, also with funding from the American Petroleum Institute, was being paraded around the journalists in Ohio as recently as September 1998.

Into the category of dirty tricks by the fossil fuel industry, is a short lived campaign run in early 1998 by a brand new organization called the Committee to Preserve American Security and Sovereignty (COMPASS). The campaign consisted of a letter that was signed by senior members of the defense establishment which was then turned into an ad with the headline: "Foreign Policy Experts Say Kyoto Is Bad For America". The letter goes on to say that the "Treaty on Global Warming signed in Kyoto, Japan threatens American security and sovereignty even if the treaty is never officially signed by the President or ratified by the Senate."

The ads, paid for by COMPASS, list the same address as the offices of Kelley Swofford Roy Helmke, Inc. This firm also represented Colombia's President Ernesto Samper, who is referred to by Assistant Secretary of State Robert Gelbard as a "clearly corrupt president" whose campaign received millions of dollars from drug interests. The phone number of Kelly Swofford, is answered as "Office of Mark Helmke." The secretary acknowledges that it is the office of COMPASS, but when asked if it is also the office of Kelly Swofford, she said no. When I pursued the question further and asked if they shared office space with Kelly Swofford, she said "I don't know how to answer that." A brass sign on the wall of the building at 1002 King St. clearly says "Kelly Swofford Roy Helmke, Inc." Many of the signatories have direct ties to fossil fuel interests including Texaco, Amoco, The Kingdom of Saudi Arabia, General Motors, Halliburton Company and Phillips Petroleum.

Industry funded misinformation campaigns reached a fever pitch just prior to the important international climate meeting in Japan in December 1997. The auto and oil companies in the U.S. were determined to keep any climate treaty from emerging in Japan. One single campaign spent $13 million to convince the American public that U.S. emissions reductions without developing country participation will lead to our competitive and economic downfall. Exxon Chairman Lee Raymond became the poster child of hypocrisy, starting his day arguing against any climate treaty that lacked developing country participation at home, and finishing with a trip to China to threaten the Chinese not to sign on if they wanted foreign investment. [For the record, the fossil fuel industry's position stands in sharp contrast with the majority of Americans, 74% of whom said they support the global warming treaty according to a December 17, 1997 Harris poll.]

The most recent smoking gun showing the fossil fuel industry's desire to distort the public debate was revealed in a memo leaked out of the American Petroleum Institute that showed up in *The New York Times* on April 26, 1998, entitled, "Industrial Group Plans to Battle Climate Treaty: Aims to Recruit Skeptics: Draft Proposal Seeks to Depict Global Warming as a Case of Bad Science." The draft plan, worked up by industry representatives including participants from Exxon, Chevron and the Southern Company, "calls for spending $5 million over two years to 'maximize' the impact of scientific views consistent with ours on Congress, the media and other key audiences."

The proposed campaign would "recruit a cadre of scientists who share the industry's views of climate science and to train them in public relations so they can help convince journalists, politicians and the public that the risk of global warming is too uncertain to justify controls on greenhouse gases like carbon dioxide that traps the sun's heat near Earth." It's the same old tricks to deny reality, similar to years of campaigns by the tobacco industry to convince us that smoking wouldn't cause cancer.[22]

Despite these ongoing heavy handed efforts, a treaty was struck in Kyoto, Japan in December 1997, the first step towards global reductions of greenhouse gas emissions. The negotiations are still on going, as details to the climate agreement are worked out. Still, it is a very weak agreement, no doubt due, in large part, to the efforts of the fossil fuel industry. Weak, as Abraham Lincoln once said, "as the soup made from the shadow of a crow that had died of starvation." Weak, in that if all nations comply with the commitments made, we will still have severe, human-induced climate change. And weaker still, considering that the relentless misinformation campaigns against this agreement continue.

Nevertheless, the Kyoto Protocol sent a signal. European carmakers recently proposed voluntarily increasing their emissions standards by 25%. GM's Chairman and CEO John F. Smith overnight became a vocal promoter of Detroit's forthcoming high-mileage vehicles. U.S. automakers are starting to see that the future belongs to the Japanese and the Europeans unless they can recover lost time — time spent lobbying and advertising against change — time they couldn't recover in the 1970s after fuel prices shot up and they only had gas guzzlers to offer the American public. In fact, Smith's chief general counsel, Thomas Gottschalk, said publicly at GM's annual meeting in Wilmington, Delaware that "They are actively looking at whether they should continue membership in the Global Climate Coalition."[23]

Industry is starting to turn. No longer can one assume large companies are against leadership to avert climate change. British Petroleum has come out for efforts to avert climate change, the Royal Dutch/Shell Group has held some dues back from the American Petroleum Institute because of the proposed misinformation campaign and Sun Oil Company recently joined a new business coalition that supports efforts to combat climate change; coalition members include Toyota, American Electric Power, Enron, BP, Boeing, Whirlpool, Maytag, 3M, Lockheed-Martin and United Technologies.

But two problems remain. First, as long as some of the major fossil fuel and auto companies that are not on the above list continue to fund efforts to hold the world back, critical time will be lost in combating this global threat. Weather will be more severe, infectious diseases will spread further and coastal communities will watch the seas climb further up their vanishing beaches. Ultimately, however, we will all use far less energy from fossil fuels in our daily lives. We will do it and we will live better for it. This leads to the second problem. If our auto and oil companies remain the last holdouts in this global conversion to a more efficient future, we Americans will be buying our cars from foreign companies that looked to the future and we'll get our renewable energy from technologies that Exxon and Texaco were too entrenched to capitalize on.

RECOMMENDED READING:

Gelbspan, Ross (1997) *The Heat is On: The High Stakes Battle Over Earth's Threatened Climate*. New York: Addison-Wesley Publishing.

THOSE WHO SUFFER ARE SILENCED

Did you know that many toxic polluting corporations decide where to build their plants or collect their waste based on the racial makeup of the surrounding community? Are you shocked? Surprised? Probably not. Racism permeates our lives on planet Earth — why should the environment be an exception?

President Clinton recognized this on February 11, 1994 when he issued Executive Order 12898, making the fight for environmental justice a federal policy. This year, the U.S. Environmental Protection Agency issued interim guidelines for investigating complaints filed under Title VI of the 1964 Civil Rights Act which bars any program receiving federal funding from discriminating on the basis of race, color or national origin.

State government programs and a series of environmentally hazardous corporate ventures in communities of color are being affected by these guidelines and pressure is on for the EPA to back off.

But this attempt by the government to do the right thing is under heavy assault from the country's worst toxic polluters who are forming associations to stop the EPA from applying civil rights law to environmental issues. One of these lobbying groups is the Business Network for Environmental Justice (BNEJ), a secretive corporate coalition under the direction of the National Association of Manufacturers. The BNEJ has demanded that the EPA issue permits for new corporate facilities "regardless of the racial composition of the surrounding community." Is there any clearer indictment of these industries?

It is unconscionable and obscene that people of color, people perceived as the powerless in our society, are dying every day from the toxic waste from corporations all over the U.S. and across the world.

No more blatant example may exist of environmental racism than "Cancer Alley," an area in Louisiana that is impacted by more accumulated toxic waste than probably anywhere in the country. In the late 1980s, the entire town of Geismar, Louisiana — with a population that is poor and African-American — was moved two miles down the road because of toxic fallout from the neighboring chemical plant. Two miles! Of course it didn't help, but you did not hear about this on the evening news, nor will you. Each member of the community signed an agreement, in exchange for $2000, saying they would never talk about the situation. This is common practice in out-of-court settlements between major polluters and people who are just trying to live their lives in small communities.

Because of the influence of groups like the BNEJ, the Chemical Manufacturers Association and the American Petroleum Institute, many state governments are protesting the implementation of Title VI of the Civil Rights Act as a tool for environmental justice. It is time for all citizens to remind the state governments who they really should be working for and to support the EPA in their efforts.

Jackie Alan Giuilano, Ph.D.
Republished with permission from the Environment News Service (ENS),
international daily newswire of the environment, online at:
http://www.ens.lycos.com

ON THE
ENVIRONMENTALIST
FRONT

POPULAR EPIDEMIOLOGY
Lay Discovery of Hazards & Causes of Illness

Phil Brown

"Technically, a disease cluster is a geographic area with an unusually high rate of a given disease. As a result of the environmental crisis affecting our society, such clusters have become commonplace in the past three decades. What has also become all too common is a pattern of denial, obfuscation and abuse from industry representatives and in many cases also from public health officials."

Dr. Joel B. Swartz

Popular epidemiology represents two related phenomena: (1) a form of citizen science in which people engage in lay ways of knowing about environmental and technological hazards, and (2) a type of social movement mobilization which increasingly plays a major part in modern political culture. This article examines these two aspects, with reference to major empirical studies.

Origins of Epidemiological Discovery

In my original formulation of popular epidemiology, I defined it in contrast to traditional epidemiology. Traditional epidemiology studies the distribution of a disease, and the factors that influence this distribution, in order to explain the etiology, and to provide preventive, public health and clinical practices. Popular epidemiology, in contrast, is a broader process whereby lay persons gather data, and also collaborate with experts. To some degree, popular epidemiology parallels scientific epidemiology, such as when lay-people conduct community health surveys. Yet it is more than public participation in traditional epidemiology since it usually emphasizes social structural factors as part of the causal disease chain. Further, it involves social movements, utilizes political and judicial approaches to remedies, and challenges basic assumptions of traditional epidemiology, risk assessment, and public health regulation.

Popular epidemiology efforts do not *require* epidemiological health studies, even though these may occur. In deed, activists generally want the hazards avoided and remediated. If the government and/or corporations involved would admit the problem and take appropriate action, activists would have no need for or interest in health studies. But the activists experience much resistance to their claims, and require the kind of public political and scientific support that often presses for more concrete evidence of causation.

From studying contaminated communities where citizens' groups have discovered toxic contamination, we observe a typical set of stages. My chief model is the Woburn, Massachusetts case, where citizens discovered a leukemia cluster, pushed for government action, collaborated with scientists to conduct a health study on leukemia, reproductive disorders and birth defects, and also filed suit against the companies they believed to be responsible.[1] In working with people at the Hanford Health Information Network, and in observing ongoing struggles in many communities, I have recently added a tenth stage. Throughout these stages, lay and professional disputes occur over lay participation— standards of proof, constraints on professional practice, disputes over the nature of risks and hazards, quality of official in different studies, and professional autonomy. These stages may vary locations, and may overlap each other:

1. Lay Observations of Health Effects and Pollutants.

Many people who live at risk of toxic hazards have access to data otherwise inaccessible to scientists. Their experiential knowledge usually precedes official and scientific awareness. They are not yet activists.

Although the first official action closing Woburn's polluted wells occurred in 1979, there was a long history of problems in the Woburn water. Residents had for decades complained about dishwasher discoloration, foul odor and bad taste. Private and public laboratory assays had indicated the presence of organic compounds. The first lay detection efforts were begun earlier by Anne Anderson, whose son, Jimmy, had been diagnosed with acute lymphocytic leukemia in 1972. Once again, it was a woman with no experience in this area, no higher education and no organizing experience who started the ball rolling. Once again, it began with knocking on doors, making intuitive connections and never stopping in the quest for the most complete information.

2. Hypothesizing Connections

Activists often make assumptions about what contaminants have caused health and environmental outcomes.

Anderson put together information during 1973 about other cases by meetings with other Woburn victims in town and at the hospital where Jimmy spent much time. Anderson hypothesized that the alarming leukemia incidence was caused by a water-borne agent. In 1975 she asked state officials to test the water but was told that testing could not be done at an individual's initiative.

3. Creating a Common Perspective

Activists begin to piece together the extent of the problem and to organize it coherently. Lay mapping of disease clusters is a typical feature of this stage. Citizens also make discoveries of actual pollution sources.

Anderson sought to convince the family minister, Bruce Young, that the water was somehow responsible, though he at first supported her husband's wish to dissuade her. The creation of a common perspective was aided by a couple of significant events. In 1979 builders found 184 empty barrels in a vacant lot; they called the police, who in turn summoned the state EPA. When water samples were then taken from a number of municipal wells, wells G and H showed high concentrations of organic compounds known to be animal carcinogens, especially trichloroethylene (TCE) and tetrachloroethylene (PCE). Well G had 40 times the EPA's maximum tolerable TCE concentration. As a result the state closed both wells.

In June 1979, just weeks after the state closed the wells, an engineer who worked for the state EPA drove past the nearby Industriplex construction site and thought he saw violations of the *Wetlands Act*. A resultant federal EPA study found dangerous levels of lead, arsenic and chromium, yet EPA told neither the town officials nor the public. The public only learned this months later, from the local newspaper. Reverend Young, initially distrustful of Anderson's theory, came to similar conclusions once the newspaper broke the story. Along with a few leukemia victims, he placed an ad in the Woburn paper seeking people who knew of childhood leukemia cases. Working with John Truman, Jimmy Anderson's doctor, Young and Anderson prepared a questionnaire and plotted the cases on a map. Six of the 12 cases were closely grouped in East Woburn. Over the years they identified more cases, claiming 28 cases over a longer period, 1965-80; 16 of those people died.

This lay mapping phenomenon is almost an instinctual public reaction. Patricia Nonnon did it in the Bronx with leukemia. A woman in Coeur d'Alene, Idaho tracked down two entire high school graduating classes when she noticed a huge cancer increase. Leon and Juanita Andrewjeski kept what they termed a 'death map' of cancer deaths and illness and of early heart attacks among farmers downwind of Hanford. In Leominster, Massachusetts, a doctor told a mother of an autistic child, 'You do so well with your children, you should talk with other parents on your block who have autistic children'. This led to lay mapping of 35 cases in a small area.

4. Looking for Answers from Government and Science

Activists expect government health agencies and their scientific colleagues to investigate the problem. Citizens are often given little or no support from officials and scientists. In-depth epidemiological studies are rarely done. Rather, officials may merely make statements about the relationship of rates in toxic communities to state or national rates.

The data convinced Dr. Truman, who called the Centers for Disease Control. The citizens persuaded the City Council in December 1979 to ask the CDC to investigate. Five days later, the Massachusetts DPH reported on adult leukemia mortality for a five-year period, finding a significant elevation only for females. This report was cited to contradict the residents' belief that there was a childhood leukemia cluster.

5. Organizing a Community Group

Citizens formalize their social action by setting up an organization to provide social support and information for toxic victims, deal with local, state and federal agencies, attract media attention, and make connections with other toxic waste groups.

In January 1980, Young, Anderson and 20 others formed For a Cleaner Environment (FACE) to solidify and expand their efforts. Community groups in contaminated communities provide many important functions. They galvanize community support, deal with government, work with professionals and engage in health studies. They are the primary information source for people living in contaminated communities, and often the only accurate source. Community groups also provide social and emotional support. Through their organization, Woburn activists report pride in learning science, in protecting and serving their community, in guaranteeing democratic processes, and in personal empowerment.

6. Official Studies are Conducted by Experts

Additional pressure leads to more detailed studies, but these tend to continue the denial of toxic waste-induced disease.

In May 1980, the Centers for Disease Control sent Dr John Cutler to collaborate with the DPH on further study. By then, the Woburn case had national visibility due to network television coverage, and then in June from Senator Kennedy having Anderson and Young testify at hearings on the Superfund. Jimmy Anderson died in January 1981, and five days later the CDC/DPH study was released, stating that there were 12 cases of childhood leukemia in East Woburn, when 5.3 were expected. Yet the DPH argued that the case-control method (12 cases, 24 controls) failed to find characteristics (e.g. medical histories, parental occupation, environmental exposures) that differentiated victims from nonvictims, and that lacking environmental data prior to 1979, no linkage could be made to the water supply. The report also stated that TCE was known as an animal, but not human, carcinogen."

7. Activists Bring in Their own Experts

At this point, more formal popular epidemiology efforts begin. Lacking official support for their claims of toxic contamination, citizens find sympathetic scientists who are willing to help them in health studies.

The activists had no 'court of appeals' to oppose the government's evidence, so FACE set out to obtain the information themselves. The conjuncture of Jimmy Anderson's death and the DPH's failure to implicate the wells led the residents to question the nature of official scientific studies. They received help when Anderson and Young presented the Woburn case to a seminar at the Harvard School of Public Health, where Marvin Zelen and Steven Lagakos of the Department of Biostatistics became interested. Working with FACE members, they designed a health study, focusing on child leukemia, birth defects and reproductive disorders. The academics and activists teamed up in a prototypical collaboration between citizens and scientists. No longer did residents have to seek scientific expertise from outside; now they were partners in scientific inquiry. The survey collected data on adverse pregnancy outcomes and childhood disorders from 5010 interviews, covering 57 percent of Woburn residences with telephones. The researchers trained 235 volunteers to conduct the survey, taking precautions to avoid bias.

8. Litigation and Confrontation

Citizens often file lawsuits against corporations they believe are responsible for contamination. They may also take other political action, such as picketing, boycotting and lobbying.

During this period, the state EPA's hydrogeological investigations found that the bedrock in the affected area was shaped like a bowl, with wells G and H in the deepest part. The contamination source was not the Industriplex site as had been believed, but rather facilities of W. R. Grace and Beatrice Foods. This led eight families of leukemia victims to file a $400 million suit against those corporations in May, 1982. A smaller company, Unifirst, was also sued but quickly settled before trial. In July 1986 a federal district court jury found that Grace had negligently dumped chemicals; Beatrice Foods was absolved. An $8 million out-of-court settlement with Grace was reached in 1986. The families filed an appeal against Beatrice, based on suppression of evidence, but the Appeals Court rejected the appeal in 1990, and the Supreme Court declined to hear the case.

The trial was a separate but contiguous struggle over facts and science. Through consultant physicians and scientists, the families accumulated further evidence of health effects. The data were not used in the trial, which never got to the point of assessing the causal chain of pollution and illness. Nevertheless, the process made the residents more scientifically informed.

9. Pressing for Official Corroboration

When community sponsored health studies are completed, citizens use them to confirm the legitimacy of their claims. At this stage they are apt to meet resistance from professional organizations, federal agencies and disease charities.

In February 1984, the FACE/Harvard data were made public. Childhood leukemia was significantly associated with exposure to water from wells G and H. Children with leukemia received an average of 21.2 percent of their yearly water supply from the wells, compared to 9.5 percent for children without leukemia. Controlling for risk factors in pregnancy, the investigators found that access to contaminated water was associated with perinatal deaths and some birth defects (deaths since 1970, eye/ear anomalies, and CNS/chromosomal/oral cleft anomalies). With regard to childhood disorders, water exposure was associated with kidney, urinary and respiratory diseases.

Though this study would not have been possible without community involvement, precisely this lay involvement led to charges of study bias from the DPH, the CDC, the American Cancer Society and the EPA. The researchers conducted extensive analyses to rule out bias, but still officials argued that interviewers and respondents knew the research question and that respondents had potential recall bias. Critics also said that the water model measured only household supply rather than individual consumption.

10. Continued vigilance

Whether or not activists have been successful in making their case, they find the need to be continually involved in cleanups, additional official surveillance, media attention, and overall coordination efforts.

Woburn activists had to keep defending their data. They were looking for confirmation from a DPH reanalysis of reproductive health effects. In 1995, activists from For a Cleaner Environment (FACE) were confronted with the long-awaited results of the Department of Public Health's reanalysis of health data. The DPH had spent years reanalyzing the data on both childhood leukemia and reproductive disorders. It was common knowledge that the results were in, yet years passed while the DPH figured out how to present the findings. One official stated publicly that she believed the leukemia data would confirm the FACE/Harvard study that implicated wells G and H in childhood leukemia, though nothing was said about the expected results of the reproductive disorders study. In 1995, a draft report was issued, for public comment that claimed no environmental basis for reproductive disorders. Upon examining the research design, FACE activists and their scientific colleagues found that the DPH had analyzed only a brief time period, which was too late to capture many of the earlier effects. FACE believes it must obtain the new DPH data set and conduct its own analyses, but that is a time-consuming task and there is no direct funding for that. Yet barring such reanalysis, FACE will be on weaker ground in arguing against the DPH. Woburn activists were at least spared the need for continual vigilance in terms of the DPH's reanalysis of the leukemia cases. There, the DPH found a dose-response relationship between childhood leukemia and maternal consumption of water from the contaminated wells G and H.

For another example of continued vigilance, we can observe the Hanford case, where activists succeeded in forcing the Department of Energy (DOE) to release the Hanford Historical Documents that showed how the Hanford Atomic Works facility, which had been engaged in the production of plutonium for weapons in World War II, had many intentional and unintentional releases of radioactivity over several decades.[2] Millions of DOE dollars were made available to develop the Hanford Health Information Network (HHIN) to educate people about this history, and to conduct health studies and surveillance. Activists were hired among the staff people of this tri-state (Washington, Oregon, Idaho) region, and appeared to have a good deal of impact. Yet, at a major 1993 conference sponsored by HHIN, it was clear that this was not a collective effort of all parties. Activists were concerned, for example, that DOE officials and numerous scientists were using restrictive definitions of radiation-caused morbidity and mortality, and that this would limit the ability of people to demonstrate health effects from Hanford releases.

The Impact of Toxic Waste Activism and of the Environmental Justice Movement

In light of the dramatic growth of the environmental justice movement and its impact on other parts of the toxic waste movement, it is necessary to discuss the applicability of the popular epidemiology approach to the toxic waste movement and to the broader environmental justice movement. My definition of popular epidemiology early in this article included the centrality of political action and social movements. Many local toxic action groups do not initially view themselves as part of a national or international movement. Rather, they are only dealing with what they perceive as local problems. But the initial groups, from the time of Love Canal, set in motion a new movement. As well, the subsequent prevalence of these groups contributes to an already existing movement which has some more consciously political centers of activity. While most of the multitude of these groups would not style themselves as popular epidemiology practitioners, this is what they are doing in their lay efforts to uncover disease and act on that discovery. The toxic waste movement represents a solidification into broader social action of popular epidemiology's critique of traditional science.

Because toxic waste activists start out from personal experience rather than political ideology, they differ from participants in the broader environmental movement in several ways: their generally lower-class background, lower levels of education, the predominance of women members and leaders among them, and a higher level of participation by minorities than in the environmental movement. Overall, activists have less general political ideology or experience than do mainstream environmentalists, though minority activists are often influenced by the civil rights legacy.[3]

Women play central roles in the toxic waste movement.[4] Women are the most frequent organizers of lay detection, partly because they are the chief health arrangers for their families,[5] and partly because their child care role makes them more concerned than men with local environmental issues.[6] Women approach the problem of toxic contamination and official response with a notion of fairness, equity and collective protection. They center their world view more on relationships than on abstract rights, and on their roles as the primary caretakers of the family. These roles lead women to be more aware of the real and potential health effects of toxic waste, and to take a more skeptical view of traditional science. They often undergo a transformation of self, based on changes noted by Belenky et al.[7] in their concept of 'women's ways of knowing'. That perspective traces the ways that women come to know things, beginning with either silence or the acceptance of established authority, progressing to a trust in subjective knowledge, and then to a synthesis of external and subjective knowledge.[8] This kind of knowledge framework makes it logical that women toxic activists would gravitate to a popular epidemiology approach.

As Mohai and Bryant noted in their review of 15 studies,[9] the wide range of geographical areas (local, regional, national) provides a combined body of knowledge which finds race and class, especially race, to be large factors in exposure to environmental hazards. Those, along with subsequent studies, provide extensive evidence for excess exposure of minorities (especially African Americans), and substantial excess exposure by class in the following categories: (1) presence of hazardous waste sites and facilities (landfills, incinerators, Superfund sites); (2) air pollution; (3) exposure to various environmental hazards, e.g. hazards in pesticides and foods, toxic releases measured by the EPA's Toxic Release Inventory; (4) actual and planned cleanups at Superfund sites; (5) fines for environmental pollution; (6) specific health statuses which are related to environmental burden (e.g. blood lead); (7) siting decisions for incinerators, hazardous waste sites, nuclear storage sites.[10]

Two Elements of Popular Epidemiology: Lay Discovery and Health Studies

The above comments on environmental justice (both the 'traditional' environmental justice focus on class and race, and the newer additional focus on women) demonstrate that epidemiological studies are not necessarily major features of toxic waste activism. This indicates that popular epidemiology describes an activist style of lay discovery of hazards and disease, more so than necessarily focusing on health studies. Some scholars have argued that popular epidemiology relies too much on scientific experts, and is not a true lay effort.[11] While I do not agree with this, I think it forces a sharper conceptualization. In tandem with the last section, this leads me to consider popular epidemiology as having two elements. The first element is the phenomenon of lay discovery of disease. Popular epidemiology is a major phenomenon even when limited to the discovery of disease clusters, overall disease excesses and the presence of hazards. The second element is the more specific participation of activists in epidemiological studies. This occurs in a relatively small number of situations where citizens are engaging in popular epidemiology approaches. It is important to note, however, that lay people who embark on a process of social discovery of toxic contamination do not typically have any idea of the extent of their future involvement. They may wind up involved in health studies even if that was never their original intent. Hence, the more 'routine' popular epidemiology (general discovery) cannot be so easily distinguished from the more 'formal' type (involvement in health studies).

Interaction of Popular Epidemiology and Critical Epidemiology

Lay epidemiological approaches have changed the nature of scientific inquiry in various ways.

1. Lay involvement identifies the many cases of 'bad science', e.g.

poor studies, secret investigations, failure to inform local health officials, fraud and cover-up.

2. Lay involvement points out that 'normal science'[12] has drawbacks, e.g. automatically opposing lay participation in health surveys demanding standards of proof that may be unobtainable or inappropriate, being slow to accept new concepts of toxic causality.

3. The combination of the above two points leads to a general public distrust of official science, thus pushing lay people to seek alternative routes of information and analysis.

4. Popular epidemiology yields valuable data that often would be unavailable to scientists. If scientists and government fail to solicit such data, and especially if they consciously oppose and devalue them, then such data may be lost. This goes against the grain of traditional scientific method.

5. Popular epidemiology has pioneered innovative approaches. For example, the Environmental Health Network is testing a new way to quickly ascertain the relationship between disease clusters and toxic wastes by using the EPA's Toxic Release Inventory, a computerized database accessible to all citizens in public libraries.

6. Popular epidemiology has had a powerful impact on public policy. The Woburn case was the major impetus for the establishment of the state cancer registry.

7. Popular epidemiology has spurred enormous amounts of research. In the case of Woburn, it led the Department of Public Health (DPH) and Centers for Disease Control (CDC) to conduct a major five-year reproductive outcome study and a case-control study of leukemia.

Toxic waste activism in Woburn has also led to several Massachusetts Institute of Technology (MIT) studies. One group of projects, totaling $3.33 million in federal support, will produce a complete hydrogeological study; a history of the tanning industry, a major polluter; and an innovative genetic toxicology study based on human cell mutation assays, which will endeavor to produce a 'unique chemical fingerprint' of a large number of known and suspected toxic substances.[13] A second MIT effort is a three-year study funded by the Agency for Toxic Substances and Disease Registry and the National Institute for Occupational Safety and Health. It will examine the scientific, ethical and legal issues in monitoring residents and cleanup workers in three Superfund sites, one of which is Woburn.

Many more examples could be cited from other locations. Critical epidemiologists understand that these developments would not have arisen from the process of normal science and routine public policy efforts.

Official and professional resistance to working with community groups may demonstrate to some epidemiologists that official scientific and governmental approaches are supporting a conservative status quo value frame. As the number of critical epidemiologists grows, we may see a greater number of well-designed health studies in which laypeople play a central role.

What Can Such Concerned Scientists Do?

Wing puts forth an alternative conceptualization of epidemiology[14] that taps the concerns of both popular epidemiology activists and critical epidemiologists:

1. It would ask not what is good or bad for health overall, but for what sectors of the population. Lay involvement has been able to show that much of routine science is actually 'bad science. For instance, lay activism has shown that government studies were frequently poorly designed, and even sometimes conducted secretly. State public health officials commonly failed to inform local health officials of environmental health problems. And some research was designed to cover up health problems rather than expose them.

2. It would look for connections between many diseases and exposures, rather than looking at merely single exposure-disease pairs.

3. It would examine unintended consequences of interventions.

4. It would utilize people's personal illness narratives.

5. It would include in research reporting the explicit discussion of assumptions, values, and the social construction of scientific knowledge.

6. It would recognize that the problem of controlling confounding factors comes from a reductionist approach that looks only for individual relations rather than a larger set of social relations. Hence, what are nuisance factors in traditional epidemiology become essential context in a new ecological epidemiology.

7. It would involve humility about scientific research, combined with a commitment to supporting broad efforts to reform society and health.

Some critical epidemiologists[15] argue that we should not merely control for central features of interest such as race and class, but should study samples of class and racial groups and use case history approaches as well. This new approach to the unit of analysis leads us to consider the significance of community. As has been so well established in recent years, the destruction or wounding of community has been a central feature of environmental disasters and hazards.[16]

Bringing the Community Back In

There are many aspects of community that we need to take into account in studying the effects of toxins. These include people's ontological sense of being, provisions of personal security and social support, the security of the home, the structures and bonds of the family, sets of relationships, physical and geographic features, racial/ethnic distributions and cultures, class distributions and cultures, economic and industrial history, sharing and control of local resources, shared political beliefs, governmental structures, civic empowerment, shared goals for social change, and community health status. This article is not the place to develop all of these elements, but I merely want to mention the complexity that is involved. The following example indicates how this might be applied.

Martha Baishem offers an alternative view of community context in her study of how a Philadelphia cancer prevention project clashed with the white, working-class neighborhood's belief system.[17] The project identified excess cancer in this area, a fact widely known by the residents and the media. The medicalized approach of the health educators focused on individual habits, especially smoking, drinking and diet. Tannerstown residents countered this worldview with their belief that the local chemical plant and other sources of contamination were responsible. Clearly, the broad-based 'blame the victim' response which has typified our nation's response to social ills and inequities in general, conditioned the health educators' tendency to treat illness as related to individual (mis)behavior, rather than to the needs and behaviors of powerful interest groups.

Baishem recounts one woman's tale of agony over her husband's death from pancreatic cancer. Jennifer.. fights for the right to see John's medical record. When she finds a line about his smoking and drinking, she asks the doctor, 'What's this on here?' He responds, 'That's not important'. Jennifer retorts, 'It's important to me. You're going to take this report saying he's an alcoholic and he smokes and this is what causes cancer. Then you wonder why we get upset because the statistics are wrong'. Jennifer struggles to get his medical record rewritten, and she demands an autopsy to show that he didn't die of lung cancer, but of metastatic pancreatic cancer, and hence is not blamed for smoking-induced cancer. To Jennifer, in her struggle to protect her deceased husband's health status, abstract ethics and value-free epidemiology simply do not make sense in a dangerous world where activists must fight against hazards and disease. Efforts such as Jennifer's are made on an individual level of experience as well as a general sense of community aggrievement. Yet there is no overwhelming drive toward epidemiological studies. Jennifer, like toxic waste activists, wants justice and fairness, and will use science if it can help. But they do not desire a science 'above' ethics and morality. That is a contradiction in terms for them.

Popular Epidemiology and the Social Construction of Disease and Illness

Popular epidemiology is, in fact, looking to gain recognition of unrecognized or under-recognized causes of diseases (as with multiple chemical sensitivity), and unrecognized or under- recognized symptoms (as with the complex of outcomes associated with trichloroethylene). Further, the lay discovery of causes of illness in contaminated communities is very akin to the lay discovery of causes of illness in general.[19] In particular, both are contingent on the social context, ranging from family structure all the way to corporate and governmental power. People recognize their problems insight of other surrounding factors, and turn their individual troubles into social problems. Their social constructionist world view shapes an understanding of personal subjectivity that is at the same time larger than personal subjectivity. By studying how illness is socially constructed, we examine how social forces shape our understanding of and actions toward health, illness and healing. We explore the effects of class, race, gender, language, technology, culture, the political economy and institutional and professional structures and norms in shaping the knowledge base which produces our assumptions about the prevalence, incidence, treatment and meaning of disease.[19] The Bashem example of Jennifer's understanding of her husband's disease that was discussed earlier is a good example of this process.

Indeed, why should we persist in calling toxics-induced illnesses, "illnesses," rather than contaminations, poisonings, etc.? While these causes might lead to specific conditions which can be regarded as illnesses, it would be wrong to regard lay persons as " discovering new diseases"; rather, they are uncovering that they have been poisoned. In illness, we are infected by a virus, a bacteria, or an organ wears out; toxics-induced illnesses, however, are situations where there has been a social intervention. There is nothing "natural" in what has happened. It has been caused by the deliberate or ignorant actions of other people. It's unnecessary, preventable, a cause for anger, and maybe for lawsuit. By focusing on the after-effect, the condition that resulted from the exposure, we use our very diction — "illness" — to disguise what is actually occurring.

As well, the lessons learned from toxic waste activism may likely be applied to other disease discovery. I wrote in an earlier article[20] that popular epidemiology was in the tradition of the women's health and occupational health movements which have been major forces in pointing to often unidentified problems and working to abolish their causes. Examples of those hazards and diseases thus uncovered are DES, Agent Orange, asbestos, pesticides, unnecessary hysterectomies, sterilization abuse, and black lung. Now, the environmental health activities of popular epidemiology practitioners have become a major force in public life, and inspire other health social movements to engage in lay discovery.

We might even add a category of 'military environmental health' to the usual categories of environmental health and occupational health. The Agent Orange episode has been recently joined by the Gulf War Syndrome episode — a situation whereby veterans recognize odd symptoms, push for diagnoses and treatment, and encounter a secretive and hostile military and espionage apparatus that hides data that could affirm the connection between veteran's illnesses and Gulf War toxic releases.

Conclusion

The current expansion of popular epidemiology efforts presents a large shift in personal and social perception of threat. In this period of widespread criticism of medicine (concerning iatrogenic disease, poor quality of clinical interaction, unnecessary surgery, and race, class and sex biases in health care), many people are more likely to challenge medical authority when it fails to recognize their environmentally induced diseases. They are also responding to a prevalent criticism of science and technology, based on growing instances of chemical, radiation and other hazards. This critique leads the public to challenge accepted notions of expertise concerning the safety of technological hazards. This combined challenge to medicine and to science and technology is not merely a power struggle between lay and professional, but a deeper challenge to the entire legacy of our unfettered poisoning industrial system.

For Beck,[21] we live in a 'risk society' where the whole world has become a place of often unseen danger. In this 'catastrophic society', 'the state of emergency threatens to become the normal state'. The new world of technological disasters and threats, Erikson[22] tells us, is unbounded, without a frame, and 'violates the rules of plot'. The totality of this threat has the potential to quiet many people, and indeed does so. Yet it also provokes growing opposition, both from highly educated people who ordinarily do become politically active, but also from less educated people for whom this toxic threat is a great rupture in the routinely accepted life they had expected.

As the world becomes more threatened by technological disasters which produce anomalous diseases, people have different illness experiences. In particular, they face distrust by experts for malingering or concocting false stories intentionally, and their suffering is not taken seriously. This can lead them to rethink the normally accepted boundaries between their bodies and their environment. The realization that one may be poisoned by the products of affluence and power leads people to suffer rents in the social fabric that previously contained them. They lose trust in many of the bonds that previously held them. If they face this on a singular level, they may suffer depersonalization and depression. But when effectively organized by natural leaders in their communities, these people can achieve political solutions, economic redress, self-esteem, self-efficacy, and a revised view of their bodies' health. And rather than only opposing official experts, they can become experts. With increased educational attainment and widened cyberspace access to resources and communication, people can formulate a very different picture of their capacities to carry out citizen science. Breast cancer activists, victims of contaminated communities, and unwitting subjects of irradiation experiments are among the many laypeople who have taken their place at the bench of popular science and popular epidemiology.

Popular epidemiology stems from the legacy of health activism, growing public recognition of problems in science and technology, and the democratic upsurge regarding science policy. Communities face difficulties in environmental risk assessment due to differing conceptions of risk, lack of scientific resources, poor access to official information, and government policies which oppose or hinder public participation. In popular epidemiology, as in other health-related movements, activism by those affected is necessary to make progress in health care and health policy. In this process there is a powerful reciprocal relationship between the social movement and new views of science. The striking awareness of new scientific knowledge, coupled with government, corporate and professional resistance to that knowledge, leads people to form social movement organizations to pursue their claims-making. In turn, the further development of social movement organizations leads to further challenges to scientific canons. The socially constructed approach of popular epidemiology is thus a result of both a social movement and a new scientific paradigm, with both continually reinforcing the other. In the last five years we have witnessed a dramatic impact of the environmental justice movement on the larger toxic waste movement, and also a growth in women's sensibilities as well as women's activism. Popular epidemiology, with some amplifications, remains a useful conceptual framework for understanding these new trends.

RECOMMENDED READING:

Brown, Phil and Edwin Mikkelsen (1997, revised edition) *No Safe Place: Toxic Waste, Leukemia, and Community Action.* University of California Press.

Bullard, Robert, ed. (1992) *Confronting Environmental Racism: Voices from the Grass-roots.* Boston: South End Press.

Couch, Steven and J. Stephen Kroll-Smith, eds. (1991) *Communities at Risk: Collective Responses to Technological Hazards.* New York: Peter Lang.

Fox, Steve (1991) *Toxic Work: Women Workers at GTE Lenkurt.* Philadelphia, Temple University Press.

Wing, Steve (1994) "Limits of Epidemiology." *Medicine and Global Survival*, No. 1, pp. 74-86.

"In California, 40% of the children working in agricultural fields have blood cholinesterase levels below normal, a strong indication of organophosphate and carbamate pesticide poisoning."

R. Repetto and S.S. Baliga
Pesticides and the Immune System: The Public Health Risks, 1996

"In the United States, the use of synthetic pesticides has grown 33-fold since 1945, to approximately 0.5 billion kg/yr. The increase in related hazards is greater than the increase in applied amounts because most modern pesticides are more than 10 times as toxic to organisms than those used in the early 1950s."

David Pimentel
"Protecting Crops," 1995

ENVIRONMENTAL ORGANIZING
& Public Service Employees[*]

Jeff DeBonis

"To dispose a soul to action, we must upset its equilibruim."
Eric Hoffer

I vividly remember the first few days of my new "dream job" as a professional Forester Trainee with the United States Department of Agriculture Forest Service. Having recently completed a two year tour with the Peace Corps in El Salvador, as a soils conservation and reforestation specialist, the frustration of watching that country's ecological health unravel was fresh in my mind. Now, though, I was in the most ecologically advanced country in the world, or so I thought. I was ready to learn how to do forestry the "right" way, working on the Kootenai National Forest, a large timber forest in northwestern Montana. I had received a degree in Forest Management from Colorado State university only four years prior. I still believed, as did many of my peers, that if one wanted to work for the good of the public trust, to practice true multiple-use management where wildlife, fisheries, soils, and long-term ecological health were considered more valuable than the industrial foresters' short-term profit motive, then one would work for a public land management agency. The good agencies were within the federal government, and the best was the USDA Forest Service.

This was an agency untarnished by political scandals; an agency with the reputation of being the most high-minded in the history of U.S. public service. It boasted nearly 100 years of true public trust management, with visionaries like Gifford Pinchot and Aldo Leopold having been part of its ranks. This image left me unprepared for the paradoxical and narrowly focused culture that was the reality.

I had spent two agonizing years in El Salvador, trying to persuade small farmers to use soil conservation strategies and replant eroded hillsides that were denuded of vegetation and soon to be depleted of soil. These conditions were due more to social circumstances than the lack of technical expertise. A few rich families held most of the country's agricultural land, while the rest of the 3.5 million people had to manage with less fertile, steep, easily erodible lands for growing the crops they needed to feed themselves and their families. As a result, farmers practiced slash and burn agriculture on steep slopes. They would clear away the vegetation, burn off the organic matter, and plant crops in the exposed soil of hillsides, typically forty-five percent slopes or higher. This combined with heavy rainfall during the warm, rainy season, resulted in massive erosion problems, leaving vast areas unfertile and often stripped to bedrock. Downstream, siltation was filling in reservoirs shortening the life expectancies of hydroelectric projects, and destroying riverine and estuarine fisheries.

Industrial forestry, as practiced in the United States, involved clear-cutting; this entailed removing all vegetation from a site and burning the cleared area, thus exposing the soil for planting. It was done in steep, mountainous terrain receiving heavy rainfall (the Kootenai National Forest had areas with 80 inches of annual precipitation). Weren't these the same policies I spent two years, at the request of a foreign government, trying to stop?

The agency replied with a rote answer explaining that this was industrial forestry at its finest. Though there may be some soil erosion, it was always within "tolerable" limits.

The expected reaction was to accept the Forest Service's explanations at face value as the truth. After all, intuition was the only thing telling me the current practices were wrong, not seasoned experience. Initially, I suppressed my intuition, suspended my questioning, and believed the agency.

This immersion into USFS culture, which continued for nearly two years, entailed arresting my concerns over apparently significant environmental damage and participating in an agency willing to allow destruction in an attempt to meet quotas. The realization that this mindset was wrong came from a seasoned "combat" biologist, Ernie Garcia. Ernie moved to my district from the Gifford Pinchot National Forest on the western side of Washington. It had a huge timber cut, along with a reputation for ruthlessly getting the cut out. Ernie was the first wildlife biologist on the Troy Ranger District of the Kootenai NF. He soon became a thorn in the side of our timber program by doing the environmental analysis and documentation required by the National Environmental Policy Act (NEPA). It was obvious that our timber program was not meeting the spirit and intent of NEPA, and Ernie began to say so.

At one particular interdisciplinary (ID) team meeting,[1] Ernie was being pressured to back down on his analysis of grizzly bear habitat needs that conflicted with timber cutting units proposed. Ernie was correct in his assessment — ecologically, morally, and professionally. I, on the other hand, was part of an effort to discredit and intimidate him into joining the "team," in order to proceed with the timber cutting at any cost. At that point, I realized I had become part of the agency's institutionalized culture without even knowing it. I had capitulated in an attempt to be part of the Forest Service family. In the process, I had also surrendered my environmental ethic, my professional ethic, and my soul. My initial intuition about the timber sale program causing unacceptable environmental damage was true. The Forest Service's true mission became clear: get the cut out regardless of the law, the extensive environmental damage, and the advice of its own biologists.

My newly expanded outlook facilitated the discovery of other like-minded employees. There were soil scientists, hydrologists, wildlife biologists, and fisheries biologists who also felt that the agency was not adhering to environmental laws, was overcutting the forests, and had become the timber industry's lackey. We remained fairly quiet and discreet, believing that working quietly within the system and being team players would be the most effective approach. The Forest Service directors only needed to hear the truth in a non-confrontational way and they would change.

My next appointment was the Nez Perce National Forest in Idaho, where the idea of internal change was close to becoming reality. The Forest Supervisor at the time was Tom Kovalicky, known internally as a maverick. In fact, he dubbed the forest the "Nez Perce Anadromous National Forest," emphasizing the importance of anadromous fisheries alongside timber. This was a very unusual attitude.

The timber shop, planning, preparing and administering timber sales were again my duties. I still believed the system could work. My apprehensions were escalating concerning the overcutting of our forests, timber primacy as a driving force, and the ensuing environmental damage. The same problems existed on the Nez Perce NF as on the Kootenai NF. For example, the Clearwater River was an important anadromous fishery. There was abundant evidence of inappropriate cutting: soil slides, washed out roads, and erosion. The agency was obviously planning to remove too much timber; and, there were no plans to correct damage done from past timber cutting and road building. In response, I formed an interdisciplinary team and started working with the 'ologists: the wildlife biologists, fisheries biologists, and hydrologists. These were, in fact, the steps required by NEPA.

Most timber sales also went through the ID team process, resulting in either an Environmental Assessment or an Environmental Impact Statement. This time, however, the timber planner no longer had a timber bias, or he at least questioned this bias constantly. Working in conjunction with the scientists, I attempted to reduce or eliminate inappropriate cutting where possible. My behavior was unusual for a forester, especially a timber planner, whose primary duty was maximizing the cut at any cost; and who typically assumed that his duty as ID team leader was to pressure the unbelievers into submission, if not conversion. This meant getting them to alter reports to allow higher cutting than they deemed appropriate. When the scientists wouldn't succumb to the pressure, the reports were usually ignored.

The planned timber sales in the John's Creek area of the Clearwater River were reduced, with the help of the hydrologists and biologists, from an estimated 14 million board feet (MBF) to a salvage opportunity of less than 3 MBF. This was significant, because under Tom Kovalicky, no pressure was made to alter the report. The system was actually working. Good documentation and science could triumph over the agency's timber primacy mandate. My internal dissent became noticeably bolder because of the open atmosphere encouraging change on the forest. Problems did exist, and timber primacy still ruled the day; but, internal dissent was not only tolerated, it was actually promoted by Tom Kovalicky.

In a private discussion once, Tom said that the difference between me and him was that I was a revolutionary, while he was a politician and would live to fight another day. But we both knew that change needed to happen quickly. Tom had designed his career to achieve the status of Forest Supervisor, a position he believed to be the first "rung" in the ladder that could truly accomplish significant change.

Once he achieved it, he did indeed push the envelope of change, and took risks many other supervisors avoided. I followed suit by sending internal messages challenging the agency's policies. An atmosphere of acceptance lulled me into believing something was about to change.

It soon became apparent that the Nez Perce NF and its supervisor were actually aberrations within the agency, not the hoped-for sign of widespread reform. That Kovalicky was not tolerated well within the agency hierarchy also came to light. His forest planning process and environmental position were under fire from the timber industry and a hostile congressional delegation, dedicated to serving the traditional timber interests of the Forest Service. Jim Overbay, Regional Forester of Region One at the time, and Kovalicky's boss, ordered him back to Missoula, Montana, to address the industry's concerns. Tom refused to compromise his position to the degree they wanted, and soon was asked to return to Washington D.C., to talk directly with the Chief of the Forest Service, and former Senator McClure from Idaho, one of the timber industry's greatest supporters. Tom Kovalicky never compromised sufficiently to meet the demands of timber industry and their allies, and eventually retired early from the agency.

In 1988, I moved to the Blue River District of the Willamette National Forest in western Oregon. At the time, this forest was cutting more timber than any other national forest in the system, and Forest Supervisor Mike Kerrick and his staff were proud of it. Shock was my initial reaction to this forest's timber management; shock at the rate and ferocity of cutting, at the steepness of the cutting units, the resulting erosion and slope failures, and at the environmental damage so nonchalantly dismissed by my superiors.

When the lack of data and models to estimate hydrological tolerance limits was questioned, I was told, "we cut until we hit puke level, and we haven't hit that yet." The Willamette NF was an accelerated look at the future of almost the entire National Forest System. The forests of western Oregon and Washington were an indicator of what was befalling most of our national forests as a result of the Forest Service's ingrained, timber-dominated mindset. There were no attempts here to meet the spirit and intent of NEPA or of other environmental laws.

My first timber sale assignment had a completed NEPA document. I was told to, "look it over; dot the i's, cross the t's, and get the decision notice signed." A field review of the proposed timber sale yielded appalling facts. An unnecessary road was being proposed for the sole purpose of lowering the costs of logging by a few dollars per thousand board feet. This road, according to the agency's own hydrologists and geotechnical engineers, had a high likelihood of failure. If this happened, it would dump tons of sediment into a tributary of the McKenzie River which contained anadromous fish and also served as a municipal water supply for the cities of Eugene, and Springfield, Oregon. The sale included cutting units in Spotted Owl Habitat Areas, which were supposed to be left intact. There were several large areas of washouts, landslides, and erosion caused by previous cutting and road building in the drainage.

The Cumulative Effects Analysis (required by NEPA) based on reports from engineers and hydrologists, and my own investigations, clearly documented the fact that this timber sale would have significant, adverse environmental impacts and, therefore, should be dropped. I submitted this analysis and was later instructed to rewrite the report. My supervisor was conveying the District Ranger's command for a new report since, "the original report might as well have written the appeal for the environmentalists." I did as instructed and wrote a new report minus the obvious references in question. I sent a copy of my draft report, with the original cumulative effects analysis, to a local environmental group who appealed the timber sale. They won the appeal.

Despite that victory, I felt that I could no longer, in good conscience, be a part of this conspiracy of silence. I decided to start speaking out, and expected to be fired.

My first action was to send an internal memo to some scientist friends in the supervisor's office. The memo documented information I had obtained after attending an Ancient Forest Seminar held at the University of Oregon. The memo was extremely critical of both the timber industry and the Forest Service. I knew the reaction from management on the Willamette NF would not be the same as on the Nez Perce NF. However, the extent of what happened was unexpected. Some members of the timber industry had apparently received a copy of the memo as quickly as I had sent it. A typical timber industry response came from Troy Reinhart, director of a local timber industry support group called Douglas Timber Operators, asking for my resignation. I alerted the local environmental constituency and the media. The controversy was commented on in the local paper and *The New York Times*.

In the meantime, I had written a letter to the chief of the Forest Service, F. Dale Robertson. My words were from the heart and extremely candid. The reality of what we were doing to our forests, and the role the Forest Service played in addressing this reality, had to be dealt with more honestly and openly. Since the media seemed so interested in the issue, *High Country News* received a copy of this letter for publication. I hoped to promote additional media coverage and get the word out to other Forest Service employees, many of whom read this particular paper.

I felt that the time had come for change, and that I had a unique perspective with which to bring it about. My international experience with deforestation and reforestation and soil conservation projects broadened my outlook, and perhaps gave me an advantage to see through the rhetoric and dogma to which so many had become obligated. I understood the necessity for beginning a transformation.

The agency's historically strong ties to resource-extraction industries, timber in particular, were now the focus of serious scrutiny as to its ramifications for the environment and future generations. In the letter to Robertson, I explained this assertion more fully, supported it with examples, and gave my input on what top management needed to do to move us into the 21st century as leaders of a new resource ethic instead of as unwilling participants being dragged along by a chain of court decisions.

I suggested we were over-cutting our National Forests at the expense of other resource values. We were incurring negative, cumulative impacts to our watersheds, fisheries, wildlife, and other non-commodity resources in our quest to meet timber targets. I provided examples, including moving Spotted Owl Habitat Area (SOHA) boundaries and allowing fragmentation of those areas to accommodate timber sales; exceeding recommended cover/forage ratios on big game winter range; ignoring nongame wildlife prescriptions, such as snag and green replacement tree guidelines; exceeding watershed/sediment "threshold values of concern" in areas with obvious, cumulative damage, etc.

At the planning level, I pointed out that Forest Plans have been built from the "top down" instead of the from the "ground up." In other words, we have taken the politically driven timber harvest level and manipulated the Forest Plan to support that amount, rather than letting the harvest level be determined by sound biological and ecological considerations mandated by our resource protection laws. I cited numerous examples of actual Forest Plans exhibiting these characteristics.

To put the issue in perspective, I posed a series of questions that the Forest Service and its employees should consider: Are we, as an agency, going to continue to support the current global epidemic of destruction of our biosphere's ecological diversity and survival, for short-sighted, short-term, economic "security"? Are we going to continue to ascribe to the timber industry's assertion that cutting our National Forests is imperative to maintain jobs, when the same industry exported three to five billion board feet of raw logs per year for the last ten years from Oregon and Washington alone? Shall we sacrifice the public's lands because the economic view of standing timber deems it "under-valued" (and a leveraged buy-out opportunity)? In essence, why are we so biased in favor of the timber industry?

I believed this last question could be partially explained by the traditional mindset within the Forest Service that has been perpetuated over the years. A combination of partial truths and an unwillingness to look too deeply into a system that was "working," led the men and women of the agency to adopt a superficial attitude towards resource management. A common belief was that good intentions and aspirations to responsible land stewardship somehow negated any detrimental effects resulting from our accelerated rates of timber harvest. Another was the acceptance that current practices were just a continuation of an historically proven, successful method of timber management. A third tenet was the idea that money could solve any problems the agency might have, that any conflict could be mitigated financially with a win/win outcome.

As the negative impacts of agency actions became more and more obvious, employees tried to pretend it was not happening. And yet at some subconscious level, we knew that we were overcutting. When I talked to coworkers about this subject, it was almost universally agreed that there was, in fact, overcutting. Most of these people, though, failed to make the connection between themselves and the agency and the direct contribution to the global, environmental onslaught. Most stopped short of admitting that the resource and their credibility with the public were being compromised.

Another contributing factor to this alliance between the Forest Service and extractive commodity-based industry involved the agency's perception of who environmental groups were and what they did. They were dismissed as "special interests" by the agency. They were not, however, to be equally weighted with the timber industry as just another special interest. The timber industry's motives were short-term, quick profits, and tended toward present-focused, economic gain. The environmental community, on the other hand, had a long-range perspective. They promoted a vision of a sustainable future, both economically and ecologically. Their motives were altruistic, not exploitative.

Environmental groups wanted change. They threatened the agency's long-held conception that it was acting in a responsible manner. That the Forest Service had traditionally allied itself with the timber industry and against the "environmentalists" only proved that it had followed a misguided purpose for some time. Again, I included a variety of specific cases to illustrate my point. The U.S. Forest Service did have the legal authority, the personnel, the research facilities, the facts and data to promote and make the needed changes to its internal value system and management practices. It was my opinion that the Forest Service must associate itself with the long range, holistic, and altruistic motives of the environmental community, which were more in line with its mission as a public resource management agency. Granted, this would involve a major confrontation with resource extraction industries, but it was something which had to be done. The time was right for this type of realignment.

I then listed specific courses of action to initiate change: The Forest Service should encourage ideological diversity and support the existing "agents for change" currently within the organization, like many wildlife biologists and other specialists. Support should be given to those within the system who try to promote the new vision. All members of the agency must insist on absolute commitment to the spirit and letter of our resource protection laws with as much energy as has been devoted to cultural diversity in the workforce. They must further promote the substantial lowering of existing and planned timber harvest levels throughout the National Forest System. By doing this, the agency could start erring on the side of resource protection instead of extraction.

The "bottom up" approach to Forest Planning must take precedence over the current "top down" strategy. We must demand realistic, specific, and meaningful Forest Plan standards and guidelines which truly protect other resource values and accurately display all effects of timber harvesting.

A new resource ethic should be encouraged by publicly endorsing alignment with the environmental community and participating in the search for a sustainable future. It is time for everyone to accept the fact that change is imperative. The unfortunate truth is that future generations will look back at the last few decades of our history with a mixture of amazement, incredulity, and disgust that we allowed such an unprecedented slaughter of our natural ecosystems during this era of massive exploitation for so little real long-term value.

I wrote this letter in hopes that it would have a maximum impact for change. Because of the import of its contents, F. Dale Robertson was unable to respond. If he had, he would have had to either concede to making changes or deny that any of the problems existed. Apparently, neither option was acceptable because I did not receive a response until eight months later (and at the request of a congressional committee), and it was only four lines acknowledging receipt of the letter and thanking me for my input.

This lack of any significant advancement left me frustrated and primed for the next opportunity to effect change. That chance came in March of 1989, hen I attended an Old Growth Symposium, sponsored by the Forest Service, in Portland, Oregon. After listening to a speech by the then Regional Forester of Region 6 at the time on how well the Forest Service was protecting wildlife, ecosystems, and the public trust, I was stunned at the degree to which this individual was out of touch with current events. On impulse, I found a computer, made some flyers, and created the Association of Forest Service Employees for Environmental Ethics (AFSEEE).

Over the next few weeks, the response was incredible, with membership requests and letters of support pouring in. I was asked to speak to agency employees, citizens, and environmental activist groups across the country. Media coverage grew, and I began traveling around the country organizing AFSEEE chapters and giving public speeches. I soon had two jobs: my usual timber planning job for the agency, and now director of a new organization of Forest Service employee dissidents. By the end of the first year, membership was over 1000 employees, approaching five percent of the total agency workforce.

The goal of AFSEEE was to foster an environmental ethic among employees that would change the culture of the Forest Service to one that was more environmentally oriented, would meet its regulatory laws, and would err on the side of resource protection.

It was a pleasant surprise that I was not fired immediately. I believe this was due to my "kamikaze" approach. I was forceful, outspoken, and not afraid of the consequences. I also had an untarnished reputation as a good forester. The element of surprise and my air of unpredictability gave me an edge.

Even so, at first, the agency response was somewhat hostile. I was told that I could not talk about old growth issues on or off the job. Though the agency had the prerogative to control my on-the-job contact with the public, news media, etc., my off-the-job activities were protected by First Amendment, free speech rights and not subject to agency or supervisory control. I informed my District Ranger and the Forest Supervisor of this, and they retreated from their initial stance. My circumstances were the impetus for a set of legal guidelines on the Willamette NF concerning free speech within and outside the workplace. Though my methods may seem to have been a little impulsive, in retrospect, I would not have changed my tactics. The time was right for my message, and the message had to be loud and clear. I continued to work for the agency as a timber sale planner and to organize AFSEEE with no interference from management. When I resigned a year later, it was a personal decision to manage the rapidly growing AFSEEE, which needed a full time director to realize its potential as a major force of change within the Forest Service.

Perhaps the most important achievement of AFSEEE to date is the enormous amount of media attention it has drawn to the issue of needed change in government agencies. We were able to speak out on topics that had before been taboo.

In fact, the results of a study on values and change in the Forest Service showed that a considerable number of employees went about their jobs differently (by speaking out, writing more responsible plans, and not ignoring pertinent data when arranging timber sales) as a direct result of the presence of AFSEEE.[1] Local chapters became very active, commenting on their Forest's Plan and serving as liaisons with the public. Meetings were held to provide information on what was going on in the National Forests and what citizens could do to participate in effecting change.

Because many of the AFSEEE members had a high degree of technical expertise in the workings of timber and other resource management, they provided testimony for several congressional hearings that further encouraged change within the Forest Service. One example was a decision made by Federal Judge Dwyer, in 1991, enjoining the agency's timber sales program in the Western Cascades and northwest California because the Forest Service still did not have a credible management plan for the Northern Spotted Owl. He was quoted as saying that:

> More is involved here than a simple failure by an agency to comply with its governing statute. The most recent violation of NFMA [National Forest Management Act] exemplifies a deliberate and systematic refusal by the Forest Service and the F'S [Fish and Wildlife Service] to comply

with laws protectin wildlife. This is not the doing of scientists, foresters, rangers, and others at the working levels of these agencies. It reflects decisions made by higher authorities in the executive branch of government.[2]

As soon as AFSEEE became a reality, it was obvious that the Forest Service was not the only agency that could benefit from such an organization. So, in January of 1993, I left as director of AFSEEE to start PEER, Public Employees for Environmental Responsibility. It was my intent to expand the AFSEEE model into other federal and state, land management and environmental protection agencies. Again, there was a tremendous response, in particular the employees of the Bureau of Land Management with regards to grazing reforms.

As we are on the cusp of the 21st century, there is a desperate need for governmental agencies to restructure in a way that will allow them to respond to the public's increasing concern over the future of our natural resources. Employee ethics groups will be instrumental in this transformation. As the current administration mandates new policy, the working levels of the agency can monitor the degree to which this policy is absorbed and implemented. They can then channel their findings outside of the agency, through AFSEEE or PEER, back to the administration. Adjustments can then be made and areas of more focused attention determined. In this manner, two-way communication and feedback can turn rhetoric into reality.

AFSEEE and PEER will hopefully inspire even more employee ethics groups in the years to come. By constantly applying pressure, adapting strategies to whatever political climate exists, and adhering to a strong commitment to resource stewardship and environmental protection, the paradigm shift will continue.

"Corporate money is wrecking popular government in the United States. The big corporations and the centimillionaires and billionaires have taken daily control of our work, our pay, our housing, our health, our pension funds, our bank and savings deposits, our public lands, our airwaves, our elections and our very governments."

"The Nature of World War III"
Rachel's Environment & Health Weekly, #567, October 9, 1997, p. 3.

A HEALTHY ENVIRONMENT
The Evolution of a
Fundamental Right

Robert V. Eye

"It should always be remembered that man's views of justice have changed as the stream of life has flowed along. It can be conceived and even hoped that the final word has not yet been said... There are men and women who have a vision of justice that, so far, the world has never yet reached."

Clarence Darrow,
Who Knows Justice?

The legal recognition of any fundamental right is the culmination of a long and difficult process. Some fundamental rights are established by mass political movements resulting in the embodiment of the right in a nation's organic law. Others are recognized by legislative bodies which either integrate such into existing law or submit the issue to voters for ratification. Still others are developed through the judicial process, generally after many cases have raised the issue unsuccessfully. Each of these processes is nearly always preceded by a widespread recognition that certain rights must be recognized as basic in order to establish or maintain a social contract.

The actual life experiences of people create the drive to establish a fundamental right. Each fundamental right established in the United States Constitution has its own unique history of controversy, failure, despair, resolve, and success. Each fundamental right is a study in how the political process is painfully slow to yield change. The right to trial by jury, the abolition of slavery, or the establishment of women's right to vote, must be considered in the light of the reasons why so many sacrificed so much to have these rights enshrined in law. Ultimately, the sacrifices were made to seek and realize justice by the recognition of specific fundamental rights.

The contemporary environmental movement has been driven by the recognition that scientific evidence indicates the biosphere has a limited capacity to support many human activities associated with the industrial and post industrial periods. The focus has been whether a particular activity poses a burden which a certain defined ecosystem cannot tolerate. This technique of analysis is clearly appropriate for a discussion of the global biological-ecological impact of human activities. However, the vernacular of this paradigm is scientific in nature, and falls short of addressing issues related to our notions of the rights and duties which define human relationships. For this paradigm, we must turn to political and legal concepts for guidance. From this paradigm has emerged the environmental justice movement.

This chapter will discuss the continuing evolution of the right to a healthy environment. This discussion will not attempt to define what constitutes a healthy environment other than to point out instances where the presence of human-created contaminants (or the prospect thereof) has forced the issue. Over time, the definition of a healthy environment will be established in the political and legal arenas based on the fluid interplay of scientific evidence and prevailing concepts of justice. The definition of a healthy environment, however, for purposes of reconciling competing and conflicting interests, becomes important only if such a right is recognized and enforced by the body politic.

The Nature of Fundamental Rights

The Supreme Court of the United States has addressed the issue of how to define fundamental rights. In *Griswold v. Connecticut*, then Justice Goldberg offered the following guidance:

> In determining which rights are fundamental, judges are not left at large to decide cases in light of their personal and private notions. Rather, they must look to the traditions and collective conscience of our people to determine whether a principle is so rooted there as to be ranked as fundamental. The inquiry is whether a right involved is of such a character that it cannot be denied without violating those fundamental principles of liberty and justice which lie at the base of all our civil and political institutions.[1]

Under this standard, it is the rare and special aspect of human conduct which is so important that to disregard it erodes the basis of liberty and justice. For example, political speech is not subject to prior restraint by the government for purposes of dictating its content. To do so would violate a specific right considered to be fundamental in a free society. While this right is subject to certain qualifications such as time, place, and manner, the liberty to criticize the government (including advocating its overthrow) is protected by law. Without this protection, it is assumed that society is not based upon principles of liberty and justice. Additionally, such a right is considered so vital that its absence or infringement has a detrimental effect on the society at large and not just on an individual or discreet identifiable group.

Lawyers often observe that determining what a particular law means is speculation until a court of competent jurisdiction has an occasion to analyze that law and announce a decision. Even then, it is subject to appeal until the court of last resort passes judgment. Then, the decision is "right" because it is final. Judicial pronouncements are neither static nor immune from social and political forces or alternative legal arguments. One only need recall that the decision in the case of *Brown v. Board of Education*, mandating school desegregation, was a reversal and repudiation of a long-standing rule of law which held that separate schools based on race were, as a matter of law, equal.

This brief discussion of the nature of fundamental rights begs the question whether a healthy environment is so important that it ought to be recognized as necessary to a free society. To those involved with the environmental movement, and who understand the relationships between ecological health and human health, the answer is clear. Unless the environment is healthy, human health will be compromised to the point that the most fundamental right, to life, will be undermined. Similarly, to the antislavery activists of the 18th and 19th centuries, it was clear that the legal protections then enjoyed by the slave trade were inimical to a free society, yet it took generations for their perspective to become recognized in constitutional law.

The Political Economy of Pollution

The major barrier to the recognition of a healthy environment as a fundamental right is economic in nature. Those who pollute the biosphere in a large scale generally have a financial interest in doing so. In terms of pure market theory, there is no incentive to prevent pollution unless a competitive advantage can be realized. Consequently, the response to the environmental crisis by the U.S. Government has been regulatory in nature. That is, legally enforceable limits on contamination are established and applied to all similarly situated polluters, thereby not permitting any one polluter to gain a competitive advantage. While rational on its face, this approach has allowed a continuous and pervasive contamination of the biosphere and the bioconcentration and bioaccumulation of toxins in the food chain. It has also preserved, by law, the right to pollute and thereby protects powerful economic interests.

This regulatory response is rooted in contemporary liberalism and the premise that science can establish contamination limits which are not harmful to human health. This approach assumes a regulatory balance can be achieved to accommodate economic interests while giving the appearance that government is doing something in response to pollution. On paper, this would seem to be a reasonable approach to environmental protection. The reality is, however, it serves neither to protect the environment nor the long term interests we all share in sustainable economic activity.

Barry Commoner observed several years ago in his book, *Making Peace With the Planet*, that the regulatory approach upon which we have relied has not worked to protect the environment.[2] The proof of his assertion is evidenced in various ways. For example, at least 70 percent of the rivers and streams in Kansas[3] are not fit for swimming due to concentrations of pesticides, herbicides and fecal coliform. Similar circumstances exist regarding the discharge of toxins into the air in numerous urban areas of the United States. After the expenditure of huge sums to "protect" water and air quality., we have achieved only modest overall improvements. (This, of course, does not account for certain areas, frequently low income, where even modest improvements have proved elusive.) Clearly, something is wrong with the regulatory approach we have adopted.

The inherent deficiency with the liberal regulatory approach is that it does not make a strong enough case linking economic prosperity with environmental health. Liberals are, too often, apologetic for taking strong stands for protecting the environment and public health. This predisposition plays into the hands of conservatives who generally see protection of the environment as a zero sum game; any gain in environmental quality comes as a direct expense to prosperity.

When protecting the environment is viewed as a luxury, something to be done if the bottom line will not suffer, advocating a healthy environment as a fundamental right is a losing proposition. In this sense, a healthy environment lacks the power of "traditions and collective conscience" necessary to call a right fundamental. As protection of the environment becomes a social tradition and part of our shared values, it becomes a fundamental right, *de facto*. Giving it force and effect, *de jure*, is usually the last and most difficult step.

The Emergence of the Right to a Healthful Environment

Most people are surprised to learn that they have no legally enforceable right to an environment free of potentially dangerous contaminants. The issue rarely is considered at all until a facility such as a landfill, incinerator, or other polluting source is proposed to be established at a particular location. Citizens then begin to become familiar with the environmental and health impacts of a proposed facility and learn that their concerns have been anticipated by the government regulators and the businesses which would build and operate the facility. The regulators and business reassure the concerned public the facility will meet all applicable standards. As citizen activists soon learn, the applicable standards still permit considerable contamination of the environment by toxins. It is no longer enough to assure people that the latest scientific evidence indicates that a state of art facility will present no unreasonable risk to health. Such an assertion simply is not adequate when toxins such as dioxin, lead, mercury, plutonium, and other known carcinogens are proposed to be discharged into the environment. Nearly always, citizens demand to know why a particular polluting facility is proposed to be built at a particular location. Enter the environmental justice movement.

The environmental justice movement asserts that dangerous facilities are located in areas where politically powerless people tend to live and work. The evidence to support this assertion is ample. Waste incinerators, hazardous and radioactive waste dumps, and other dangerous facilities have traditionally been located in low income areas. The demographic characteristics of an area play a disproportionate role in site selection. In short, hazardous facilities are not proposed for geographic areas dominated by middle and upper income people.

The U.S. Environmental Protection Agency decided to look at the evidence to determine whether there is a relationship between location of polluting facilities and low income. In 1992, the EPA concluded what had been known for years by many in the civil rights and environmental movements: minorities and low income people are much more likely to suffer exposures from air pollution, waste facilities, pesticides, and contaminated water than their upper income and Euro-American counterparts.[4] From a legal/political perspective, the issue which emerges is whether low income and racial minorities have the same right to a healthy environment as their Euro-American and upper income counterparts.

This finding by the EPA raised the stakes to constitutional proportions. It also put even a greater premium on utilizing scientific evidence to persuade low income and minority opponents that dangerous facilities present no undue risk. When science could not carry the day, financial incentives were often introduced into the equation to neutralize opponents. However, science and money (characterized as bribes by many) do not equate, per se, with justice. Concentrations of dangerous polluting facilities in low-income areas is no longer coincidental, but rather calculated to exploit the politically less powerful.

The Clinton administration issued Executive Order No. 12898 in 1993 which directed federal agencies to consider whether their programs have adverse environmental and health impacts on low income and racial minorities. This was a small but potentially important step. It was small because executive orders are not rules of law, and this order stops short of directing that specific actions be taken by federal agencies to prevent environmental inequities. It is potentially important because it is at least an implicit recognition that the assertions of environmental racism and inequity are based on sound evidence.

Whether this Executive Order actually results in preventing environmental injustice or inequity is, at best, problematic. If the administration is really serious about this issue, it will take the initiative by proposing legislation to establish enforceable protections for low income areas to prevent their exploitation. Additionally, for example, the administration should propose the repeal of the provisions of the *Atomic Energy Act* which offer financial incentives to communities to establish radioactive waste facilities. These and similar measures would send a clear signal that the right to a healthy environment will not be contingent on economic status or geographic location.

On another front, civil rights laws are being utilized to bring claims based on equal protection theories. The jurisprudence related to this theory in the environmental context is in the developmental phase. The EPA's Office of Civil Rights has begun to recognize that certain permits for waste facilities and incinerators may violate Title VI of the Civil Rights Act of 1964 due to a disproportionate impact on minority and low income communities.[5]

Louisiana has also seen a major victory for a predominantly poor, African American community against the nuclear industry. On April 22, 1998, plans were abandoned to obtain an operating license from the Nuclear

Regulatory Commission for a uranium enrichment plant planned to be built near Homer, Louisiana — even though powerful forces, including the nuclear industry and its allies in Congress, strongly advocated the plant be built. Residents of Homer, Louisiana organized and activated the anti-nuclear networks in Washington, D.C.. Together, they waged a fierce struggle which caused, in the final analysis, the project to be abandoned. This case is a model for other low income communities to utilize in resisting facilities which threaten the environment and public health.

Responses to the Environmental Justice Movement

There are, however, countervailing developments which could have the effect of continuing the practice of preying on low income areas to establish dangerous facilities. The Clinton administration and many in Congress appear to advocate utilizing risk-based analysis to make many environmental decisions. In effect, this technique relies upon an assessment of technical information to determine whether a particular activity or facility presents any unreasonable risk to human health. This method, when examined carefully, nearly always reveals that people will be exposed to toxins from the proposed facility. However, in order to justify the site of a dangerous facility, the exposure is proposed to be "managed" or "regulated" to present an "acceptable" risk. This approach has the support of large segments of the scientific community and is a major growth industry for environmental consultants.

Risk analysis has many shortcomings, including reliance on incomplete data, incapacity to thoroughly consider the synergistic effects of combinations of toxins, and reliance on assumptions about the state of the public's health which may be overly optimistic. In short, risk analysis is mostly used by the polluting community to justify further contamination.

The use of risk analysis has nothing to do with fairness and justice. It is, in essence, a rationalization for the continued concentration of dangerous facilities in low income areas. Yet, the current political leadership finds its appeal irresistible and is determined to use it to provide underpinnings for decisions in the areas of environment and public health.

It will be most interesting to observe how Executive Order No. 12898 mentioned earlier, and the use of risk analysis will be reconciled. Proponents of dangerous facilities will utilize risk analyses to show that poor communities are at no more risk from incinerators/dumps/nuclear facilities than their counterparts in affluent neighborhoods are from being hit by an errant golf shot at the local country club. The interplay of environmental justice principles and risk analyses is likely to highlight the disparities between various income groups related to toxic exposures.

Another threat to the emergence of the right to a healthy environment is the so-called wise use movement. This idea is variously described as a means to protect private property rights and to constrain unreasonable government regulation. In essence, the wise use movement presumes any government action which diminishes an owner's right to use property constitutes a "taking." This theory then asserts that such takings should be compensable by the government. Hence, if an owner of property located next to a hospital wanted to utilize it to operate a hazardous waste incinerator, the government could not prohibit that use without compensating the owner for his loss of use value. If allowed to become law, this policy would open the door to gutting environmental regulations now in effect. Government regulators would become even more reluctant to confront dangerous projects because to do so might subject them to liability for interfering with the use of private property.

This so-called wise use movement is a throwback to the days when there was little or no awareness about the relationships between human health and environmental contamination. Taken to its logical conclusion, this movement would likely result in, among other things, further concentration of polluting facilities in low income areas. Property values are lower in these areas and investors would, therefore, have an incentive to locate there. Siting decisions would become simple matters of real estate acquisition with no regard for the current state of the public's health in the area or consideration of future environmental impacts.

The environmental justice movement has been successful in large part due to the exercise of another fundamental right: free speech. The capacity to effectively organize citizens is in direct proportion to the ability to disseminate information about environmental threats. In response, a legal ploy has emerged in an attempt to silence critics of polluting facilities. The tactic is known as strategic lawsuits against public participation (SLAPP). SLAPP cases are usually filed by businesses against the most vocal critics and threaten that money damages will be sought if these critics do not censor themselves. The mere threat of litigation is often sufficient to chill effective communication and organization. Actual litigation diverts scarce resources away from community organizing activities to legal defense. In this regard SLAPP suits do not have to be successful to be effective. Generally, SLAPP suits are not successful in the judicial context.[6] Courts are reluctant to silence citizens who have organized to resist threats to the environment and their health. On occasion, SLAPP suits backfire and actually help the environmental justice movement by pointing out the extent polluters will go to have their way.

Prospects for Legal Recognition
of a Fundamental Right to a Healthy Environment

Despite these threats to the establishment of the right to a healthy environment, there is room for guarded optimism. The public expects that the huge expenditures of resources over the past thirty years will result in a healthy environment. Though these expenditures have yet to yield the expected outcomes, such have had the effect of establishing a climate of rising expectations. Additionally, more evidence is compiled daily concerning the links between environmental contamination and the adverse impacts on public health. As these relationships become a part of the public's consciousness, there will be added pressure to have policy address the links between environmental contamination and health consequences.

Moreover, evidence will continue to accumulate showing economic health increasingly dependent on environmental quality. Pollution for the sake of profits will be seen as counterproductive to long-term economic interests. Whether manifested in terms of cleanup costs, increased health costs or a realization that scarce resources increase costs of production, pollution will eventually be a luxury that most societies will not be able to afford. Moreover, pollution is generally a function of inefficiency. The more inefficient a process or machine is, the more likely it is to pollute. As more emphasis is placed on efficiency, the rates of pollution should, in theory, fall.

The environmental movement has progressed now to the realm of international law and diplomacy. The international agreements concerning ozone depletion (Montreal Convention), global warming (Kyoto Agreement), and other environmental issues (Rio Treaties) recognize implicitly that a healthy environment is fundamental to human existence. However, whether the lofty ideals set forth in these agreements have any substantive enforceability under domestic law is linked to where we, as a society, rank environmental health as a priority.

These agreements, however, would not have had any chance of becoming reality absent a widespread recognition that environmental health is fundamental to life. The agreements are not a guarantee of enforcement. Enforcement is tied to the willingness of the domestic body politic to accept limitations on sovereignty imposed by international agreements.

Conclusion

One school of thought argues that fundamental rights become recognized only when the power elite decide it is time to do so. Recognizing that there is frequently a vast difference between rhetoric and action, consider this statement in 1970 by then President Richard Nixon:

> Environment is not an abstract concern, or simply a matter of esthetics, or of personal taste — although it can and should involve these as well. Man is shaped to a great extent by his surroundings. Our physical nature, our mental health, our culture and institutions, our opportunities for challenge and fulfillment, our very survival — all of these are directly related to and affected by the environment in which we live. They depend upon the continued healthy functioning of the natural systems of the Earth.

> The newly aroused concern with our natural environment embraces old and young alike, in all walks of life. For the young, it has a special urgency. They know that it involves not only our own lives now, but the future of mankind. For their parents, it has a special poignancy —because ours is the first generation to feel the pangs of concern for the environmental legacy we leave our children.

At the heart of this concern for the environment lies our concern for the human condition, for the welfare of man himself, now and in the future. As we look ahead to the end of this new decade of heightened environmental awareness, therefor, we should set ourselves a higher goal than merely remedying the damage wrought in decades past. We should strive for an environment that not only sustains life, but enriches life, harmonizing the works of man and nature for the greater good of all.[7]

This statement, made in a time when dioxin-laden Agent Orange was being widely used in Vietnam, during a building binge of nuclear plants and weapons, and other environmental atrocities, is so idealistic that it would make one believe establishing a right to a healthful environment would be done because our collective conscience recognized the damage we were doing to the earth and ourselves. The contradiction between reality and the ideal evidenced in Nixon's statement and actions as President should not deter us from continuing struggles to establish a fundamental right to a healthful environment. Rather, it should serve as a reminder that the madness of destruction is always close at hand and it is that which we must struggle against as we struggle for environmental justice as a fundamental right. When environmental justice is part of our collective conscience and tradition, a healthy environment will inevitably follow.

AGRICULTURE OF THE FUTURE
Conceptually Rich or a Technological Tour de Force

Wes Jackson

"Nature bats last." **Dr. Paul Ehrlich**

I have two stories to tell. The first consists of an ecological comparison between a never-plowed Nebraska prairie and an adjacent wheat field. It is based on a study done in the early 1930s. The second story is about two tropical rain forest regions where natural succession is allowed to occur after forest clearing, and another nearby area, also cleared, where scientists have mimicked natures succession using plant species introduced by humans.

From September 1933 to September 1934, over an 11-month period, William C. Noll, a graduate student at the University of Nebraska, carried out a research project under the direction of the famous J.E. Weaver. Near Lincoln, Nebraska, Noll compared an upland, climax prairie with an adjacent field of winter wheat.

That year was the driest and hottest ever recorded. Precipitation was below normal for every month except December. The average precipitation over several decades for that area was a little over 24 inches for an 11-month period. That year only 10.5 inches fell. The soil type and drainage conditions for both the wheat field and the prairie was Carrington silt loam.

Nearly 9 percent of the water that fell on the wheat field ran off. Only a little over 1 percent ran off the prairie. The water available for plant growth was usually several percent higher in the prairie to a depth of 2 feet, but slightly lower than in the field in the third and fourth foot. The prairie held its water at a higher level, not allowing it to go down to the third and fourth foot. As one might expect, shallow-rooted plants died on the prairie; those of moderate depths suffered greatly and only deeply-rooted species functioned normally. The point is, at least some part of the prairie survived. The wheat plants, on the other hand, were all dwarfed due to the drought and died early in June before the half-filled kernels could ripen.

The maximum weekly average temperatures were 100 °F for the prairie but 102 for the field during the first week in June. During the hot spring and summer, the average weekly humidity was 20 percent or less. It was higher in the wheat field but only during the short period until the wheat dried. From then on humidity was usually 5 or more percent higher in the prairie. In other words, it was more evenly distributed on the prairie.

There was 2-2.4 miles per hour greater wind movement in the prairie until harvest. After harvest, however, it was from 0.5-5 miles per hour greater in the field. Before harvest, evaporation was 3-20 cubic centimeters (per square what?) per day greater in prairie; thereafter and for a much longer period of time, 4-13 cubic centimeters greater in the field.

The maximum surface soil temperature, at 10 a.m. was 107 °F on the prairie and 111 °F in the field. Even at 3 inches in depth, in July the maximum soil temperature on the prairie was 104 °F but 112 °F in the field. Winter and summer extremes of temperature were greater in the field to a depth of 3 feet.

The water losses per square foot per day from phytometers after the first of April were 2-2.5 times greater in the field, where a maximum of 1.97 pounds per square foot was reached. Water used by vegetation while producing a pound of dry matter was less for the wheat field-2,400 pounds compared to 5,584 pounds in the prairie. Once the wheat stalk was dead, it no longer required water. The leaves and stems of the prairie plants were still alive and in need of water. But the prairie had saved more water, too. The water losses from March 13 to June 11, calculated from weighed phytometers, were 20.8 tons per acre per day in the field and 13.3 tons in the prairie.

End of story one. Now for the second story.

At a Land Institute-sponsored symposium entitled "The Marriage of Ecology and Agriculture," Jack Ewel, from the University of Florida, described some investigations he and some colleagues had undertaken in a tropical rain forest environment. They attempted to mimic the natural vegetation there and discover standards for designing farming schemes for the humid tropics. The plant community they were trying to imitate was the forest.

Since the mid-70s, Jack and his co-workers had studied successional vegetation on deforested rain forest sites that had been abandoned. They chose to study successional vegetation as a model for the design of sustainable agriculture in the tropics because, for one thing, it is notoriously fast-growing. They reasoned that if they could channel some of that productivity into food, fuel, fiber, fodder, or cash, they might be able to develop economically viable alternatives to the destructive slash and burn methods settlers use.

A second reason is that, unlike most agricultural crops, successional vegetation is rarely devastated by pests. They wanted to unravel the mechanisms that endow these natural communities with resistance and resilience to pest attack.

Third, they reasoned that successional vegetation seemed to afford excellent soil protection. Many kinds of cropping schemes, on the other hand, are all too often accompanied by nutrient leaching and soil erosion in the high-rainfall tropics. At best, these processes create an addiction to expensive chemical fertilizers; at worst, they lead to semi-permanent impoverishment of the soil. After studying this diverse, fast-growing creation of nature, they set out to build and study a parallel community, one that contained as many species of plants and in the same proportions of life forms — herbs, shrubs, trees, vines, palms, and so forth — as the natural community. Their mimic, however, was to contain only species chosen by them, none chosen by nature.

Some of the species in the mimic were cultivars. Some were wild. Only two criteria were used in their selection. First, they had to resemble some plant that occurred naturally in the successional vegetation. They substituted vine for vine, herb for herb, tree for tree, shrub for shrub. The second criterion was that the plants selected could not have found their way to the site without the help of Jack and his colleagues — that is, they had to be aliens. Both communities, the natural and the mimic, in other words were species-rich and structurally complex.

The first year they perfected their horticultural skills. The composition of the successional community was rapidly changing, and the natural successional vegetation out-paced the mimic in biomass production by 1,800 grams per square meter compared with 1,150 grams per square meter. By the fifth year, however, their efforts at imitating nature paid off. The productivity of the mimic rose to more than 90 percent of that of the model. That's the production story. As for pests, the two communities lost leaf tissue to insects at about the same rate, but in the face of a pest outbreak the mimic fared worse than the natural-vegetation model.

The successional vegetation was only slightly more complex than the mimic. Nevertheless — and here is an important point — early on, nature's system pumped much more water back into the atmosphere than did the imitation. For a year or two the mimic suffered from having 23 percent more water moving downward than did the successional vegetation, which meant more leaching, for with that water went nutrients essential for plant growth — three grams of nitrate-nitrogen per square meter, for example. And that nitrate washed out along with a gram each of calcium, magnesium, and potassium. But the good news is, within a couple of years leaching rates in the two communities were indistinguishable. The mimic proved as watertight and as nutrient- tight as the model. As one might expect, an adjacent plot of bare soil lost water to deep infiltration 50 percent faster than those that were vegetated. Nutrient losses from bare soil were even more dramatic. Losses of nitrogen and cations were ten times faster than either the mimic or the model. Now here is an important point, as Jack reminds us: "Bare soil is the nemesis of fertility in the humid tropics. Without the return of water to the atmosphere through transpiration, and in the absence of nutrient uptake by plants, leaching is rampant and soil impoverishment is inevitable."

What lessons did they learn? They verified the fact that nature's way worked pretty well. They found that if they mimicked the structure, they were usually, but not always, in Jack's words, "granted the function-high productivity, responsiveness to pests, and good protection of the soil."

Though the complex systems they worked with were essential for the kinds of basic understanding they sought, as scientists they knew they were not working with a technological package they could hand over to tropical farmers. All that diversity was a horticultural nightmare. As Jack said, "A tropical farmer would have to be out of her mind to attempt anything so complex."

Jack left us with the good news that the immense richness of species they worked with was likely unnecessary agronomically and ecologically. Much simpler communities might provide many of the same benefits, provided the plants they contain are capable of high productivity, either resistant or resilient to pest attack, and able to use the soil thoroughly and continuously.

End of story two.

Consider the relationship of water to each ecosystem. The native prairie retains water. It "struggles" to keep it from the atmosphere by storing it in the soil and thatch, just what you would expect from a Great Plains ecosystem. The wheat fields more readily discharge water to the atmosphere.

The tropical rain forest, the opposite of the prairie, has evolved to get rid of water fast. It keeps water from the soil and discharges it to the atmosphere. Nearly all agricultural efforts in this part of the tropics lead to reduced fertility because water moves nutrients downward. While Jack's mimic eventually came close to nature's succession, it is important to remember that when we humans interfere, we tend to run what local nature wants to do in the opposite direction no matter where we are — Lincoln, Nebraska, or Costa Rica.

There is another point worth considering. Between these two extremes of water retention and water discharge to the atmosphere, where nature is at work, is an ecological mosaic, a world of diverse ecosystems. If nature is to be our standard for agriculture and culture, and sustainability our watchword, how we handle this mosaic between and including extremes should determine our agenda for the next century. A tension arises. We are Homo the homogenizer with a generalizing technological capability and relatively narrow enzymatic demands. Whether it is a Nebraska prairie community or the tropical rain forest or anything else in between, problems arise because we humans tend to invert what nature does well. If nature is our standard or measure for sustainable culture, then we should expect a mosaic of human cultures or communities across our land. Our problem becomes: how much of this mosaic do we dare aggregate if we are to do it sustainably?

In the words of John Todd, co-founder of the New Alchemy Institute, we want to come up with "elegant solutions predicated on the uniqueness of place." At a conference held at the University of Minnesota, a soil expert cited a 54-acre field which had 7 mappable units — not uncommon, especially for the glaciated agricultural soils of North America. For corn alone, the yield potential of these 7 units varied from 112 to 162 bushels per acre. A fertilizer expert could logically recommend anywhere from 88 to 132 pounds of nitrogen per acre per year for that 54-acre field, depending on the unit. The phosphorus need, based on the soil tests, ranged from 10 to 35 pounds of P205 per acre. Here is why it is important to think of small scale, or a high eyes-to-acres ratio in a world of scarce resources. Farmers usually apply a fertilizer level their best soil can handle, which leads to a waste of the resource and pollution of the ground water. These seven soil types are even smaller units of that ecological mosaic I mentioned. To some degree, this is the mosaic for which we must provide the "elegant solutions," the mosaic which has to be considered as we think of those countless mini-economies.

This very field was described to illustrate that information from soil surveys and production data is becoming available on floppy disks and that this new technology can accommodate a new but still infant fertilizer spreader equipped with a microprocessor designed to control the appropriate distribution of fertilizer as the machine proceeds across the field. This machine is predicted to have a major future role in reducing pollution. So here, where cultural intelligence could play a major role and justify small farm size, technological cleverness is introduced as a future likelihood to take care of what is regarded as the main problem: pollution. This is why a conceptually rich solution requiring cultural intelligence should be given a chance to override any technological tour de force. In other words, "elegant solutions being predicated on the uniqueness of place" can easily become a battle cry for an extension of the old industrial mind into a new technological frontier. Two perceptions of the world are on the line. Those who propose a cultural solution to an environmental problem will always be faced with a proposed technological solution promoted by the technological zealot.

A primarily cultural solution requiring our wits in the management of the ecological mosaic is more likely to be more ecologically and economically correct than the technological solution. The former requires thought. The latter requires expertise. The former has the opportunity for furthering the cause of economic justice. The latter has the potential to be sold on the stock market. The former will look to the sort of possibilities that Nebraska prairie suggests for agronomic arrangements. The latter will attempt to overcome the problems with an expanding technological array including the employment of satellites. The former will feature small loops of influence away from the farm; the latter features larger loops widening away from the farm and into the corporate industrial realm where, if the past is any indication of the future, ecological and financial problems will continue to multiply.

SPENDING OUR CAPITAL*

David Ehrenfeld

> The rich man's wealth is his strong city. The ruin of the poor is their poverty.
>
> **Proverbs 10:15**

A deeply felt aversion to spending capital is an ancient part of the heritage of most societies. My father was a doctor, not a businessman, but he taught this to me. He had lived through the bankruptcy of his own parents during the Great Depression, watching as they gradually sacrificed the inventory of their store in Passaic, New Jersey, to keep the family in food and clothing. On one side of this store, his father sold records, phonographs, and sewing machines, and repaired the appliances that he sold, while on the other side his mother, a brilliant dress designer, prepared bridal gowns for customers who came from as far away as New York City. She would dress the brides on the wedding day too, and was celebrated for her ability to make the plainest bride look beautiful. But as the Depression wore on, business fell off, the customers stopped coming from New York, and their stock of goods dwindled away. There was no choice but to close the store. My grandparents livelihood was gone forever. Later my father and the oldest of his four brothers made it a priority to pay their parents' creditors in full, a decision that entailed its own sacrifices. And the lesson was handed down to me: you can spend your earned income and any interest you may have received, providing you first set aside a portion to increase your savings, but *never* spend the principal, your capital, except as an act of desperation.

To most of us, capital is associated with business, yet the habit of preserving capital and handing it on to the next generation likely started not as an economic or financial practice, but as an agricultural one. It must have begun in neolithic societies when farming replaced hunting and gathering as the main source of food. From Anatolia to North Africa to Peru, the staple grains, including wheat, rice, corn, millet, oats, barley, and rye, the legumes, such as peas and beans, and the other vegetables, from squashes to eggplants, were almost all annual crops. Every year, our ancestors needed new seeds and tubers for planting; without new seed they would have starved. So every year a portion of the harvest, including seeds from the finest and most productive plants, was saved for the next planting. This seed would have been considered inviolable — to eat it was to risk starvation and death. The idea of preserving capital, the saved-up goods and assets upon which future earnings depend, is a simple and

*Reprinted from *Orion*, Autumn 1996, Vol. 1, No. 2, pp. 7-8. with permission from The Orion Society, 195 Main St., Great Barrington, MA. 01230.

inevitable extension of the ancient rule of saving a part of the harvest for use next year.

Even in this profligate century, there have been those who have paid the ultimate price rather than consume the necessities of future generations. The Russian plant biologist Nikolai Vavilov, father of modern crop plant protection, organized and directed the gathering and classification of viable samples of crop seeds from all over the world in his Institute of Plant Industry in Leningrad. By 1941, while Vavilov, chief preserver of agricultural diversity, was dying of malnutrition somewhere in the Siberian Gulag, and while the German armies were approaching Leningrad, more than 187,000 varieties of wheat, rice, peas, and other crops were represented in the Institute's genebank. Vavilov's deputies rushed to preserve the priceless collection, preparing duplicates of the most valuable samples, evacuating some, and regrowing many of the older seeds on a small plot of land near the front lines. As starvation grew rampant in blockaded Leningrad, the Institute's scientists did their best to protect the seeds from pillagers, rats, and exploding shells. Dimitry Ivanov, a rice specialist, died of starvation while preserving the several thousand packs of rice in his keeping. Georgy Kriyer, in charge of the medicinal plants, L. M. Rodina, a keeper of the oat collection, Alexander Stuchkin, a specialist in groundnuts, and other extraordinary men and women slowly starved to death rather than eat the agricultural capital that surrounded them. The continuing, heroic efforts of those who survived saved the genebank for a largely oblivious posterity.

Saving capital has gone out of style in the postwar era. We have developed a flair for squandering that would be the envy of even the most accomplished wastrel. Show us something that our children and grandchildren will need for their survival, and we will find a way to spend it or ruin it, even if the spending and ruining require much effort. Some of these efforts are remarkably ingenious: take, for example, the draping of a massive government building or an entire island in petroleum-based plastic sheeting as — I suppose some would call it — an artistic statement about light and form. Is one a grinch for asking how many barrels of our children's vanishing oil were used in this bizarre public display of reckless waste? Is it far fetched to see a similar heedlessness in the political effort to lower the federal tax on what is already the most dangerously underpriced gasoline in the world? Is it unreasonable to question the decisions to sell the nation's emergency oil and helium reserves merely to hold down the prices of these limited commodities? I naively thought that the whole point of supply-and-demand capitalism was that high prices would serve to limit the consumption of scarce goods.

One of the insidious causes of our pervasive thoughtlessness is that stocks of capital are no longer as easy to assess as they were when all we needed to know was how much seed was stored in the shed or how many sewing machines and phonographs were still on the shelves. Now, the cars are in the driveway in Massachusetts or Montana, but the oil wells are in Kuwait, Nigeria, Iran, and the Gulf of Mexico. Or the commodities themselves, such as helium, are so remote from our daily experience that most of us have no reason ever to give them a thought.

Another set of causes of our waste of capital is related to the problem of exponential growth. When the practice of saving capital evolved, population and consumption were comparatively stable or were growing slowly by today's standards. In the second half of the twentieth century, a rapid increase in both has occurred. This has confounded our efforts to save capital in three related ways that were described in 1978 by an American physicist, Albert A. Bartlett. Bartlett applied a well-known formula concerning growth rates to examine the question of how rapidly resources are being used up. From this equation, he developed principles that he thought every educated person should know, and illustrated them with a series of vivid examples. First, if we assume an annual growth rate in consumption of seven percent, the amount of a resource used during the next ten years will equal the total of all preceding consumption of that resource in history. Our conventional thinking, based on millennia of slow growth or no growth, does not prepare us for such a dramatic depletion of our resources. *Growth makes us overestimate our future needs for capital.*

Second, with exponential growth, the time at which the diminishing stocks of a resource still appear sufficient and the time at which the resource runs out occur surprisingly close together. Imagine, says Bartlett, a single bacterium put in a bottle at 11:00 PM. The bacterium doubles every minute. By midnight, the bottle is full. When was the bottle only half full? The chilling answer: at one minute to midnight. By the time the more perceptive and activist bacteria first notice that they are running out of space and call a public meeting, it may already be too late to do anything about it. *Growth makes us exaggerate the time we have left in which to take action to save capital.*

And third, counter to all of our intuitions, in exponential growth, the lifespan of a finite resource is only weakly related to the size of the remaining stock. The number of oil fields or copper deposits left to be found makes little difference in calculating the date when reserves of oil and copper will be exhausted. To return to the bacteria, let us assume that at 11:59 P.M. an enterprising bacterium discovers three empty bottles off the north slope of Alaska. This quadruples the space resources available to the crowded population. How long will this new space last? The calculation is very simple: only two more minutes! *Growth makes a mockery of our hope for salvation by as yet undiscovered resources of capital.*

The list of dwindling capital resources is not confined to obvious commodities such as oil, coal, helium, and scarce metals: it includes items that we used to take for granted — soil, fresh water, clean air, open space, trees, and fish in the sea. Moreover, there is another kind of capital besides natural resources — human capital — and we are wasting that, too. By human capital, I mean the skills in agriculture, crafts, academic learning, mechanics, the arts, and all the other necessary human occupations. Also,

I mean the community-fostering skills of judgment, patience, consideration, and knowledge. Human capital, like crop varieties, but unlike iron ore in the earth or intact forests, does not preserve itself automatically until it is used or damaged. It demands an ongoing, active, proficient effort to prevent its inexorable decay. It can disappear in one generation — ten generations of work may not bring it back. Only a stable society can provide the continuity needed to preserve human capital.

Parents, teachers and elders are the primary caretakers of human capital, and the low esteem that modern society accords to these civilization-preserving people helps explain why human capital is now melting away. In a world dominated by huge corporations and global trade, the need of humans and their societies for continuity has become subservient to a non-human, economic demand for cheap production, with its ever-changing technologies and job definitions. In this world, stability is considered stagnation, and human capital is not considered at all.

During the past year or two, the job placement advisers have begun to say to graduating seniors, "You will each have to get used to the idea that you should not expect to have one job that lasts your lifetime — -you will have to be flexible about what you do and where you live." In other words, stability is an undeserved luxury; traditions of wisdom and workmanship are not worth the effort to acquire, let alone pass on; skill is short-lived or at best utterly generic; and saving and planning are the quaint activities of characters in history books and old novels. We have come a long way from the farmer who took comfort during the dark winter night in knowing that the seeds for spring planting were safely stored.

Today, even as "conservative values" are blasted from loudspeakers everywhere, the venerable custom of protecting our assets is rejected and vilified. Everything is topsy turvy, as self-proclaimed conservative capitalists lead the assault against the conservation of capital. Here is an evil that no easy remedies will erase. Saving capital is hard work, and our quick-profit culture imposes extra social and economic punishments on those who undertake the effort. We are not yet asked to make the extreme sacrifice of a Vavilov or an Ivanov. But pulling back from the seductive attractions of corporate culture — the Super Marts, the television, the deadly conveniences — which are eating away at our true and only wealth like a cancer, calls for a quiet heroism and a daring creativity of its own. Heroism in choosing the more difficult and sometimes more expensive path; creativity in finding new ways of saving capital in an environment that despises prudence and loathes the future. Our task has been described concisely and elegantly in Proverbs (13:22): *A good man leaves an inheritance to his children's children.*

NONE SO BLIND
The Problem of Ecological Denial*

David Orr & David Ehrenfeld

"None so blind as those that will not see."

Matthew Henry

Willful blindness has reached epidemic proportions in our time. Nowhere is this more evident than in recent actions by the U.S. Congress to deny outright the massive and growing body of scientific data about the deterioration of the Earth's vital signs while attempting to dismantle environmental laws and regulations. But the problem of ecological denial is bigger than recent events in Congress. It is flourishing in the "wise use" movement and extremist groups in the U.S., among executives of global corporations, media tycoons, and on Main Street. Denial is in the air. Those who believe that humans are, or ought to be, something better than ecological vandals need to understand how and why some people choose to shun reality.

Denial, however, must be distinguished from honest disagreement about matters of fact, logic, data, and evidence that is a normal part of the ongoing struggle to establish scientific truth. It is, rather, the willful dismissal or distortion of fact, logic, and data in the service of ideology and self-interest. The churchmen of the 17th century who refused to look through Galileo's telescope, for example, engaged in denial. Their blind obedience to worn-out dogma was expedient, serving to protect ecclesiastical authority. Denial is apparent in every historical epoch as a willing blindness to the events, trends, and evidence that threaten one established interest or another.

In our time, great effort is being made to deny that there are any physical limits to our use of the Earth or to the legitimacy of human wants. On the face of it, the case is absurd. Most physical laws, as Herman Daly has pointed out, define the limits of what it is possible to do. And all of the authentic moral teachings of three thousand years have been consistent about the dangers and futility of unfettered desire. Rather than confront these things directly, however, denial is manifested indirectly.

A particularly powerful form of denial in U.S. culture begins with the insistence on the supremacy over all other considerations of human economic freedom manifest in the market economy. If one chooses to believe that economies so dominated by lavishly subsidized corporations are free, then the next assumption is easier: the religious belief that the

* Excerpted with permission from *Conservation Biology*, Vol. 9, No. 5, October 1995, pp. 985-987.

market will solve all problems. The power of competition and the ingenuity of technology to find substitutes for scarce materials, it is believed, will surmount physical limits. Markets are powerful institutions that, properly harnessed, can accomplish a great deal. But they cannot substitute for healthy communities, good government, and farsighted public policies. Nor can they displace the laws, both physical and moral, that bound human actions.

A second indirect manifestation of ecological denial occurs when unreasonable standards of proof are required to establish the existence of environmental threats. Is the loss of species a problem? Well, if you think so, name one species that went extinct today! The strategy is clear: Focus on nits, avoid larger issues, and always demand an unattainable level of proof for the existence of any possible problem before agreeing to any action to forestall potential catastrophe. True, no such standards of proof of likely Soviet aggression were required to commit the United States to a $300 billion defense budget. But denial always works by establishing double standards for proof.

Third, denial is manifest when unwarranted inferences are drawn from disconnected pieces of information. For example, prices of raw materials have declined over the past century. From this, Julian Simon for one, draws the conclusion that there can be no such thing as resource scarcity. But the prices of resources are the result of complex interactions between resource stocks/reserves, government subsidies, unpriced ecological and social costs of extraction, processing and transportation, the discount rate, and the level of industrial growth (which turned down in the 1980s). This is why prices alone do not give us accurate information about depletion, nor do they tell us that the planetary sinks, including the atmosphere and oceans, are filling up with wastes they cannot assimilate.

Moreover, the argument from prices and other economic indicators does not take into account the sudden discontinuities that often occur when limits are reached. A typical example from physics is provided by Hooke's law: "Stress is proportional to *strain, within the elastic limit.*"   The length of an elastic band is proportional to the stretching force exerted on it — until the band snaps. In biology, the population crashes that sometimes occur when carrying capacity is reached provide another example. There are many more.

Fourth, denial is manifest in ridicule and *ad hominem* attacks. People inclined to think that present trends are not entirely positive are labeled doomsayers, romantics, apocalyptics, Malthusians, dread-mongers, and wackos. In a book that dominated environmental discussion on Earth Day 1995, *Newsweek* writer Gregg Easterbrook, for example, says that such people (whom he calls 'enviros') "pine for bad news." They suffer from a "primal urge to decree a crisis,"[1] and "subconscious motives to be alone with nature."[2] Pessimism, for them, is "stylish." This example is characteristic of ridiculous people with nonsensical views who do not deserve a serious response; this relieves those doing the name calling and denying from having to think through complex and long-term issues.

Fifth, denial is manifest in confusion over time scales. Again Gregg Easterbrook spends the first 157 pages of his 698 page opus explaining why in the long view things such as climatic changes and soil erosion are minor events. Shifting continents, glaciation, and collision with asteroids have wreaked far greater havoc than human-caused degradation. "Nature, " he says,"has for millions of centuries been generating worse problems than any created by people."[3] We do not for a moment doubt the truth of this assertion. Nor do we doubt that from, say, Alpha Centauri, a nuclear war on Earth would scarcely make the mid-day farm report. Easterbrook enjoins us to place our ecological woes in the perspective of geologic time, and from a sufficient distance they do indeed look like a quibble. The Earth is a "fortress" he says, capable of withstanding all manner of insult and technological assault. But we don't live on Alpha Centauri, and events that may be trivial in a million years loom very large to us, with our 75-year life spans, our few-hundred-year-old countries, and our 8,000-year-old agricultural civilization.

Denial is manifest, sixth, when large and messy questions about the partisan politics of environmental issues are ignored. In the fall of 1994, about the same time that Gregg Easterbrook would have been working over the galley pages for *A Moment on the Earth*, agents of the Republican party were drafting the final version of *The Contract With America*, a major goal of which was to dismantle all of the environmental laws and regulations so painstakingly erected over the past 25 years. Ecological optimism was blindsided by political reality.

So much for how denial is manifested, but why is it happening? It is happening, first, because in the face of serious problems such as the increasing gap between the rich and everyone else, and the related problems caused by unrestrained corporate power, we look for scapegoats rather than confront problems directly. Historian Richard Hofstadter once called this the "paranoid style of politics." Practitioners of paranoid politics use conspiracy theories to explain why things are not as good as they ought to be. Since the collapse of the Soviet Union, dependably loathsome enemies are more difficult to find. Accordingly, environmentalists, bureaucrats, gays and ethnic minorities have replaced communists as the enemies of choice.

Second, and perhaps most obvious, denial is a defense against anxiety. Many of the environmental changes that are now happening are deeply disturbing, but they constitute only a part of the assaults on our well being that most of us face daily. It is natural to want to lighten our load of troubles by jettisoning a few. Environmental problems are rarely as personally pressing as sickness or loss of a job, so out they go. This kind of denial can provide some immediate relief of anxiety. However, it merely delays the confrontation with ecological reality until the time when environmental events, breaking through the screen of denial, force themselves upon us. When that occurs, our ecological troubles will be far more painful and far less tractable to deal with than they are now.

Ecological denial is happening, third, because it seems plausible to the ill-informed. A recent poll conducted by the National Opinion Research Center at the University of Chicago found that only 44% of Americans believed that "human beings developed from earlier species," while only

63% were aware that "human beings are the main cause of plant and animal species dying out." This was the lowest response among the citizens of 20 counties surveyed. People so ignorant are mere fodder for those who would harness denial for their own purposes.

Fourth, it may be fair to say that ecological denial is happening because environmental advocates often appear to be elitist and overly focused on an ideal of pristine nature, to the exclusion of real people. We have not bridged the gap between environmental quality and class as imaginatively and aggressively as we ought to have. As a result, many people see conservation biologists and environmental activists as members of yet another special interest group, not working for the general good. It is clear that we will have to do a better job explaining to the public why environment is not an expendable concern unrelated to real prosperity and community.

How is this to be done?

We would like to recommend the following steps. We members of the conservation community must not deny that we live in a society that desperately needs fixing and for which denial is seductively easy and cheap, at least for a time. We must acknowledge and seek to understand the connection between poverty, social injustice, and environmental degradation. We must acknowledge and seek to understand the connection between rootlessness and environmental irresponsibility. We must acknowledge and seek to understand the connection between the loss of functional human communities and the inexorable decline in the state of the Earth.

We should take our critics seriously enough to read what they have to say. We should take words more seriously than we have in the past. Without much of a fight we have abandoned words such as "progress," "prosperity," and "patriotism" to those who have cheapened and distorted their meanings beyond recognition. We need to take back the linguistic and symbolic high ground from the deniers. At the same time, however, some of us need to be much more careful about using apocalyptic words such as "crisis." "Crisis," a word taken from the field of medicine, implies a specific time in an illness when the patient hovers between life and death. But few environmental problems conform closely to that model. We do not doubt for a second that we now face some genuine crises and that we will face others in the future. But for the most part, ecological deterioration will be a gradual wasting away of possibilities and potentials, more like the original medical meaning of the word "consumption."

Finally, we should all learn to recognize the signs of ecological denial so that when we see it in operation we can expose it for what it is and force an honest discussion of the real issues that deserve our immediate and full concern.

RECOMMENDED READING:

Ehrlich, Paul and Anne (1996) *Betrayal of Science and Reason*. Washington, DC: Island Press,

COMPLICITY WITH EVIL

The Very Rev. Thomas E. Punzo

"Unless there is within us that which is above us, we shall soon yield to that which is around us."

P. T. Forsyth

The Rev. Richard Viladesau wrote a reflection about the banality, the ordinariness, of evil., which began with a focus upon the Polish director Agnieska Holland's film, *Europa, Europa*. The film tells the harrowing and true story of Solomon Perel, a teenager in a German- Jewish family that fled to Poland just before the outbreak of WWII. When the Nazis invaded Poland, the boy — by a mixture of good luck and charm — managed to pass himself off as an orphaned and displaced "Aryan" and even to be sent to an exclusive school for Hitler youth, where he spent most of the war years. In this setting, with his Jewish identity hidden, he found and gave friendship and love. At one point in the film, he wonders how this can be: how can the same people who would despise and destroy him if they knew he was a Jew, be so good and kind to him? And, for his part, how can he find them lovable?

Viladesau then goes on to comment:

The paradox the boy faces reveals something about the nature of evil. The majority of those who collaborated with and even perpetrated some of human history's most terrible crimes and falsehoods were not inhuman monsters; they were like us. They were capable of giving and receiving sympathy and love; they could have beautiful feelings and noble ideas; family love, loyalty, and culture could all co-exist with almost unbelievable evil...

A German woman who grew up in the Nazi years recently wrote an article reflecting on this fact. 'Americans,' she wrote, ' look with horror at the German society that accepted Hitler's leadership and ask, "How could they?" But for someone who lived through it, it is quite easy to see how they could. There were no apparently momentous decisions to be made for or against evil. People were concerned with their daily affairs, and on that level, Nazism seemed good: it brought prosperity, it made things work, it allowed people to feel good about themselves and their country. The larger issues — the ones that we now see as having been central — did not enter into most people's minds.'

After giving some thought to Viladesau's reflections, I would concur that there is, in fact, a certain ordinariness about evil. It does not necessarily present itself in monstrous and dramatic shape. It can take the form of simple conformity to what everyone is doing, to what the leadership says is right. It can be something we do not even notice because we are not paying attention, something barely on the periphery of our consciousness. When the full horror of Nazism was revealed at the end of the war, the German people responded — with basic truth — "We did not know."

Could they have known? *Should* they have known?

And what about ourselves? Are we immune to cooperating with evil in our own time? Are we collaborating in something — the destruction of the environment; the exploitation of the poor; the inability to bring to a halt the outrageous evil of "ethnic cleansing" in former Yugoslavia of which future generations will ask, "But how 'could' they?" Keep in mind this haunting truism:

> ALL THAT IS NEEDED FOR GREAT EVIL TO TRIUMPH
> IS FOR THE GOOD TO BE COMPLACENT
> AND NOT ASK QUESTIONS.

(That may be a good paraphrase of Edmund Burke's observation along these same lines.) It will weigh heavily upon the future memory and conscience (hopefully!) of the Serbian Orthodox Church — leaders and faithful alike — (not to mention the worldwide Christian Church as a whole) that too much "deafening silence" has muted outright condemnation and abhorrence against the warfare, human torture and related atrocities afflicted by the Serbian military and citizenry on their Muslim fellow country men, women and children. Isn't this the same ghost of shameful silence haunting the memory and heritage of German Christians and others beyond Germany (even perhaps high churchmen such as Pope Pius XII) in Hitler's reign of terror?

"Religious leadership where is your voice of outrage and condemnation?" And while we're at it... could the same not be addressed to the contemporary Christian leadership (church) of Northern Ireland — Protestant and Catholic? White Christians of our own country will always bear the scar-tissue of racism toward Black Americans, Native Americans and other minority peoples in our modern era.

Cooperating with evil is manifested beyond question when would-be "good Samaritans" stand by as observers when a woman is being raped or a man is being beaten to death, or when forests are being destroyed, rivers polluted, and the air we breathe injested with billions of tons of toxic chemicals each year. The shallow plea of "not wanting to get involved for fear of one's own personal safety" holds not one point of justification no matter how loud the plea. Once again silence condones, and complicity with evil is beyond question.

Often, the justification note is sounded when workers refuse to challenge management in the production of heinous weapons of torture (e.g. stun guns, etc.) and anti-personal weaponry — the justification being that the job is needed to provide for one's family. Coupled with this is the end result of much of this activity, which is to peddle these products in the arms bazaar, an activity which includes our own country as one of the largest (if not the largest) exporters of such products. Something more — much, much more — has to be offered than merely meeting economic bottom lines. We shall all have to answer for this; and "I didn't know" or "I had no part in the commerce component of this activity" or "it's a matter of free enterprise" will have no validity as self-justification. Either we are on the side of affirming human dignity or we are not.

Hiding behind the disguise of "free speech" lurks some who exhibit very little if any sense of connecting "free speech" with "responsible speech." I refer to many talk show hosts whose villainous, hate-filled jargon and conversation work their very effective corrosive ways into the minds and psyches of many more than just the vulnerable and unsuspecting. We are indeed shaped by what we hear and see. We are enculturated by both the good and the bad as human beings. Like the song from *South Pacific* says, "You've got to be taught how to hate." Words and signs have effective power. Both can be used to uphold and strengthen the good; they can both become the propaganda that instills and fans the fires of hatred and destruction. The right to "free speech" is lost or surrendered the moment it becomes dislodged from "responsible speech." Both those who take the low road and those who refuse to challenge the negative and destructive voices and messages are duly locked into a complicity with evil, and both bear responsibility when the message turns into destructive, degrading, and demoralizing actions and attitudes toward others and the environment. That the stewardship of our human speech is of divine importance, let there be no doubt!

One of the functions of religion, as it confronts us with God as our final horizon, is to make us pay attention to the ultimate issues, the deeper questions, to keep us from being complacent, sometimes to shake us out of our comfortable habits and perspectives.

The late Dr. Karl Menninger wrote of evil in his book entitled *Whatever Became of Sin?*. He spoke to environmental sins when he said:

> The worldwide threat to the survival of human life springing from uncontrolled technology, unregulated industry, unlimited population growth, and ruthless wastage and pollution, is a moral problem. It goes back to our basic beliefs regarding the world and ourselves. Religion today repudiates and even challenges such once-held beliefs as that the universe exists for man's exclusive and unconditional use, that production and consumption must increase endlessly, that the earth's resources are unlimited and that a major purpose of government is to make it easy for individuals and corporations to exploit the environment for the amassing of wealth and power for a few.[1]

We see daily stories in the media about just this behavior and what's worse, much of it is supported by elected officials. We are told that the United States and developed nations of the world are the largest contributors to conditions producing climate change and yet, certain industries in the United States are reluctant to cut back on profits in order to prevent what are erroneously being called "natural" disasters. President Clinton has agreed to sign the treaty drawn in Buenos Aires and certain senators are immediately predicting that the Senate will not approve because to do so, they say, will be bad for the economy.[2] Certain senators have vowed to defeat its passage.

As long as we blame "nature" and "God" for the climatic changes that result in personal death and horrendous destruction, we can avoid looking at what we are doing to exacerbate the conditions which produce these disasters. We thereby neglect our duty to God's earth, which He has put under our stewardship. Matthew uses the familiar image of the Lord's vineyard.

> Listen to another parable. There was a man, a landowner, who planted a vineyard; he fenced it around, dug a winepress in it and built a tower; then he leased it to tenants and went abroad. When vintage time drew near he sent his servants to the tenants to collect his produce. But the tenants seized his servants, thrashed one, killed another and stoned a third. Next he sent some more servants, this time a larger number, and they dealt with them in the same way. Finally he sent his son to them. "They will respect my son," he said. But when the tenants saw the son, they said to each other, "this is the heir. Come on, let us kill him and take over his inheritance." So they seized him and threw him out of the vineyard and killed them. Now when the owner of the vineyard comes, what will he do to those tenants? They answered, "He will bring those wretches to a wretched end and lease the vineyard to other tenants who will deliver the produce to him when the season arrives." Jesus said to them, "Have you never read in the scriptures: It was the stone rejected by the builders that became the keystone. This was the Lord's doing. And it is wonderful to see. I tell you then, that the kingdom of God will be taken from you and given to a people who will produce its fruit."
> *Matthew 21:33-43*

Here, it will be taken away from the unjust tenants and given to new ones who "yield a rich harvest" for the Lord. Matthew has allegorized this parable to make it apply more exactly to the rejection of Jesus by his own people and the admission of Gentiles into the Church. It would be a mistake, however, to think that this means that the vineyard is now safely in the hands of good tenants, i.e. "us." Often churches or congregations enjoy an occasion whereby a particular member makes available what at first glance seems to be an extremely generous gift (e.g. $50,000 for a new organ). Once it is received or accepted by the pastor and congregational leadership, the donor then begins to make demands about how the "gift" is to be used. In other words, this was not a gift at all but a subtle (and not so subtle) way to control or manipulate others to the extent that the "strings-attached behavior" finally works to the total disintegration of the congregation and the painful resignation of the pastor. "Mrs. Got Bucks" behavior is but a contemporary example of how evil can disguise itself in the disguise of generosity and giving within the very household of faith. The imperative and warning of the parable still apply. We are reminded that we are responsible for God's vineyard and must give to God, God's due: namely, the harvest of goodness, love and care for all God's creation inclusive of one another and all the others.

The Eucharist that we celebrate continually reminds us of this imperative and what it means. It challenges us to break out of the narrowness and complacency that are a constant temptation. Instead of insisting that we are right, we begin by confessing that we all fall short, that we are sinners. Instead of grasping at life, we remember the life and death of Jesus. Instead of restricting our perspective to the small world of our ego and its extensions, we affirm the whole of humanity as our vineyard, our family. Instead of accepting the world on its own terms, we commit ourselves to bringing about the true harvest — God's kingdom. And that, dear Fellow Tenants, is our singular vocation!

May God prosper the work of our hands!

RESOURCES FOR ACTIVISTS

Judith Scherff

"The politics of restoration will start, not in Washington, but in many other places, separately and together, when people decide to close the gap between what they believe and what is. People may begin this work by understanding what they are up against."

William Greider
Who Will Tell the People: The Betrayal of American Democracy

For anyone interested in helping to snatch democracy from the jaws of Political Action Committees, the industries and the special interest groups they represent, there are many and varied platforms from which to work, with good company. The issues discussed in the preceding chapters are not unsolvable — even though a return to the virtues of tolerance, consideration for the "least of these," an appreciation for diversity, caring for the creation — these religious values — is going to be very hard work. In our current incredibly entrenched zeal for the market economy, these virtues are remarkably unpopular. Anyone who wishes for change will have plenty of resources, but the process is going to take a lot of work, because there will be considerable opposition.

The Constitutional injunction "to promote the general welfare" is an uncontrovertible pillar of liberal value. "Promote" is an active verb which means "to raise to a more important or responsible rank or job... further... advocate." It means that elected representatives who have taken the oath of office promising to uphold the Constitution of the United States are compelled to see to it that all residents of this nation have their basic needs met. But it is up to us to see to it we elect candidates who take their commitment to the Constitution seriously, not simply as a rote process required to be seated in whatever governing body. Gratefully, there are still some elected officials whose goal is to promote the general welfare. Changing a society that has a death grip on material values (taking) over value systems based on integrity and concern for other life (giving) is not going to be easy.

The government is a public trust, not a rubber stamp for special interests. So when elected officials act on behalf of special interests rather then their constituency, they are violating the public trust. The popular (and racist) image of welfare recipient is that of an unmarried woman (usually black) living in a tenement with her illegitimate children. However in reality, approximately $50 billion a year is spent on individuals who are in need of government assistance but — a whopping $167 billion a year goes to support corporations,[1] many of whom are pork barrel friends of elected officials. Corporations are being given over three times the dollars received by individuals — the long time scapegoat for welfare criticism and the main target of welfare " reform".

So what do we begin sweeping in our search for means to protest our air, our rivers, our mineral resources, and even our climate, from further despoilation at the hands of corporations? Political reform to get corporations out of politics would be an important first step. Corporations are public bodies created by public charter to serve a public interest. It is their proper role to follow the rules, not make them. If those rules restrict the freedom of corporate action, that is one of their essential purposes. If corporations have more power than democratically elected governments, the appropriate response for citizens is not to abandon democracy. It is to reclaim that power and restore democracy.[2]

Select candidates of integrity and support them. And in your selection process, follow the money trail — always follow the money trail — and that course will tell you just about all you need to know about the candidates. In the spring of 1998, a story appeared on CNN entitled "Gore Slows EPA Pesticide Review" because "the possible loss of widely used pesticides was causing an uproar in key political states."[3] The petrochemical companies and farm organizations are controlling an issue, through politics, that is most assuredly a "general welfare" issue. This is where tolerance needs to side with and promote the activities of organizations such as those listed at the end of this chapter who are holding the line against the malfeasance of greed.

Pay attention to what your representatives are doing and how they're voting. For example: a really old ruse to watch for is the elected official who refuses to vote for a popular public measure by stating that s/he doesn't care for the bill as it is being presented, so has offered their own bill. That is merely a disgusting means by which the politician begs off voting for what the public is shown to want, and s/he knows full well her/his particular bill will never even be brought to a vote. Whether the issue is campaign finance reform, stronger environmental regulations, or any other measure beneficial to the whole public, keep a watch and see who pulls this loathsome trick.

A CBS television news analyst said that Congressional politics has reached new lows and will not get any better until campaign finance reform is passed.[4] Even as this book is being readied for publication, a story on CNN about campaign finance reform clearly distinguishes the pro from the con forces on that issue, stating that League of Women Voters, Common Cause, AARP, National Council of Churches and Union of American Hebrew Congregations in favor of the reform are up against organizations such as the National Rifle Association and Christian Coalition who oppose campaign finance reform.[5] Isn't it clear to the public who is manipulating our political system?

Churches

Mainstream — the leading denominations — of organized Christianity have lost 25% of their members in the last two decades. "The chief reason for this is that they have accommodated the culture."[6] Some local churches have weighted secular values to a point of dominance in their congregations; they have chosen culture over Christ. The culture v. Christ pressures are undoubtedly playing in most congregations these days. And the future of the organized Christian church depends upon which set of "values" is selected.

Information

Again, follow the money trail. A regular feature of NBC's "Evening News" with Tom Brokaw is called "The Fleecing of America". It is this kind of journalism that tells the other side of the story. If it weren't for segments such as "Fleecing", and CNN's "Earth Matters", the public would have to dig much harder to learn what else is going on and in reality, probably never know.

Read "alternative" literature, publications fewer than in the volume of *Time* or *Newsweek*. So you won't be able to get these publications at grocery stores, but they're readily available at book stores. They are written and published by individuals who believe that alternative news (as opposed to "official" statements) should be available to the public. They print the other side of the story, in essence.

Public Health

Pay attention to any illness that seems to be prevalent in your community. Perhaps a city along a river has a higher than national average incidence of kidney and bladder cancer. If you are researching in this area, you must be proactive. For example, as simple as it should be, misinformation continues among physicians on the determination of lyme disease in blood tests even after a plethora of information has stated blood tests are not reliable. So in dealing with toxic chemicals, a more complex issue, there is even greater professional ambivalence. Nevertheless, stay with your feelings about what you think is going on in your community in regard to a cluster illness caused by any number of impacts on the environment.

Consumer Activism

Americans must search for more ways to democratically impact our economic system than through our roles as consumers. Nonetheless, it remains an inescapable fact that we vote with our dollars every time we spend them. When you buy a product, you confirm its value to the corporation that manufactured or produced it. Look into what you're buying; expand your shopping horizons and look into products from companies such as Seventh Generation and Real Goods who sell environmentally benign products. Example: If you have to have a redwood deck, insist that the lumber company provide you with proof that the redwood lumber you're about to purchase is from a tree farm, not the old growth forests. If he can't do it, go elsewhere.

Demand organic — start lobbying grocery stores if the source isn't in your community. Plant as much as you can yourself.

Contact or join the Council on Economic Priorities. This organization researches and reports on corporations with respect to ten social values. There are many good corporations which deserve our economic support. Find out which they are. Invest in them; purchase their products. And there are several socially conscious investment groups such as Working Assets, Calvert Social Fund, PAX International and others that will invest your money in corporations with positive values. You can contact the Council on Economic Priorities (address below). Learn what to boycott and what to buy. There is a web site for social investing.

Universal goodness will not rain down on us unsolicited. Marking the 20th anniversary of Love Canal, Lois Gibbs stated that government is not voluntarily going to help people solve their problems. She said, "Nobody [government] is going to give it to them. Nobody is going to do it for them. They have to make it happen."[7]

Whatever you wish were better, you are simply going to have to help bring about! You will have to work. You will have to go outside of your comfortable self. You will have to take a stand and get others to take that stand. You may even have to take certain risks. But imagine how you're going to feel when you learn of situations which are out of bounds with justice and compassion and tolerance and ecosystem integrity, and you've done nothing. Feelings associated with turning your back are going to be even more uncomfortable. There are certainly no shortages of issues — so pick one, or two, and get started. Some issue, somebody, some bodies, need your help. You won't be alone.

Mainstream Religious

Interfaith Alliance. An organization "dedicated to protecting America's basic freedoms of speech, press and religion from the fringe groups who cloak in religious garb their challenge to these principles". They are located at 1012 14th Street, NW, Suite 700, Washington, DC 20005. Phone: 202/639-6370. Web site: http://www.tialliance.org.

Politics and Government

Center for Public Integrity. 1634 Eye St., NW, Ste. 902, Washington, DC 20006. Phone: 202/783-3900. Web: http://www.publicintegrity.org. Nonprofit, nonpartisan research organization concentrating on ethics and public service issues. They are best known for their project, the International Consortium of Investigative Journalists. Their homepage features a quote by Abraham Lincoln: "I am a firm believer in the people. If given the truth, they can be depended upon to meet any national crisis. The great point is to bring them the real facts."

Common Cause. 1250 Connecticut Ave., NW, Washington, DC 20036. Web: http://www.commoncause.org/. They have considerable information on campaign finance and feature information on political payoffs called "follow the money trail."

League of Women Voters. 1730 M St., NW, Washington, DC 20036-4508. Web: http://www.lwv.org. Their mission is "to encourage the informed and active participation of citizens in government and to influence public polity through education and advocacy."

Project Vote Smart. 129 NW 4th St., Ste 204, Corgallis, OR 97330. Phone: 541/754-2746. Voters Research Hotline toll free: 1-800-622-SMART. Web: http://www.vote-smart.org. Project Vote Smart tracks the performance of over 13,000 political leaders — they are nonpartisan.

Public Citizen. Web: http://www.citizen.org. Founded by Ralph Nader, this is a multiple issue organization with staff working on Congress Watch and the Health Research Group. Their issues involve government, corporations and the public welfare. It is related to U.S. **PIRG (Public Interest Research Group)** 215 Pennsylvania Ave., SE, Washington, DC 20003. Phone is 202/546-9707. Web site: http://www.uspirg.org.

Center for Responsive Politics. This organization publishes information that the interested voter should not go without! They list each member of Congress and the money they have accepted from PACS; publish information on soft money, loopholes for corporate taxes and many other issues defined by their following the money trail. Located at 1320 19th St., NW, Ste. 620, Washington, DC 10036, their phone is 202/857-0044 and their web site, filled with interesting facts, is http://www.crp.org.

League of Conservation Voters. This League tracks the voting records of members of Congress in relation to environmental bills that are voted on. It also makes the connection between PAC money and the way members of Congress vote. Their address is 1707 L Street, NW, Suite 750, Washington, DC 20036. Phone 202/785-8683 and web: http://www.lcv.org.

CCHW Center for Health, Environment and Justice. Founded and currently managed by Lois Gibbs whose motto is: "First, Do No Harm." Located in Falls Church, Virginia, (P.O. Box 6806, zip: 22040), their phone is 793/237-2249 and their web site is: http://www.essential.org/cchw.

The National Institute for Money in State Politics. Provides information about state level candidates. They have searchable databases of campaign finance information for state legislators and gubernatorial races. Their web site is http://followthemoney.org. They are located at 25 S. Ewing St., Ste. 506, Helena, MT 59601 and their phone is 406/449-2480.

Economic Organizations

Council on Economic Priorities. 30 Irving Place, New York, NY 10003. Phone: 212/420-1133. For membership: 1-800-729-4CEP. Email: cep@echonyc.com. For a number of years, the Council has been tracking corporations according to ten parameters of social consciousness. They have a directory of corporations rated by them for socially conscious investment and purchasing.

Environment

Natural Resources Defense Council An organization that specializes in gathering facts encompassing environmental issues and lobbying for environmental protection legislation. NRDC publishes *Amicus,* an informative, in depth report of their research on their issues they are covering. Located at 40 W. 20th St., New York, New York, 10001. Their phone is 212/727-2700; web site is http://www.nrdc.org.

Union of Concerned Scientists. As close to the truth as we're going to get with the information that is currently available, this is the organization whose members are among the leading, most respected scientists of the world; scientists who support the data that is not influenced by corporate money. They are located at 26 Church St., in Cambridge, MA 02138; phone is 617/547-5552 and web site is http://www.usc.usa.org. They publish a newsletter.

National Wildlife Federation. Their motto: "People and nature: our future is in the balance". Head office is at 8925 Leasburg Pike, Vienna, Va 22814. Phone is 703/790-4000. One of the major environmental organizations in this country with an excellent educational program for children. On the Net: http://www.nwf.org.

Earthjustice Legal Defense Fund. This group calls itself "Lawyers for the environment," with nine regional offices they safeguard natural resources, strive to reduce pollution and achieve environmental justice. Web site: http://www.earthjustice.org. The Earthjustice central office is located at 180 Montgomery, San Francisco, CA. 94100, and their phone is 415/627-6700.

Ozone Action. Specializing in issues regarding global warming and stratospheric ozone depletion, Ozone issues fact sheets on companies which add to both issues. They can be found on the Internet at http://www.ozone.org, for a complete reference to the information they provided in this book.

Sierra Club. One of the oldest U.S. environmental groups; Sierra Club does in depth studies of environmental issues which are published in their magazine, *Sierra,* and on their web site at www.sierraclub.org. Their main office is 730 Polk St., San Francisco, 94109. Phone is 415/776-2211.

PEER. Public Employees for Environmental Responsibility (PEER) is a Washington, DC-based employee group which promotes ethics and integrity in natural resource and environmental management agencies. PEER works with government scientists, law enforcement officers and other field specialists who implement and enforce environmental regulations and agency policies. As the final line of defense for our natural heritage, these dedicated professionals are making a difference for a better government and a healthier environment. Located at 2001 S. Street, NW, Ste. 570, Washington, DC 20009. Phone: 202/205-7337. Web site: http://www.peer.org.

Environmental Working Group. This organization publishes air and water pollution statistics state by state and even in your local area. Just click on their web site: http://www.ewg.org. They have sections such as:

campaign contributions and environment; tap water; air pollution; pesticides; anti-environmental groups; farm subsidies; toxics and transportation policy. This research organization has a very impressive presentation for general information but most especially, for local communities. EWG is located at 1718 Connecticut Ave, NW, Suite 600, Washington, DC 20009. Phone: 202/667-6982.

Greenpeace USA. Greenpeace has published a book entitled *The Greenpeace Guide to Anti-Environmental Organizations*, listing groups that have an anti-environmental agenda, some which have deliberately disguised themselves by their names as pro-environment. The ISBN number is 1-878825-05-4, Odonian Press, Box 7776, Berkeley CA 94707. The address of Greenpeace itself is 1436 U Street NW, Washington, DC 20009; their web site is at http://www.Greenpeaceusa.org.

Environmental Defense Fund Longtime watchdog and litigator of environmental issues, their web site contains a number of reports that may be of interest. Electronically connected at http://www.edf.org, they are located at 257 Park Avenue South, New York, NY 10010.

RECOMMENDED READING:

Hawken, Paul (1993) *The Ecology of Commerce: A Declaration of Sustainability.* New York: Harper Collins.

Korten, David C. (1995) *When Corporations Rule the World.* West Hartford, CT: Kumarian Press Books.

Meadows, Donella H. *The Global Citizen,* on the Internet at: http://iisdl.iisd.ca pcdf/meadows.

Meeks, Rev. M. Douglas (1989) *God the Economist.* Minneapolis: Fortress Press.

Menninger, Dr. Karl (1977) *Whatever Became of Sin.* New York: Hawthorne Books.

Mokhiber, Russell (1988) *Corporate Crime and Violence.* San Francisco: Sierra Club Books.

Nollman, Jim (1990) *Spiritual Ecology.* New York: Bantam Books.

Rachel. Environmental Research Foundation, P.O. Box 5036, Annapolis, MD 21403. Fax (410) 263-8944. Web: http://www.monitor.net/rachel. Email: erf@rachel.org. A great deal of information written and posted by Peter Montague.

Rolston, Holmes (1988) *Enviromental Ethics.* Philadelphia: Temple University Press.

Stern, Philip (1988) *The Best Congress Money Can Buy.* New York: Pantheon Press.

UN WORLD CHARTER FOR NATURE

A/RES/37/7
28 October 1982

The General Assembly,

Reaffirming the fundamental purposes of the United Nations, in particular the maintenance of international peace and security, the development of friendly relations among nations and the achievement of international cooperation in solving international problems of an economic, social, cultural, technical, intellectual or humanitarian character,

Aware that:

(a) Mankind is a part of nature and life depends on the uninterrupted functioning of natural systems which ensure the supply of energy and nutrients,

(b) Civilization is rooted in nature, which has shaped human culture and influenced all artistic and scientific achievement, and living in harmony with nature gives man the best opportunities for the development of his creativity, and for rest and recreation,

Convinced that:

(a) Every form of life is unique, warranting respect regardless of its worth to man, and, to accord other organisms such recognition, man must be guided by a moral code of action,

(b) Man can alter nature and exhaust natural resources by his action or its consequences and, therefore, must fully recognize the urgency of maintaining the stability and quality of nature and of conserving natural resources,

Persuaded that:

(a) Lasting benefits from nature depend upon the maintenance of essential ecological processes and life support systems, and upon the diversity of life forms, which are jeopardized through excessive exploitation and habitat destruction by man,

(b) The degradation of natural systems owing to excessive consumption and misuse of natural resources, as well as to failure to establish an appropriate economic order among peoples and among States, leads to the breakdown of the economic, social and political framework of civilization,

(c) Competition for scarce resources creates conflicts, whereas the conservation of nature and natural resources contributes to justice and the maintenance of peace and cannot be achieved until mankind learns to live in peace and to forsake war and armaments,

Reaffirming that man must acquire the knowledge to maintain and enhance his ability to use natural resources in a manner which ensures the preservation of the species and ecosystems for the benefit of present and future generations,

Firmly convinced of the need for appropriate measures, at the national and international, individual and collective, and private and public levels, to protect nature and promote international cooperation in this field,

Adopts, to these ends, the present *World Charter for Nature*, which proclaims the following principles of conservation by which all human conduct affecting nature is to be guided and judged.

I. GENERAL PRINCIPLES

1. Nature shall be respected and its essential processes shall not be impaired.

2. The genetic viability on the earth shall not be compromised; the population levels of all life forms, wild and domesticated, must be at least sufficient for their survival, and to this end necessary habitats shall be safeguarded.

3. All areas of the earth, both land and sea, shall be subject to these principles of conservation; special protection shall be given to unique areas, to representative samples of all the different types of ecosystems and to the habitats of rare or endangered species.

4. Ecosystems and organisms, as well as the land, marine and atmospheric resources that are utilized by man, shall be managed to achieve and maintain optimum sustainable productivity, but not in such a way as to endanger the integrity of those other ecosystems or species with which they coexist.

5. Nature shall be secured against degradation caused by warfare or other hostile activities.

II. FUNCTIONS

6. In the decision-making process it shall be recognized that man's needs can be met only by ensuring the proper functioning of natural systems and by respecting the principles set forth in the present Charter.

7. In the planning and implementation of social and economic development activities, due account shall be taken of the fact that the conservation of nature is an integral part of those activities.

8. In formulating long-term plans for economic development, population growth and the improvement of standards of living, due account shall be taken of the long-term capacity of natural systems to ensure the subsistence and settlement of the populations concerned, recognizing that this capacity may be enhanced through science and technology.

9. The allocation of areas of the earth to various uses shall be planned, and due account shall be taken of the physical constraints, the biological productivity and diversity and the natural beauty of the areas concerned.

10. Natural resources shall not be wasted, but used with a restraint appropriate to the principles set forth in the present Charter, in accordance with the following rules:

(a) Living resources shall not be utilized in excess of their natural capacity for regeneration;

(b) The productivity of soils shall be maintained or enhanced through measures which safeguard their long-term fertility and the process of organic decomposition, and prevent erosion and all other forms of degradation;

(c) Resources, including water, which are not consumed as they are used shall be reused or recycled;

(d) Non-renewable resources which are consumed as they are used shall be exploited with restraint, taking into account their abundance, the rational possibilities of converting them for consumption, and the compatibility of their exploitation with the functioning of natural systems.

11. Activities which might have an impact on nature shall be controlled, and the best available technologies that minimize significant risks to nature or other adverse effects shall be used; in particular:

(a) Activities which are likely to cause irreversible damage to nature shall be avoided;

(b) Activities which are likely to pose a significant risk to nature shall be preceded by an exhaustive examination; their proponents shall demonstrate that expected benefits outweigh potential damage to nature, and where potential adverse effects are not fully understood, the activities should not proceed;

(c) Activities which may disturb nature shall be preceded by assessment of their consequences, and environmental impact studies of development projects shall be conducted sufficiently in advance, and if they are to be undertaken, such activities shall be planned and carried out so as to minimize potential adverse effects;

(d) Agriculture, grazing, forestry and fisheries practices shall be adapted to the natural characteristics and constraints of given areas;

(e) Areas degraded by human activities shall be rehabilitated for purposes in accord with their natural potential and compatible with the well-being of affected populations.

12. Discharge of pollutants into natural systems shall be avoided and:

(a) Where this is not feasible, such pollutants shall be treated at the source, using the best practicable means available;

(b) Special precautions shall be taken to prevent discharge of radioactive or toxic wastes.

13. Measures intended to prevent, control or limit natural disasters, infestations and diseases shall be specifically directed to the causes of these scourges and shall avoid adverse side-effects on nature.

III. IMPLEMENTATION

14. The principles set forth in the present Charter shall be reflected in the law and

practice of each State, as well as at the international level.

15. Knowledge of nature shall be broadly disseminated by all possible means, particularly by ecological education as an integral part of general education.

16. All planning shall include, among its essential elements, the formulation of strategies for the conservation of nature, the establishment of inventories of ecosystems and assessments of the effects on nature of proposed policies and activities; all of these elements shall be disclosed to the public by appropriate means in time to permit effective consultation and participation.

17. Funds, programs and administrative structures necessary to achieve the objective of the conservation of nature shall be provided.

18. Constant efforts shall be made to increase knowledge of nature by scientific research and to disseminate such knowledge unimpeded by restrictions of any kind.

19. The status of natural processes, ecosystems and species shall be closely monitored to enable early detection of degradation or threat, ensure timely intervention and facilitate the evaluation of conservation policies and methods.

20. Military activities damaging to nature shall be avoided.

21. States and, to the extent they are able, other public authorities, international organizations, individuals, groups and corporations shall:

(a) Co-operate in the task of conserving nature through common activities and other relevant actions, including information exchange and consultations;

(b) Establish standards for products and manufacturing processes that may have adverse effects on nature, as well as agreed methodologies for assessing these effects;

(c) Implement the applicable international legal provisions for the conservation of nature and the protection of the environment;

(d) Ensure that activities within their jurisdictions or control do not cause damage to the natural systems located within other States or in the areas beyond the limits of national jurisdiction;

(e) Safeguard and conserve nature in areas beyond national jurisdiction.

22. Taking fully into account the sovereignty of States over their natural resources, each State shall give effect to the provisions of the present Charter through its competent organs and in co-operation with other States.

23. All persons, in accordance with their national legislation, shall have the opportunity to participate, individually or with others, in the formulation of decisions of direct concern to their environment, and shall have access to means of redress when their environment has suffered damage or degradation.

24. Each person has a duty to act in accordance with the provisions of the present Charter; acting individually, in association with others or through participation in the political process, each person shall strive to ensure that the objectives and requirements of the present Charter are met.

CONTRIBUTORS

Thomas Berry is author of *The Dream of the Earth* (Sierra Club Books) and numerous articles. He was founder and director for twenty-five years of the Riverdale Center for Religious Research in the Bronx, a research institute devoted to the study of a viable mode of human presence upon the earth.

Conger Beasley, Jr. is a prolific writer on environmentalist issues. His latest book is the text, *Colorado Closeup*, with photographs by J. C. Leacock.

Phil Brown is Professor of Sociology at Brown University and holds a Ph.D. in sociology from Brandeis University. He has published articles on community response to toxic wastes, social stratification and toxic exposures, environmental social movements, physician involvement in identifying toxic waste-induced disease, the politics of diagnosis, and clinician-patient interaction. He is author of *The Transfer of Care: Psychiatric Deinstitutionalization and its Aftermath*, and has edited *Mental Health Care and Social Policy and Perspectives in Medical Sociology* His book, *No Safe Place: Toxic Waste, Leukemia, and Community Action*, has recently been reprinted.

Jeff DeBonis is a former federal government employee who "blew the whistle" on illegal timber sales in the Willamette National Forest in Oregon. *The Economist* referred to DeBonis as the "spark plug" of the internal revolt in the U.S. Forest Service. While still with the Forest Service in 1989, DeBonis formed the Association of Forest Service Employees for Environmental Ethics, an internal employee dissident group, and later PEER (Public Employees for Environmental Responsibility), both for encouragement and protection of employees who attempt to support conservation laws.

Dr. David Ehrenfeld teaches ecology at Rutgers, and is author of *The Arrogance of Humanism*, and founding editor of *Conservation Biology*.

Robert V. Eye, formerly Chief Counsel, Kansas Dept. of Health & Environment, was an independent candidate for governor of Kansas in 1994.

Dan Fagin is an environmental writer for *Newsday*, and was a Pulitzer finalist in 1994 for his articles documenting links between pesticides and breast cancer.

Wes Jackson is founder of The Land Institute in Salina, Kansas, and a well known advocate for agricultural reform.

Very Rev. James Parks Morton was Dean of Cathedral of St. John the Divine, New York City for 25 years. He is now President of the Interfaith Council of New York., which he founded in 1997.

Marianne Lavelle is a writer for the *National Law Journal* and author of a 1992 investigation into racial bias in environmental protection that won the Polk Award and 6 other national prizes.

David Orr is professor of environmental studies at Oberlin college and the author of *Earth in Mind* (1994) and *Ecological Literacy* (1992).

John Passacantando is co-founder and Executive Director of Ozone Action. He served as an Executive Director of the Florence & John Schumann Foundation. He holds an M.A. in economics from New York University.

Very Rev. Thomas Punzo is an Episcopal priest, Dean of Northeast Convocation, Diocese of Kansas.

Tarso Luís Ramos is program director at the Western States Center, a regional non-profit research and training institute based in Portland, Oregon. He heads the Center's Wise Use Public Exposure Project, a leading source for information and analysis of anti-environmental backlash since 1992.

Judith Scherff is an ecologist and longtime activist in environmental issues and editor of the anthology, *The Mother Earth Handbook*. She holds graduate degrees in political science and environmental studies.

John Stauber and Sheldon Rampton work for the non-profit Center for Media & Democracy in Madison, Wisconsin, and edit the investigative quarterly, *PR Watch*. They are co-authors of two books published by Common Courage Press, *Toxic Sludge is Good for You: Lies, Damn Lies and the Public Relations Industry*, and *Mad Cow U.S.A.: Could the Nightmare Happen Here?* They can be reached at 608-233-3346 or via their web site: www.prwatch.org.

Dr. Sandra Steingraber is coauthor of a report on ecology and human rights in Africa, "The Spoils of Famine". She has taught biology for several years at Columbia College, Chicago, and held visiting fellowships at the University of Illinois, Radcliffe College, and Northeastern University. She was recently appointed to serve on the National Action Plan on Breast Cancer administered by the U.S. Department of Health and Human Services. She is author of *Living Downstream: An Ecologist Looks at Cancer and the Environment*.

Glenn P. Sugameli is Senior Counsel for the National Wildlife Federation, Washington.

Douglas Trent, owner of an ecotourism company, once lived in Brazil and saw the decimation of tropical rain forests first hand. A graduate of the University of Kansas Environmental Studies Program, he lives in Minnesota where his tour company, Focus Tours, is based. He can be reached at FocusTours@aol.com.

Ann Vileisis is a writer, researcher, and activist.

Johanna H. Wald is a senior attorney with the Natural Resources Defense Council (NRDC), in its San Francisco, California office, and a Pew Scholar in Conservation & Environment. **Susannah French**, an attorney with the San Francisco firm of Shute, Mihaly and Weinberger, was a student at the University of California School of Law, Berkeley, when this chapter was originally drafted.

ENDNOTES

Publisher's Foreword:

[1] *Rachel's Environment & Health Weekly* #569, 23 October 1997, "Trends in Corporate Accountability," WW III, Pt. 3. Electronic version.

Berry: Foreword

[1] Morton Horowitz, *The Transformation of American Law 1780-1860,* p. 253.

Scherff: Introduction

[1] Genesis 1:1.
[2] Matthew Fox, an address to COMISS, an organization of hospital chaplains, Minneapolis, 1988.
[3] William K. Stevens, "How Much is Nature Worth? For You, $33 Trillion," *The New York Times,* May 20, 1997, Internet.
[4] Letter from PEER (Public Employees for Environmental Responsibility) Executive Director, Jeff Ruch, March 25, 1998.
[5] Peter Montague, "Trends in Corporate Accountability--WW III, Pt. 3," *Rachel's Environment & Health Weekly,* #569, October 23, 1997, Electronic Edition.
[6] Peter Montague, "The Major Causes of Illness," *Rachel's Environment & Health Weekly* #584. February 5, 1998, Electronic Edition.
[7] Federal Election Commission, Internet: www.fec.gov. April 22, 1997
[8] Matthew 25:31-46.

Beasley: The Killing Ground: The Destruction of Native American Species

[1] A.W. Schorger, *The Passenger Pigeon,* University of Oklahoma Press, 1973, p. 225.
[2] *Id.,* p. 193.
[3] John Perrin, *A Forest Journey: The Role of Wood in the Development of Civilization,* Harvard University Press, Cambridge, 1989, p. 151.
[4] *Id.,* p. 173.
[5] Frederick Marryat, *A Diary in America,* Alfred A. Knopf, New York, 1962, p. 97.
[6] Schorger, *supra* endnote 1, p. 29.
[7] E. Douglas Branch, *The Hunting of the Buffalo,* University of Nebraska Press, Lincoln, 1962, p. 45.
[8] *Id.,* p. 86.
[9] Nicholas Lapham, *A History of Predator Control in the Western States* unpublished senior essay, p. 5.
[10] *Id.,* p. 11.
[11] *Id.,* p. 17.
[12] Michael Milstein, "A Federal Killing Machine Rolls On," *High Country News,* January 28, 1991.
[13] J. Frank Dobie, *The Ben Lilly Legend,* University of Texas Press, 1985, p. 217.
[14] Lapham, *supra* endnote 9, p. 11.
[15] *Id.,* p. 12.
[16] *Id.,* p. 13.
[17] *Wildlife Management: Effects of Animal Damage Control,* GAO Report to Senator Alan Cranston, August 1990, p. 16.
[18] Milstein, *supra* endnote 12, p. 13.
[19] "Animal Damage Control Program Highlights," United States Department of Agriculture, Publication No. 1501, 1991.
[20] "Dances With Coyotes," *People Magazine,* June 15, 1992, p. 67.

[23] Lapham, *supra* endnote 9, p. 44
[24] Telephone conversation with Coleen Ault, Comanche National Grasslands, La Juntga, CO, April 1998.
[25] Telephone conversation with Conger Beasley, Jr., November 1992.
[26] Ed Abbey,"Even the Bad Guys Wear White Hats," *Harpers*, January 1986, p. 55.
[27] Milstein, *supra* endnote 12, p. 12.
[28] David Plowden, *Floor of the Sky*, Sierra Club Books, 1972, p. 56.
[29] Milstein, *supra* endnote 12, p. 14.

Trent: American Forests

[1] Jeff DeBonis, "On Speaking Out, Conflict of Interest, And Being An Embarrassment To The Agency, " *Inner Voice*, Vol. 1, No. 1, Summer 1989, p. 2.
[2] Native Forest Council "Some Important Facts about Native and National Forests," a single page fact sheet, no date.
[3] Timothy Hermach,"The Great Tree Robbery," *The New York Times* editorial, April 20, 1992.
[4] Cheri Brooks, Association of Forest Service Employees for Environmental Ethics (AFSEEE). Personal comment, February 4, 1993.
[5] Richard Ober, "Past as Prologue: The Story of the Weeks Act," *Inner Voice*, December 1991, Vol. 3, No. 6, p. 4.
[6] Randal O'Toole, "Political Power Rears Its Ugly Head...Again," Forest Watch Supplement to the *Inner Voice*, December 1991, Vol. 3, No. 6, p. S-1.
[7] Jim Stiak, "A Voice From Within: Jeff DeBonis's Quest to Save His Forest Service," *E Magazine*, Premier Issue, p. 55.
[8] Comments from the Willamette National Forest Public Affairs Office, dated April 7, 1992.
[9] Richard Nilsen, " Reforming the Forest Service From Within: The Crusade of Jeff DeBonis," *Whole Earth Review*, Fall 1989, p. 76.
[10] Paul Roberts, "Zero Cut,"*Seattle Weekly*, November 4, 1992.
[11] John Baden, "Spare that tree!" *Forbes*, Dec. 9, 1991, p. 229.
[12] *Id.*
[13] *Id.* p. 230.
[14] *Id.*
[15] Tim Hermach. Native Forest Council. Personal comment, February 15, 1993.
[16] T.H. Watkins and Staff, Wilderness Society, "The Deforestation of America," in Judith S. Scherff, ed., *The Mother Earth Handbook*, Continuum, New York, p. 65.
[17] *Id.* p. 66.
[18] John Baden, *supra* endnote 11.
[19] *Id.* p. 230-231.
[20] Native Forest Council. The U.S. Government subsidizes logging on National Forests. *Forest Voice*, Vol. 1 No. 4, 1992. p. 5.
[21] Timothy Hermach, Personal comment, February 15, 1993
[22] Native Forest Council. Behind their Ads. *Forest Voice*, Vol. 1, No. 4, 1992. p. 10.
[23] "Behind The Lines: Stories You Won't See Here, " *Reader's Digest*, November 1992, p. 7.
[24] Mark Lear, Forest Conservation Council. Personal Comment, February 15, 1993.
[25] Randy Fitzgerald, "The Great Spotted Owl War," *Reader's Digest*, November 1992, p. 91.
[26] Deanne Kolepfer, The Wilderness Society, Portland, Oregon office. Personal comment, February 12, 1993.
[27] Randy Fitzgerald, *supra* footnote 25.
[28] "The Giant Conifers," *The Oregonian*, October 15, 1990, p. 27 of Special Report : Forests in Distress.
[29] Paul Koberstein, "Private forests face critical log shortages," *The Oregonian*, October 15, 1990, p. 3 of Special Report: Forests in Distress.
[30] Russel Sadler, "Truth and Weyerhaeuser Truths," *Inner Voice*, Summer 1989, p. 5.
[31] Native Forest Council brochure "Stop the Chainsaw Massacre."
[32] Paul Koberstein, *supra* endnote 29, p. 3.
[33] Jeff DeBonis, (AFSEEE). Personal comment, January 14, 1993.
[34] Timothy Hermach, *supra* endnote 3.
[35] Paul Roberts, "Zero Cut,"*Seattle Weekly*, November 4, 1992, p. 21.
[36] Russel Sadler, *supra* endnote 30.

[35] Paul Roberts, "Zero Cut,"*Seattle Weekly*, November 4, 1992, p. 21.

[36] Russel Sadler, *supra* endnote 30.

[37] Paul Koberstein, *supra*. endnote 29, p. 4.

[38] *Id.*

[39] Rick Hoppe, The Wilderness Society. Personal comment, February 15, 1993.

[40] Jim Pissot, *Northwest Timber: An Industry in Transition*, National Audubon Society, Washington State Office, Olympia, Washington, November 1992, p.2.

[41] Timothy Hermach, Personal comment, February 15, 1993.

[42] U.S. Forest Service, Region 10. Statements of Obligations. Juneau, Alaska.

[43] The Wilderness Society, *Americas's Vanishing Rain Forest; A Report On Federal Timber Management In Southeast Alaska*, April, 1986.

[44] Randy Fitzgerald, *supra* endnote 25.

[45] Timothy Hermach, *supra* endtnote 3.

[46] Jimmy Carter, "Salmon swimming against logging tide," *USA Today*, June 22, 1992.

[47] Deanne Kloepfer, The Wilderness Society, Portland Office. Personal comment, February 15, 1993.

[48] The Wilderness Society, *Fact Sheet: Ancient Forests and Taxol*. June 1992.

[49] Timothy Egan, "Trees That Yield a Drug For Cancer Are Wasted," *The New York Times*, January 29, 1992.

[50] *Id.*

[51] Kathie Durbin, "Watershed logging ignites fight over clean rivers," *The Oregonian*, Northwest Forests: Day of Reckoning special insert, October 15, 1990, p. 13.

[52] John Kilpatrick and Richard Isaacson, The Minnesota Landscape Arboretum. Personal comment, March 1, 1993.

[53] Michael Frome, "In Celebration of Wilderness," *Inner Voice*, Vol. 1, No. 1, Summer 1989.

[54] Karen Heiman, "Protecting the Rich Heritage of the Southeast," *Inner Voice*, Vol. 3, No. 6, 1991, p. 12.

[55] Scott Sonner, "Workers tell tales of pain, retaliation," *The Oregonian*, Associated Press, March 27, 1992.

[56] Stan Boicourt, "No Friends at the Dixie," *Inner Voice*, Vol. 3 Issue 6, December 1991, p. 19.

[57] Jimmie L. Hickman, Director of Aviation and Fire Management, Forest Service, Southwest Region. Personal comment, November 23, 1992.

[58] "Owls, Trees and Loggers," *The Washington Post* editorial appearing June 25, 1991, p. A18.

[59] Jeff DeBonis, "Speaking Out: A Letter to the Chief of the US Forest Service," reprinted in *Inner Voice*, Vol. 1, No. 1, Summer 1989, p. 4-5.

[60] Douglas B. Trent, "Tropical Forests," in Judith S. Scherff, ed., *The Mother Earth Handbook*, *supra* p. 42-62.

[61] Jose A. Lutzenberger, A letter to President George Bush. Secretario do Meio Ambiente, Brasilia, Brazil. October 15, 1991.

Vileisis: Cash Register Rivers

[1] U.S. Dept. of the Interior, National Park Service, "Nationwide Rivers Inventory," Washington, D.C. , 1982.

[2] U.S. Environmental Protection Agency, "The Quality of the Nation's Water," Washington, D.C., 1982.

[3] Tim Palmer, *Endangered Rivers and the Conservation Movement*, University of California Press, 1986, p. 177.

[4] Jeffrey Stine, "The Tennessee-Tombigbee Waterway," *Environmental Review*, 1991, pp. 5-7.

[5] U.S. Environmental Protection Agency, Office of Water, *National Water Quality Inventory 1990*, Report to Congress, April 1993, p. 3.

[6] Steve Coffel, *But Not a Drop to Drink: The Lifesaving Guide to Good Water*, Ballentine Books, New York, 1989, p. 312.

[7] *Id.*, p. 92; John Cronin and Robert F. Kennedy, Jr., *The River Keepers: Two Activists Fight to Reclaim our Environment as Basic Human Right*, Scribner, New York, 1997, pp. 58-60; Christopher Thomas Freeburn, "The Hudson Still Hurts," *Audubon*, May-June, 1997, p. 19.

[8] Tim Palmer, "Going to the Heart of America: Our Landscape and Our Future," pre-publication manuscript to be published in 1999, p. 8; Merrimack River Watershed Council, *Midstream News*, Summer 1991, p. 1.

⁹ National Wildlife Federation Leader, April 1991, p. 6; U.S. Environmental Protection Agency, Office of Water, *National Water Quality Inventory 1998, Report to Congress, April 1990*, p. 109; U.S. Environmental Protection Agency, 1992, p. 94; Palmer, *supra* endnote 6, p. 11; Tim Palmer, *Lifelines: The Case for River Conservation*, Island Press, Washington, D.C., 1994, p. 106.

¹⁰ Palmer, *Id.*

¹¹ *High Country News*, 4 May 1992, p. 8.

¹² U.S. Environmental Protection Agency, 1992, p. 10.

¹³ Palmer, *supra* endnote 1, p. 8; Palmer, *supra* endnote 7, p. 106.

¹⁴ U.S. Environmental Protection Agency, 1992, p. 99; *Sacramento Bee*, 31 July 1991.

¹⁵ *Id*, p. 85.

¹⁶ Palmer, *supra* endnote 1, p. 105.

¹⁷ National Wildlife Federation Leader, June 1990, p. 9.

¹⁸ *Free Flow*, Journal of the Oregon Rivers Council, Fall, 1990.

¹⁹ *60 Minutes*, CBS Television Network, December 1996.

²⁰ Phillip Wallin and Rita Haberman, *People Protecting Rivers*, River Network, Portland Oregon, April, 1992; Dean Miller, "Making a Difference on the Clark Fork,"*High Country News*, 4 Dec. 1989, pp. 13-14.

²¹ *High Country News*, 20 November 1989, p. 11.

²² U.S. Environmental Protection Agency, 1992, p. 9.

²³ *Id.*

²⁴ Pat Ford, "Idaho Points the Way to Stream Quality," *High Country News*, 4 Dec. 1989, pp. 15-16.

²⁵ U.S. Environmental Protection Agency, 1992, p. 9.

²⁶ U.S. Environmental Protection Agency, 1992, p. 162; Palmer, *supra* endnote 6, p. 18.

²⁷ Palmer, *supra* endnote 1, p.15.

²⁸ *Id*, p.172.

²⁹ Greg Hanscom, "Reclaiming a Lost Canyon," *High Country News*, 10 Nov. 1997, p. 11.

³⁰ Cheryl Bradley and Derald G. Smith, "Plains cottonwood recruitment and survival on a prairie meandering river floodplain, Milk River, southern Alberta and northern Montana," *Canadian Journal of Botany*, 64, 1986, p.1433.

³¹ Tim Palmer, "Columbia: Sustaining a Modern Resource," *The Mountaineers*, Seattle, 1997, p. 43-71; "Saving Idaho's Salmon," *Fish and Game News*, Autumn 1991, p. 4.

³² Palmer, *supra* endnote 29, p. 87.

³³ Tim Palmer, *The Snake River: Window to the West*, Island Press, Washington, DC, 1991, pp. 220-224.

³⁴ Palmer, *supra* endnote 29, p. 62; Palmer, *supra* endnote 7, p 80; Marc Reisner, "Power, Profit and Preservation: The Invasion of Small-Scale Hydropower," *Wilderness*, Fall 1984, p. 27.

³⁵ Palmer, *supra* endnote 29, p. 62.

³⁶ Palmer, *supra* endnote 7, p. 82.

³⁷ John D. Echeverria, Pope Barrow, and Richard Roos-Collins, *Rivers at Risk: The Concerned Citizen's Guide to Hydropower*, Island Press, Washington, DC, 1989, pp. 17-22.

³⁸ Palmer, *Lifelines*, p. 82.

³⁹ Echeverria, *supra* endnote 35, p. 130.

⁴⁰ Palmer, *supra* endnote 7, p. 75.

⁴¹ "Supreme Court," *American Rivers Newsletter*, Fall 1990, p. 11.

⁴² Tim Egan,"Dams May Be Razed So the Salmon Can Pass, " *The New York Times*, 15 July 1990.

⁴³ Peter Kirsh, "Maine Dam Decision Reverberated in the West," *Denver Post*, 29 Jan. 1998, p. 7B; Palmer, *supra* endnote 7, pp. 44-45.

⁴⁴ Bruce Farling, "Drained Rivers Rouse Montana," *High Country News*, 20 Nov. 1989, p. 16.

⁴⁵ Palmer, *supra* endnote 1, p. 34.

⁴⁶ Farling, *supra* endnote 42, p. 16.

⁴⁷ Palmer, *supra* endnote 1, p. 139.

⁴⁸ U.S. General Accounting Office, Comptroller General, "Federal Charges for Irrigation Projects Reviewed do not Cover Costs," *Report to Congress*, 18 March 1981; James P. Degan, "The Desert Shall Rejoice and be Made to Blossom as the Rose..."*The Living Wilderness*, Summer 1982, p. 24.

⁴⁹ Lass Peterson, "Cut Rate water, Surplus Crops," *The Washington Post*, March 1988; Palmer, *supra* endnote 7, p. 148.

⁵⁰ Peterson, *Id.*; Palmer, *supra* endnote 7, p. 148.

⁵¹ Luna Leopold,"Ethos, Equity and the Water Resource," The Abel Wolman Distinguished Lecture, University of California at Berkeley, National Research Council, February 1990, referring

to Congressional Hearings on HR 1443.

[52] Ed Osann, "Water Use and Abuse in California," *National Wildlife Federation Leader*, Sept. 1991.

[53] Hundley Norris, *The Great Thirst*, University of California Press, Berkeley, 1992, p. 315; Degnan, p. 24.

[54] Palmer, *supra* endnote 1, pp. 180-194; Degnan, 24.

[55] Hundley, *supra* endnote 51, pp. 385, 417.

[56] Degnan, p. 25.

[57] Palmer, *supra* endnote 7, p. 146.

[58] Eliot Diringer, "President Signs Water Bill," *San Francisco Chronicle*, 31 Oct. 1992; Palmer, *Lifelines*, p. 146.

[59] Tim Palmer, *Wild and Scenic Rivers of America*, Island Press, Washington, DC, 1993, p. 6.

[60] "Congress Passes Landmark Grand Canyon Protection Legislation," *American Rivers Newsletter*, Fall 1992, p. 5.

[61] Palmer, *supra* endnote 6, p. 24; Personal interview with Tim Palmer by author, 20 March 1998.

Wald & French: The Mining Law of 1872

[1] Dan Hirschman & Debby Tipton, "The Wearing of the Gold," *Clementine*, Spring/Summer 1996.

[2] Stewart L. Udall, "A letter from MPC Chairman," *Clementine*, Autumn 1991.

[3] George C. Coggins, *Federal Public Land and Resources Law*, University Casebook Series: The Foundation Press, Mineola, NY, 2d ed., 1987.

[4] George C. Coggins, "Public Natural Resources Law," Environmental Law Series: Clark Boardman Company Ltd., New York, 1990.

[5] Vardis Fisher, *Gold Rushes and Mining Camps of the Early American West*, Caxton Printers, Ltd. Caldwell, Idaho, 1968.

[6] Congressional Research Service, "CRS Issue Brief – The 1872 Mining Law: Time for Reform," updated February 21, 1997.

[7] Southern Utah Wilderness Alliance, "'Miners' Law Needs Major Reform," *Newsletter*, Vol. 9, No. 1, Spring 1992.

[8] Todd Wilkinson, "Undermining the Parks," *National Parks*, January/February 1991.

[9] George C. Coggins, *supra* endnote 4.

[10] *Id.*

[11] *Id.*

[12] John Horning, "Selling Our Land Cheaply," *National Wildlife EnviroAction*, National Wildlife Federation, June 1992.

[13] Thomas J. Hilliard, with James S. Lyon & Beverly A. Reece, *Golden Patents, Empty Pockets – A 19th Century Law Gives Miners Billions, the Public Pennies*, Mineral Policy Center, June ,1994.

[14] U.S General Accounting Office, "Interior Should Ensure Against Abuses from Hardrock Mining", GAO/RCED 86-48, March 1986.

[15] U. S. General Accounting Office, "The Mining Law of 1872 Needs Revision," GAO/RCED-89-72, March 1989.

[16] George C. Coggins, *supra* endnote 4.

[17] Michael Zielenziger, "Interior Secretary Fuzzy About Policy," *Miami Herald*, March 23, 1989.

[18] *Green Scissors '98 – Cutting Wasteful and Environmentally Harmful Spending*, Friends of the Earth *et al.*, January 1998.

[19] Thomas J. Hilliard *et al.*, *supra* endnote 13.

[20] Philip M. Hocker, President, Mineral Policy Center, Statement beforethe Subcommittee on Mineral Resources Development and Production, Committee on Energy and Natural Resources, U.S. Senate, regarding S. 433 and S. 785, June 11, 1991.

[21] Green Scissors '98, *supra* endnote 18.

[22] U. S. General Accounting Office, *supra* endnote 15.

[23] *Id.*

[24] John D. Leshy, Statement before the Subcommittee on Mineral Resources Development and Production, Committee on Energy and Natural Resources, U.S. Senate, regarding S. 433 and S. 785, June 11, 1991.

[25] U.S. General Accounting Office, "Unauthorized Activities Occurring on Hardrock Mining Claims," GAO/RCED-90-111, August 1990.

[26] George C. Coggins, *supra* endnote 4.

[27] U. S. General Accounting Office, *supra* endnote 15.

[28] *Id.*

[29] U.S. General Accounting Office, "Increased Attention Being Given to Cyanide Operations," GAO/RCED-91-145, June 1991.

[30] Peter Applebome, "Striking Gold and Fear in S. Carolina," *The New YorkTimes*, June 5, 1991.

[31] U.S. Department of the Interior, Bureau of Mines, Mineral Commodity Summaries, 1991.

[32] Dan Hirschman & Debby Tipton, *supra* endnote 1.

[33] U.S. General Accounting Office, *supra* endnote 29.

[34] Dan Hirschman & Debby Tipton, *supra* endnote 1.

[35] Anne Kersten, "Wildlife Mortalities Resulting from Plastic Mine Claim Markers," Univ. of Nevada, Reno, Political Science 782, April 19, 1991; Lawrence F. LaPre, Ph.D., "Wildlife Deathtraps," *Clementine*, Spring/Summer1990.

[36] Dan Hirschman & Debby Tipton, *supra* endnote 1.

[37] U.S. General Accounting Office, *supra* endnote 29.

[38] *Id.*

[39] *Id.*

[40] *Id.*

[41] Mining Waste Study Team of the University of California, Berkeley, *Mining Waste Study — Final Report*, July 1, 1988.

[42] Thomas J. Hilliard, *supra* endnote 13.

[43] Ann S. Maest, Ph.D., Statement before the Committee on Interior and Insular Affairs, U.S. House of Representatives, regarding H.R. 918, June 20, 1991.

[44] U.S. General Accounting Office, *supra* endnote 29.

[45] Mining Waste Study Team, *supra* endnote 41.

[46] Lynn A. Greenwalt, Vice President for International Affairs, National Wildlife Federation, Statement before the Subcommittee on Mineral Resources Development and Production, Committee on Energy and Natural Resources, U.S. Senate, regarding S. 433, June 11, 1991.

[47] *Id.*

[48] Dirk Johnson, "Digging for Ore Still Pays, Should Miners Pay, Too?" *The New York Times*, February 12, 1992.

[49] U.S. Department of the Interior, Bureau of Land Management, Final Environmental Impact Statement: Betze Project, Barrick Goldstrike Mines, Inc., June 1991.

[50] *Id.*

[51] *Id.*

[52] *Id.*

[53] *Id.*

[54] The Wilderness Society & Mineral Policy Center, "The General Mining Law of 1872: An Idea Whose Time Has Gone," 1992.

[55] Johnnie N. Moore, , Ph.D. & Samuel N. Luoma, Ph.D., "Mining's Hazardous Waste," *Clementine*, Spring 1991.

[56] Tom Knudson, "Mining's Grim Ecology," *Clementine*, Spring/Summer 1990.

[57] Robert L.P. Kleinmann, Research Supervisor, U.S. Bureau of Mines, "Acid Mine Drainage in the United States," 1990.

[58] U.S. Department of Agriculture, Forest Service, *Acid Drainage FromMines on the National Forests – A Management Challenge*, March 1993.

[59] Peter Enticknap, & Katya Kirch, "Report from Windy Craggy," *Clementine*, Winter 1991.

[60] Tom Knudson, *supra* endnote 56.

[61] Edmund T. Piasecki, "Adding Insult To Injury," *Clementine*, Winter 1996-97.

[62] Peter Nielsen & Bruce Farling, "Mining Catastrophe in Clark Fork,"*Clementine*, Autumn 1991.

[63] *Id.*

[64] Lynn A. Greenwalt, *supra* endnote 46.

[65] Johnnie N. Moore, *supra* endnote 54.

[66] Jim Mayer, "Mine's Toxics Worsen; Foes Blame State Panel," *Sacramento Bee*, December 2, 1991.

[67] Jane Kay, "Iron Mountain – a study in perpetual pollution," *San Francisco Examiner*, November 2, 1997.

[68] Tom Knudson, *supra* endnote 56.

[69] Steve Shirley, "After the Gold Work," *Clementine*, Autumn 1989.

[70] U.S. Department of Agriculture, *supra* endnote 58.

[71] Mining Waste Study Team, *supra* endnote 41.

[72] *Id.*

[73] U.S. House of Representatives, Interior and Natural ResourcesCommittee, Subcommittee on Mining and Natural Resources, Majority Staff Report, "GAP STUDY – Western State Non-coal Mining Reclamation Requirements and Proposed Reclamation Standards of HR 918 Substitute Bill," June 24, 1992.

[74] Thomas J. Hilliard, "States' Rights, Miners' Wrongs," Mineral Policy Center, July 1994.

[75] L. Thomas Galloway & Karen Perry, "The Orange Paper," Mineral Policy Center, March 1995.

[76] Western Interstate Energy Board, *Inactive and Abandoned Noncoal Mines,* Volume I, "A Scoping Study," August 1991.

[77] U.S. General Accounting Office, *supra* endnote 29.

[78] John S. Fitzpatrick, Director, Community and Governmental Affairs, Pegasus Gold Corporation, Statement before the Subcommittee on Mineral Resources Development and Production, Committee on Energy and Natural Resources, U.S. Senate, regarding reclamation and bonding practices associated with hard rock mining on federal and state lands, April 19, 1990.

[79] Cy Jamison, Director, Bureau of Land Management, U.S. Department of the Interior, "Responses to Additional Questions" submitted for the record to the Subcommittee on Mineral Resources Development and Production, Committee on Energy and Natural Resources, U.S. Senate, regarding reclamation and bonding practices associated with hard rock mining on federal and state lands, April 19, 1990.

[80] U.S. General Accounting Office, "Limited Action Taken to Reclaim Hardrock Mine Sites," GAO/RCED-88-21, October 1987.

[81] U.S. Department of the Interior, Office of Inspector General, Audit Report No. 91-I-654, "Survey of Selected Activities of the California State Office, Bureau of Land Management," March 1991.

[82] U.S. General Accounting Office, *supra* endnote 14.

[83] U.S. General Accounting Office, "An Assessment of Hardrock Mining Damage," GAO/RCED-88-123BR, April 1988.

[84] Western Interstate Energy Board, *supra* endnote 76.

[85] Green Scissors '98, *supra* endnote 18.

[86] "Grassroots-Community Action on Mining," *Clementine,* Autumn 1991.

[87] Dirk Johnson, *supra* endnote 48.

[88] Ray Ring, "All the king's horses, and all the king's men …," *High Country News,* Vol. 30, No. 1, January 19, 1998.

[89] David A. Mullon, Jr., "Rescuing Cave Creek Canyon," *Clementine,* Autumn 1991.

[90] *Id.*

[91] Robert Reinhold, "Unusual Accord Opens the Way for a Gold Mine," *The New York Times,* November 25, 1990.

[92] Public Land Law Review Commission, *One Third of Our Nation's Land: A Report to the President and to the Congress,* U.S. Government Printing Office, Washington, DC, 1970; Council on Environmental Quality, *Hard Rock Mining on the Public Land,* 1977; U. S. General Accounting Office, *supra.,* endnote 15.

[93] John Lancaster, "Marketing a Public Image: From Mail Delivery to Mining Law," *The Washington Post,* March 25, 1991.

[94] Dan Hirschman & Debby Tipton, *supra* endnote 1.

[95] U.S. Department of the Interior, *supra* endnote 31.

[96] *Id.*

[97] Lynn A. Greenwalt, *supra* endnote 46.

[98] Southern Utah Wilderness Alliance, *supra* endnote 76.

[99] Lynn A. Greenwalt, *supra* endnote 46.

Steingraber: Living Dangerously

Editor's note: Dr. Steingraber used a format for documentation that consisted of listing of sources by chapter and within the chapter, by page --- but not by a method of citation tied to specific statements. The material that we selected can be found between pages 159 and 210 of the original text, and is densely documented. See Dr. Sandra Steingraber, *Living Downstream: An Ecologist Looks at Cancer and the Environment,"* Addison-Wesley Publishing Company, Inc., New York, 1997.

Fagin & Lavelle: Science For Sale

[1] Nicholas A. Ashford, "A Framework for Examining the Effects of Industrial Funding on Academic Freedom and the Integrity of the University," *Science, Technology, & Human Values*, 8, no.2 (Spring 1983): 16-23.

[2] The authors used the MEDLINE database, the online version of the IndexMedicus, to analyze 346 articles on alachlor, atrazine, perchloroethylene, or formaldehyde that appeared in major biomedical journals from 1989 through 1994 and were indexed as of February 1995. That set was designed to be comprehensive — containing every article in the index concerning alachlor (57), atrazine (100), perchloroethylene(62), or the toxicity of formaldehyde (127). The search for formaldehyde articles was narrowed to those that concerned toxicity because the word 'formaldehyde" appears in thousands of abstracts unrelated to study of its effects on health because of the chemical's use as a biomedical preservative. The more common current name "perchloroethylene" was used in the search instead of the chemical's older name, although some studies may have been overlooked because of this choice. Sixteen studies dealt with both alachlor and atrazine and were analyzed separately for each chemical. Forty-five studies could not be analyzed because the full article or an English translation was not readily available. Of the remaining articles, the analysis concluded that four on alachlor, 24 on atrazine, 32 on formaldehyde, and 21 on perc were irrelevant to the question of the chemical's effects on health. These included studies on how to remediate groundwater or waste sites contaminated with such chemicals and studies in which the chemicals were used or mentioned that did not focus on the toxicity of the chemicals themselves — for example, the common test of the effect of medical anesthetics through irritation of rats with formaldehyde. All of the remaining 209 studies were analyzed to determine whether their results that tended to be favorable or unfavorable to the continued use of the chemicals or were ambivalent or too difficult to characterize. Study sponsors were determined from the acknowledgments sections included in the studies themselves. In cases where no sponsor was acknowledged but the study was conducted at a research institute, that institute was deemed the sponsor. In cases where multiple sponsors were listed, but only one could be characterized as either industry or independent, it was deemed the sponsor. In a few cases, a study's sponsor was not explicitly stated but could be reasonably inferred. No sponsor could be determined for 48 of the studies (five onalachlor, nine on atrazine, 27 on formaldehyde, and seven on perchloroethylene), and these were excluded from the final analysis. The full list of studies, including how they were classified by the authors, is accessible through the Center for Public Integrity's site on the World Wide Web (http://www.essential.org/cpi).

[3] Orvin C. Burnside, "Weed Science-The Step Child," *Weed Technology* 7, No. 2, April-June 1993, pp. 515-18.

[4] Numbers from a database created by the Center for Public Integrity using Internal Revenue Service Form 990 reports for the charitable foundations or other not-for-profit organizations associated with the companies.

[5] EPA formaldehyde documents, 16 February 1982.

[6] American Cancer Society, "Cancer Risk Assessment: Learning to Live With Cancer Risk," *Research News* 4, No. 3, December 1995, p. 1.

[7] Chemical Industry Institute of Toxicology 1994 Annual Report, p. 17.

[8] EPA alachlor public docket, March 1985.

[9] University Extension, University of Missouri-Columbia, press release, "'Farmers are not the bad guys': Crop chemical faces ban; found in drinking water," 24 February 1995.

[10] University of Missouri-Columbia press release, "Pesticides-in-Water Scare Is 'Unfounded' Researchers Say," 1995.

[11] Nancy Mays, College of Agriculture, Food, and Natural Resources, University of Missouri-Columbia, memorandum to authors, n.d.

[12] David B. Baker *et al*.," Some Characteristics of Pesticide Transport in Lake Erie Tributaries," *Proceedings of the 26th Conference on Great Lakes Research*, May 23-22,1983 (Summary of conference), p. 22.

[13] R. Don Wauchope *et al*., "Pesticides in Surface Water and Ground Water," *CAST Issue Paper*, No. 2 , April 1994, pp. 1, 6.

[14] David B. Baker, letter to the reader in "A Review of the Science, Methods of Risk Communication and Policy Recommendations in *Tap Water Blues*: Herbicides in Drinking Water," by David B. Baker, R. Peter Richards, and Kenneth N. Baker, of Hiedelburg College Water

Resources Program, 1994, n.p.
[15] R. P. Richards *et al.*, "Atrazine Exposures Through Drinking Water: Exposure Assessments for Ohio, Illinois, and Iowa," *Environmental Science Technology* 29, No. 2 , 1995, pp. 406-12.

[16] Water Quality Laboratory, Heidelberg College, 1994 Annual Report, 16.
[17] Fl documents, 12 August ,1982.
[18] Fl documents, 15 January 1981.
[19] *Gulf South Insulation v. CPSC*, 701 F.2d 1137 (1983), 1145.
[20] Fl documents, 25 July 1979; Gary M. Marsh, "Proportional Mortality Among Chemical Workers Exposed to Formaldehyde," prepared for the Monsanto Company, 15 May 1981, submitted to the Office of the Surgeon General; and J.Donald Miller, Assistant Surgeon General, Director, to William R. Caffey, Manager,   Epidemiology, Monsanto Company, 23 March 1982.
[21] EPA atrazine public docket, n.d.
[22] EPA atrazine public docket, 17 May 1985.
[23] EPA alachlor public docket, 14 March 1996.

Stauber & Rampton: News Pollution

[1] Ben Bagdikian, *The Media Monopoly*, 4th Edition, Beacon Press, 1992, p.xxvii.
[2] Interview with Ben Bagdikian.
[3] Buck Dunham, "All the Criticism of Journalism,"Internet posting to alt.journalism.criticism, March 3 1995
[4] Jeff and Marie Blyskal, *PR: How the Public Relations Industry Wntes tbe News*, William Morrow & Co., New York, 1985, p.28.
[5] Interview with Pam Berns.
[6] PR Newswire promotional material, 1994.
[7] North American Precis Syndicate promotional material, 1994.
[8] *Radio USA* promotional material, 1994.
[9] Interview with Bob Goldberg, president of Feature Photo Service.
[10] Susan B. Trento, *The Powerhouse: Robert Keith Grey and the Selling of Access and Influence in Washington*, St. Martin's Press, New York, 1992, p. 245.
[11] David Lieberman,"Fake News," *TVGuide*, Feb. 22-28, 1992, p.10.
[12] George Glazer, "Let's Settle the Question of VNRs," *Public Relations Quarterly*.
[13] Trento, *supra* endnote 10, pp. 231, 233.
[14] Speech by Rotbart at Nov. 1993 PRSA conference.
[15] TJFR promotional material.
[16] TJFR Environmental News Reporter; Feb.
[17] Rowan and Blewitt report to National Dairy Board, July 13, 1989.
[18] CARMA report to National Dairy Board, May-Aug. 1989.
[19] "12 Reporters Help Shape Pesticides PR Policies,"*Environment Writer;* Vol.6, No. 11, National Safety Council, Washington, DC, Feb. 1995, pp. I, 4-5.
[20] *Id.*
[21] Dashka Slater, "Dress Rehearsal for Disaster," *Sierra*, May / June 1994, p.53.
[22] Promotional information, Video Monitoring Services, 1994.
[23] Jonathan Rabinovitz, "Computer Network Helps Journalists Find Academic Experts," *The New York Times*, May 23, 1994.
[24] Howard Kurtz, "Dr. Whelan's Media Operation," Re *Columbia Journalism Review*, March / April 1990.
[25] *Id*. See also Ann Reilly Dowd. "Environmentalists Are on the Run," *Fortune*, September 19, 1994, p. 92.
[26] Rhys Roth, *No Sweat News*, Olympia, WA, Fall 1992.
[27] David Shaw, "Feeling Bombarded by Bad News," *Los Angeles Times*, Sept. 11,1994.
[28] Samuel S. Epstein, "Evaluation of the National Cancer Program and Proposed Reforms," *American Journal of Independent Medicine*, No.24, 1993, pp. 102-133.
[29] David Steinman, "Brainwashing Greenwashers: Polluting Industries Are Waging a Long Term Disinformation Campaign to Attack the Environmentalist Agenda,"*LA Village Vioice*, Nov. 18-23, 1994, pp. 11-12.
[30] *Measures of Progress Against Cancer, Cancer Prevention, Significant Accomplishments 1982-1992*, The National Cancer Institute.

[31] Rick Weiss, "How Goes the War on Cancer? Are Cases Going Up? Are Death Rates Going Down?" *The Washington Post*, February 14,1995.

[32] *Cancer at a Crossroads: A Report to Congress for the Nation*, National Cancer Advisory Board, September, 1994.

[33] Blyskal, *supra* endnote 4, p.34.

[34] Kim Goldberg, *This Magazine*, Toronto, August, 1993.

[35] Ben Parfitt, "PR Giants. President's Men, and B.C. Trees," *The Georgia Straight*, Vancouver, BC. , February 21-28, 1991, p. 7.

[36] Robert L. Dilenschneider, *Power and Influence: Mastering the Art of Persuasion*, Prentice-Hall, New York, 1990, p. 177.

[37] Ronald K.L. Collins, *Dictating Content*, Center for the Study of Commercialism, Washington, D.C., 1992.

[38] *National Journal*, October 9,1993.

[39] "Resisting Disclosure," *Political Finance & Lobby Reporter*, Vol. XVI, No.12, June 28, 1995, p. 12.

[40] John Dillon, "Poisoning the Grassroots," *Covert Action*, No. 44, Spring 1993, p.36

[41] Trento, *supra* endnote 10, p. xi.

[42] John Keane, *The Media and Democracy*, Polity Press, Cambridge, UK, 1991, p. 63.

[43] Robert W. McChesney, "Information Superhighway Robbery," *In These Times*, July 10, 1995, p. 14.

[44] Kirk Hallahan, "Public Relations and Circumvention of the Press," *Public Relations Quarterly*, Summer 1994, pp. 17-19.

Sugameli: Legalized Extortion

[1] Counsel, National Office of Conservation Programs, National Wildlife Federation, 1400 16th Street, N.W., Suite 501, Washington, D.C. 20036-2266 (202) 797-6865; sugameli@nwf.org. Portions of this chapter are adapted from the author's writings, "Takings Bills Threaten Private Property, People, and the Environment," *Fordham Envtl. L.J.* , Vol. 8, 1997-98, p. 521 [hereinafter "Takings Bills"]; "Environmentalism: The Real Movement to Protect Property Rights," in Philip D. Brick and R. McGreggor Cawley, eds., *A Wolf in the Garden: The Land Rights Movement and the New Environmental Debate*, Rowman & Littlefield Publishers, Inc., Lanham, MD.,1996, p. 59; and "Takings Issues in Light of *Lucas v. South Carolina Coastal Council*," *Va. Envtl. L.J.*, Vol. 12, 1993, p. 439 (reproduced in Kenneth Young, ed., *Zoning and Planning Law Handbook*, Clark Boardman Callaghan, Webster, NY, 1994). NWF intern Carrie Noteboom and legal intern Colleen Kennedy helped prepare this chapter.

[2] *Guildford County Dept. of Emergency Services v. Seaboard Chemical Corp.*, 441 S.E.2d 177 (N.C. App.), *review denied*, 447 S.E.2d 390 (N.C. 1994); *M & J Coal Co. v. United States*, 47 F.3d 1148, 1154 (Fed. Cir. 1994), *cert. denied*, 116 S. Ct. 53 (1995) (coal company should have known it "could not mine in such a way as to endanger public health or safety"); *Clajon Production Corp. v. Petera*, 70 F.3d 1566, 1574-79 (10th Cir. 1995) (state hunting license and bag limit regulations of wildlife on private lands was not a regulatory or physical taking; *United States ex rel. Bergen v. Lawrence*, 848 F.2d 1502, 1507 (10th Cir.), *cert. denied*, 488 U.S. 980 (1988).

[3] *Pennsylvania Coal Co. v. Mahon*, 260 U.S. 393, 413 (1922).

[4] *Private Property Rights and Environmental Laws: Hearings Before the Senate Comm. on Env't and Pub. Works*, Sen. Hrg. No. 104-299, at 218 (1996) [hereinafter Sen. Hrg. No. 104-299].

[5] See *Alves v. United States*, 133 F.3d 1454, 1457-58 (Fed. Cir. 1998) (grazing allotments are not a compensable property right under the Fifth Amendment).

[6] See *Kunkes v. United States*, 78 F.3d 1549 (Fed. Cir. 1996) (plaintiffs did not have a valid takings claim because they did not follow the statutory requirements to maintain their mining claim; it was their inaction that caused the forfeiture of the unpatented claim under the Mining Law of 1872). See also *United States v. Locke*, 471 U.S. 84 (1985); *Freese v. United States*, 639 F. 2d 754 (Ct. Cl.), *cert. denied*, 454 U.S. 827 (1981).

[7] Senate Comm. on the Judiciary, *The Omnibus Property Rights Act of 1995 — S. 605*, Sen. Rep. No. 104-239, at 82 (1966) [hereinafter Sen. Rep. No. 104-239](describing how S. 605's definition of property to include "the right to use or the right to receive water" and contractual water rights could require "fair market value" payments of $100 to $250 per acre foot if federal reclamation project reforms reduced the subsidized water for which users pay from $3.50 to $7.50 per acre foot) (additional views of Sen. Feingold).

[8] As Professor Richard J. Lazarus testified about a takings bill before the Senate Environment

and Public Works Committee: "Perverse incentives will abound. Property owners will propose activities not because of any real interest in their undertaking, but rather simply so that the holder of the property right can be denied permission and thus be entitled to compensation. The law would create an economic incentive for land owners to engage in the most environmentally destructive activities possible, short of a classic common law nuisance, in order to force government to pay the land owner not to do so." Sen. Hrg. No. 104-299, *supra* endnote 4, at 220.

[9] Sen. Rep. No. 104-239, *supra* endnote 7, at 79 (additional views of Sen. Leahy).

[10] Douglas T. Kendall & Charles P. Lord, "The Takings Project: A Critical Analysis and an Assessment of the Progress So Far," *B.C. Envtl. L.J.*, Vol. 25, 1998, pp. 509, 539-45.

[11] 505 U.S. 1003 (1992).

[12] Petitioner's Brief on the Merits, p. 19-35; Brief of Amici Curiae the American Mining Congress, the National Coal Association, the National Forest Products Association, the American Forest Council, and the American Forest Resource Alliance in Support of Petitioner, p. 7-11; Brief of Amicus Curiae the American Farm Bureau Federation and the South Carolina Farm Bureau Federation in Support of Petitioner, p. 16-20.

[13] The author represented NWF in these cases: *K & K Construction, Inc. v. Dep't of Natural Resources*, 575 N.W.2d 531 (Mich. 1998); *Zealy v. City of Waukesha*, 548 N.W.2d 528 (Wis. 1996); *Florida Game and Fresh Water Fish Commission v. Flotilla*, 636 So.2d 761 (Fla. App. 2 Dist. 1994).

[14] Charles Fried, *Order and Law: Arguing the Reagan Revolution: A Firsthand Account*, Simon & Schuster, New York, 1991, p. 183.

[15] See Glenn P. Sugameli, "Takings Bills," *supra* endnote 1, p. 529 n.30.

[16] *Id.* p. 528.

[17] Nancie G. Marzulla and Roger J. Marzulla, *Property Rights: Understanding Government Takings and Environmental Regulation*, Government Institutes, Rockville, MD, 1997, p. 174.

[18] 141 Cong. Rec. H2606-07 (Mar. 3, 1995).

[19] Letter from President Bill Clinton to Orrin Hatch, Chairman of the Senate Judiciary Committee (Dec. 13, 1995), quoted in Sen. Rep. No. 104-239, *supra* endnote 7, at 55.

[20] Glenn P. Sugameli, "Takings Bills," *supra* endnote 1, p. 533 and n. 16.

[21] *Concrete Pipe & Products v. Construction Laborers Pension Trust*, 508 U.S. 602, 645 (1993).

[22] *Id.* p. 644.

[23] *Id.* pp. 642-45.

[24] *The Right to Own Property: Hearings on S. 605 Before the Senate Comm. on the Judiciary* , Sen. Hrg. No. 104-535, at 139, 226 (1995) [hereinafter Sen. Hrg. No. 104-535].

[25] Keith Bagwell, "Property Rights Proposal Rejected By Wide Margin," *Ariz. Daily Star*, November 9, 1994, p. 13A (voters rejected both referendums by a 60-40% margin); Doug Conner, "Property Vote Losers to Keep Fighting," *L.A. Times*, November 9, 1995, p. 35; see also Bruce Rushton & Heath Foster, "Property Rights Initiative Swamped/ Prop. 48 Boosters Blame the Weather," *News Trib.* (Tacoma, Wash.), November 8, 1995, p. A1; Dennis Wagner, "'War' In Wings As Voters Reject Property Rights Issue," *Phoenix Gazette*, November 9, 1994, p. A13 (opponents of takings bills outspent by 2-1).

[26] Eric Pryne & David Postman, "Ref. 48 Defeat Has Louder Echoes: A Property Rights Stall in Congress, Too?", *Seattle Times*, November 9, 1995, p. A1.

[27] For example, Senate Committee on the Judiciary Chairman Orrin G. Hatch (R-UT), in his opening statement at a Committee hearing on S. 605, the unsuccessful Omnibus Property Rights Act of 1995, declared that: "excesses in land use regulations collectivize property by prohibiting the owners of their property the ability to use productively their property. S. 605 was written to fulfill the promise of the fifth amendment that no property shall be taken by the Government except for public use and with just compensation to the property owner... It codifies recent Supreme Court decisions and clarifies the meanings of sometimes confusing case law." Sen. Hrg. No. 104-535, *supra* endnote 24, at 139 (statement of Sen. Orrin G. Hatch). Sen. Hatch essentially repeated his assertion in describing a slightly revised bill that tracked the S. 605 standards. See 142 Cong. Rec. S7888 (July 16, 1996).

[28] See Glenn P. Sugameli, "Takings Bills," *supra* endnote 1, p. 521; Frank I. Michelman, "A Skeptical View of 'Property Rights' Legislation," *Fordham Envtl. L.J.* Vol. 6, 1995, p. 409; John A. Humbach, "Should Taxpayers Pay People to Obey Environmental Laws?", *Fordham Envtl. L.J.*, Vol. 6, 1995, p. 423.

[29] See Glenn P. Sugameli, "Takings Bills," *supra* endnote 1, p. 522.

[30] Statement On The New "Takings Bill," S. 1256— National Press Club Morning Newsmaker

Event, October 23, 1997 — http://www.senate.gov/member/vt/leahy/general/s971023.html (visited September 21, 1998).

[31] See, e.g., endnote 1 and the author's articles "Courts Reject Wetland Takings Claims," *Natl. Wetlands Newsl.* , Envtl. L. Inst., 1994; "Species Protection and Fifth Amendment Takings of Private Property," in Kelly G. Wadsworth and Richard E. McCabe, eds., *Transactions of the Sixtieth North American Wildlife and Natural Resources Conference,* Wildlife Management Inst., 1995; and Colloquium: "The Fifth Amendment's Just Compensation Clause: Implications for Regulatory Policy," *Admin. L.J. Am. U.* , Vol. 6, 1993, p. 676 ; Glenn P. Sugameli, "A Most Extreme Takings Bill," letter to the editor, *The Washington Post,* December 3, 1995; Glenn P. Sugameli, "Takings Exception (Cont'd)," letter to the editor, *The Washington Post,* August 18, 1998.

[32] See, e.g., Editorial, "Balancing Private Rights and Public Well-Being," *National Wildlife,* October-November 1994, p. 3; Doug Harbrecht, A Question of Property Rights and Wrongs, *id.* p. 4.

[33] See, e.g., Nancie G. Marzulla, "Property Rights Movement: How It Began and Where It Is Headed," in *A Wolf in the Garden, supra* endnote 1, p. 39.

[34] See *National Association of Home Builders v. Babbitt,* 130 F.3d 1041, 1052 (D.C. Cir. 1997) (value of endangered plants and animals as sources of medicine and genes); EPA Off. of Water, "Liquid Assets: A Summertime Perspective on the Importance of Clean Water to the Nation's Economy," 800-R-96-002 (May 1996).

[35] *Sammy's of Mobile, LTD and Sammy's Management Company, Inc. v. City of Mobile,* 928 F. Supp. 1116 (S.D. Ala.1996); *Specialty Malls of Tampa, Inc. v. City of Tampa, Florida,* 916 F. Supp. 1222 (M.D. Fla 1996).

[36] *Guildford County Dept. of Emergency Services v. Seaboard Chemical Corp.,* 441 S.E.2d 177 (N.C. App.), *review denied,* 447 S.E.2d 390 (N.C. 1994).

[37] *Triple G Landfills v. Board of Commissioners,* 977 F.2d 287 (7th Cir. 1992).

[38] *Get Away Club, Inc. v. Coleman,* 969 F.2d 664 (8th Cir. 1992).

[39] *Naegele Outdoor Advertising, Inc. v. City of Durham,* 803 F. Supp. 1068 (M.D.N.C. 1992), *affirmed without opinion,* (4th Cir. Mar. 1, 1994), *cert. denied,* 115 S. Ct. 317 (1994).

[40] *Goldblatt v. Hempstead,* 369 U.S. 590 (1962).

[41] *City of Minneapolis v. Fisher,* 504 N.W.2d 520 (Minn. App. 1993).

[42] *Midnight Sessions, Ltd. v. City of Philadelphia,* 945 F.2d 667 (3d Cir. 1991), *cert. denied,* 112 S. Ct. 1668 (1992).

[43] *Hunziker v. State of Iowa,* 519 N.W.2d 367 (Iowa 1994), *cert. denied,* 115 S. Ct. 1313 (1995); *Department of Natural Resources v. Indiana Coal Council, Inc.,* 542 N.E.2d 1000 (Ind. 1989), *cert. denied,* 493 U.S. 1078 (1990).

[44] *Glasheen v. City of Austin,* 840 F. Supp. 62 (W.D. Tex. 1993).

[45] See generally *Virginia Surface Mining and Reclamation Ass'n, Inc., et al. v. Andrus,* 483 F. Supp. 425 (W.D. Va. 1980); Secretary of the Interior Bruce Babbitt and Office of Surface Mining Reclamation and Enforcement Director Robert Uram, "GOP bill puts King Coal back on throne," *Charleston Gazette,* July 17, 1996. ("Buffalo Creek was one of the worst manmade disasters in U.S. history. More than any other event, Buffalo Creek led to the passage of the surface mining law by showing that control of surface coal mining operations is a matter of life and death, not mere landscape aesthetics.")

[46] Gerald M. Stern, Prologue, *The Buffalo Creek Disaster,*Vintage Books, New York, 1976.

[47] See, e.g., *Hodel v. Indiana,* 452 U.S. 314 (1981).

[48] See *id.* at 329 ("Congress adopted [SMCRA] in order to insure that production of coal for interstate commerce would not be at the expense of agriculture, the environment, or public health and safety..."); *Keystone Bituminous Coal Ass'n v. DeBenedictis,* 480 U.S. 470, 488 (1987) (rejecting takings challenge to Pennsylvania law).

[49] *M & J Coal Co. v. United States,* 30 Fed. Cl. 360 (1994), *affirmed,* 47 F.3d 1148 (Fed. Cir.), *cert. denied,* 116 S. Ct. 53 (1995).

[50] "Valuing Wetlands: The Cost of Destroying America's Wetlands," National Audubon Society, 1994.

[51] See Michael Allan Wolf, "Overtaking the Fifth Amendment: The Legislative Backlash Against Environmentalism,"*Fordham Envtl. L.J.,* Vol. 6, 1995, p. 637 (rebutting hyperbolic arguments and horror stories of private property proponents); see also Patrick A. Parenteau, "Who's Taking What? Property Rights, Endangered Species, and the Constitution," *Fordham Envtl. L.J.* , Vol. 6, 1995, p. 619 .

[52] "Campaign '96: Transcript of the Vice Presidential Debate," *The Washington Post,* October 10, 1996, p. A25.

[53] See Al Kamen, "Next Time, Perhaps, a Shorter Speech," *The Washinton Post*, October 23, 1996, p. A21. The Fish and Wildlife Service and the Democratic National Committee checked, but "no one has been able to find anything in their files that remotely resembles this incident." *Id.*

[54] *Lac du Flambeau Band of Lake Superior Chippewa Indians v. Stop Treaty Abuse-Wisconsin, Inc.*, 781 F. Supp. 1385, 1394 n.7 (W.D. Wis. 1992), *modified on other grounds*, 991 F.2d 1249 (7th Cir. 1993).

[55] *Christy v. Hodel*, 857 F.2d 1324, 1335 (9th Cir. 1988) (rancher fined for killing grizzly bears that were eating sheep); *United States v. Kepler*, 531 F.2d 796 (6th Cir. 1976) (ban on interstate or foreign transport of endangered species as applied to species lawfully possessed before passage of the ESA); *United States v. Hill*, 896 F. Supp. 1057 (D. Colo. 1995) (no taking from ESA prohibition on sale of animal parts; defendant was not denied all value as other uses remained, defendant had never applied for an ESA sale permit and defendant had no property right in animal parts he obtained after they were subject to ESA proscriptions); *Good v. United States*, 39 Fed. Cl. 81 (1997), *appeal pending*.

[56] See *U.S. Army Corps of Engineers, Fiscal Year 1995 Regulatory Programs Statistics*, November 8, 1995.

[57] 141 *Cong. Rec.* H2532 (March 2, 1995).

[58] In analyzing the results of S. 605, Senator Robert Dole's 1995 takings bill, the Congressional Budget Office concluded that "CBO expects that the majority of the new suits would involve relatively large claims against agencies that regulate the use of land or water, particularly the U.S. Army Corps of Engineers and the Department of the Interior (DOI)." Sen. Rep. No. 104-239, *supra* endnote 7, at 43 (Congressional Budget Office Cost Estimate, enclosure to March 8, 1996 letter from June E. O'Neill, Director, CBO, to Sen. Orrin G. Hatch, Chairman, Committee on the Judiciary, U.S. Senate).

[59] James Gerstenzang, "Tanker Seeking Return to Alaskan Waters," *L.A. Times*, May 7, 1996, p. A16; David Whitney, "Exxon wants notorious tanker back on duty in Alaska waters," *Anchorage Daily News*, April 5, 1996, p. A-1. See also Margaret Kriz, "Taking Issue," *National Journal*, June 1, 1996, p. 1202 ("for the environmental lobbyists fighting the Republican property-rights legislation, the lawsuit might just as well have come wrapped in pretty paper and tied with a big bow."); Joan Claybrook (president of Public Citizen), "'Takings' legislation would reward polluters," *San Francisco Examiner*, May 29, 1996, p. A-17 (citing Exxon Valdez takings claim).

[60] 33 U.S.C. 2701.

[61] See *supra* endnote 59.

[62] "Maritrans to sue over spill law losses," *The Journal of Commerce*, August 26, 1996.

[63] Sen. Hrg. No. 104-299, *supra* endnote 5, at 205 (written statement of C. Ford Runge, Professor, Dept. of Agricultural and Applied Economics, Univ. of Minn.).

[64] *Id.*

[65] *Id.* at 207-08.

[66] "Legislature Should Quickly Kill Measure to Scrap Land-Use Controls," *Tampa Tribune*, March 9, 1993, p. 6.

[67] Sen. Hrg. No. 104-299, *supra* endnote 4, at 142 (June 7, 1995 letter from Alice M. Rivlin, Director, Office of Management and Budget to Sen. Orrin G. Hatch, Chairman, Committee on the Judiciary, U.S. Senate), *accord id.* at 134-46, 181-83 (testimony and written statement of Alice M. Rivlin).

[68] Taxpayers for Common \$ense, "Blank Check: The Cost to U.S. Taxpayers of Creating a New 'Takings' Entitlement," May 1996. See Margaret Kriz, "Taking Issue," *National Journal*, pp. 1200, 1203, June 1, 1996 ("Opponents of the Senate bill assert that the full price could rise to \$100 billion over seven years—an estimate developed by Taxpayers for Common Sense, a Washington-based public-interest group.").

[69] Sen. Hrg. No. 104-299, *supra* endnote 4, at 146, 183.

[70] Sen. Rep. No. 104-239, *supra* endnote 7, at 70 n.17 (citing *Referendum 48—Economic Impact Study of the Property Rights Initiative*, University of Washington, Institute for Public Policy Management 1995).

[71] Glenn Sugameli, National Wildlife Federation, quoted in Margaret Kriz, "Taking Issue," *National Journal*, June 1, 1996, pp. 1200, 1202.

[72] Sen. Rep. No. 104-239, *supra* endnote 7, at 60, n.9.

[73] *Id.* at 124.

[74] Nathan Arbitman, "Takings Proponents At It Again— Developer Bill passes House; action moves to Senate," *EnviroAction*, NWF, February 1998, pp. 11, 13.

[75] *Congressional Quarterly* stated "[e]xamples of the home builders' work abound. Early drafts

of the property rights bill were written by Linowes and Blocher, a Silver Spring, Md., law firm retained by the association." Allan Freedman, "Property Rights Advocates Climb the Hill to Success: After tasting failure in the 104th, home builders come back with a narrower bill and a lobbying blitz to win their case in the House," October 25, 1997, p. 2591. A front page story in the October 23, 1997 Capitol Hill *Roll Call* newspaper stated "John Delaney, a lawyer at Linowes & Blocher who helped write the bill for the Home Builders, acknowledged that Gallegly's bill, which would expedite the process for property owners to pursue claims against the government for taking land 'seems to be close to what we proposed.'"

[76] John Brinkley, "Lobby gave Gallegly landowner measure: 'Unacceptable': politician blasted for not disclosing source of 'home builders' bill,"*Ventura County Star*, November 5, 1997, p. A1.

[77] Jonathan Weisman, "Narrow failure of land-use bill proves object lesson in lobbying: Home builders vs. environmentalists and local governments,"*The Baltimore Sun*, July 14, 1998, p. 3A.

[78] See, e.g., *NAHB v. Babbitt*, 130 F.3d 1041 (D.C. Cir. 1997) (unsuccessful NAHB claim that it is unconstitutional to enforce the ESA to protect a species that was only found in one state).

[79] Katherine Salant, "Home Fire Sprinklers Have Merit Despite Opposition by Builders," [Housewatch: Safety First Commentary], *The Washington Post*, February 14, 1998, p. E1.

[80] Allan Freedman, *Congressional Quarterly, supra* endnote 75, p. 2591.

[81] Jonathan Weisman, *The Baltimore Sun, supra* endnote 77, p. 3A; Jennifer Sheeter, *Center for Responsive Politics Money in Politics Alert*, Vol. 3, No. 42, November 17, 1997. The Center for Responsive Politics reported that, according to August 1, 1998 FEC data, the NAHB PAC contributions to federal candidates during 1997-1998 totalled $955,249. http://www.crp.org/pacs/pacs/00000901.htm. The same industries that support takings bills also oppose fundamental environmental protection laws with substantial PAC contributions. A 1998 report by the U.S. Public Interest Research Group found that 228 PACs representing real estate, mining, petrochemical, timber, and agribusiness industries have contributed over $100 million to political candidates since 1989. Candidates who opposed the Endangered Species Act (ESA) received twice as much money in contributions than supporters of the ESA. The study found that the Realtors and Home Builders lobbies "are the largest contributors among anti-wildlife PACs with a combined total of over $28.6 million in PAC contributions to political candidates." Sims Weymuller and Kim Delfino, *The Price of Extinction: Political Contributions and the Attack on the Endangered Species Act*, 1998.

[82] Eliza Newlin Carney, "Power Grab," *National Journal*, April 11, 1998, p. 798, 800.

[83] Weisman, *The Baltimore Sun, supra* endnote 77, p. 3A.

[84] Peter S. Goodman and Scott Wilson, "Builders Giving Big in Maryland Campaigns," *The Washington Post*, August 20, 1998, p. D7.

[85] Paul Nussbaum, "Pa. panel calls suburban sprawl main threat to the environment: Ridge's commission recommended giving local governments more control over development," *The Philadelphia Inquirer*, September 17, 1998.

[86] Justin Blum, "Loudoun Seeks the Power to Handle Growth," *The Washington Post*, September 17, 1998, p. B01.

[87] Senate Comm. on the Judiciary, *The Private Property Rights Implementation Act of 1998*, Sen. Rep. No. 105-242, at 45 (1998) (minority views).

[88] February 20, 1998 Dear Senator letter from the U.S. Catholic Conference, National Council of Churches, Coalition on the Environment and Jewish Life, and Evangelical Environmental Network (on file with author).

[89] "All this work was done out of the limelight, catching the opposition off guard. The association eschewed flashy news conferences. Elton Gallegly, R-Calif., the bill's top sponsor, quietly signed up hundreds of cosponsors." Freedman, *Congressional Quarterly, supra* endnote 75, p. 2591.

[90] Nathan Arbitman, "Takings Proponents At It Again— Developer Bill passes House; action moves to Senate," *EnviroAction*, NWF, February 1998, p. 11.

[91] 143 Cong. Rec. H8952 (October 22, 1997).

[92] *Id*. at H8951.

[93] *Id*. at H8963-64.

[94] *The State Journal-Register*, Springfield, IL, October 31, 1997.

[95] 143 Cong. Rec. at H8962.

[96] Sen. Patrick Leahy (D-VT) described this opposition in debate on the Senate Floor. 144 Cong. Rec. 58029 (July 13, 1998).

[97] See "Conservative Flip-Flop: Let's Not Federalize Local Zoning Disputes," *The Union Leader*,

(Manchester, New Hampshire), November 3, 1997; "Not a Federal Case," *The News & Observer* (Raleigh, North Carolina), November 1, 1997; "Property Rights: Senate Should Let Takings Bill Die Without a Vote," *The Dallas Morning News*, November 3, 1997; "Son of Takings," *The Washington Post*, October 22, 1997; "War on Local Control," *The Arizona Daily Star*, October 23, 1997; "Zoning Bill Tilts the Scales," *The Wisconsin State Journal* (Madison, Wisconsin), January 4, 1998.

[98] "Undermining Local Government," *The New York Times*, October 22, 1997.

[99] Statement of Administration Policy, S. 2271—*Property Rights Implementation Act of 1998*, reprinted in Senate debate at 144 Cong. Rec. S8030 (July 13, 1998). See also Editorial, "Takings Exception," *The Washington Post*, July 12, 1998, p. C6.

[100] Weisman, *The Baltimore Sun*, *supra* endnote 77, p. 3A.

[101] *Id.*; NAHB News, "Senate Denies Property Owners on Private Property Rights Bill." http://www.nahb.com/update/story4.html.

[102] John Echeverria, "The Politics of Property Rights," *Oklahoma Law Review*, Vol. 50, 1998, pp. 351, 368-69.

[103] Sen. Hrg. No. 104-299, *supra* endnote 4, at 163-65, 203-09 (testimony of Dr. C. Ford Runge before the Senate Environment and Public Works Committee).

[104] Minority Staff of Senate Comm. on Agriculture, Nutrition & Forestry, *An Overview of Animal Waste Pollution in America: Environmental Risks of Livestock & Poultry Production*, 1997.

[105] Peter S. Goodman and Joby Warrick, "U.S. Plans Rules to Curb Livestock Waste Pollution: Livestock Waste to Be Regulated," *The Washington Post*, September 14, 1998, p. A1.

Ramos: Mobilizing Against Environmentalism

[1] The "Sagebrush Rebellion" refers to the efforts of conservative western state legislators during the late 1970s and early '80s to transfer control of federally managed public lands to state government. Many had privatization as their ultimate goal. The most outspoken champion of these efforts at the federal level was Reagan administration Interior Secretary James Watt.

[2] For instance, Craig's staff organized workshops for the 1994 Alliance for America conference in Washington, D.C.

[3] Republished as *Ecology Wars*.

[4] Ron Arnold, "Defeating Environmentalism," *Logging Management Magazine*, April 1980, p.39. In this article, Arnold misspells Saul Alinsky's last name as "A linski," probably a deliberate attempt to discredit Alinsky by associating him with Soviet communism.

[5] *Id.*, pp. 40-41. This formula has been followed very closely by industry funders of "wise use" organizations and campaigns, who often are more likely to make "in kind" contributions of training seminars, transportation, printed materials, paid leave time for employees, public relations assistance, etc., than direct financial contributions. Among the incentives for this approach, such indirect contributions are considerably more difficult for wise use opponents to document. On a different note, it is interesting that Arnold uses the collective "our" in speaking about the role of industry in developing activist groups.

[6] Claude Emery, *Share Groups in British Columbia*. Library of Parliament Research Branch, Canada, 10 December 1991, p. 12.

[7] Ron Arnold, "Defeating Environmentalism," *supra* endnote 4, p.41. The publications program was to include classroom-targeted materials.

[8] See Ron Arnold, *At the Eye of the Storm: James Watt and the Environmentalists*, Regnery-Gateway, Chicago, 1982. An Associated Press article on the publication of the book notes that "Arnold conceded the book is ideological rather than objective, written from a conservative viewpoint sympathetic to Watt's goals." William Kronholm, "Conservative Book Portrays 'Real James Watt'," Associated Press, 9 November 1982. The Free Congress Foundation was created, in part, by Paul Weyrich, known as the "godfather" of the New Right. Weyrich helped to create a number of other prominent right-wing institutions, including the Heritage Foundation and the American Legislative Exchange Council, and also is one of the key architects of the Religious Right. According to Arnold's acknowledgments, the book about Watt was Weyrich's idea, and it was Weyrich who offered Arnold a contract to write it. Arnold also notes that Watt helped to edit the volume.

[9] In addition to his reputation as the best right-wing direct-mail fundraiser outside of Washington, D.C., Gottlieb is a convicted tax felon. [See Loretta Callahan, "Organizations, financing murky," *The Columbian* (Vancouver, Wash.), 17 May 1992, p. A11.] Gottlieb also has been involved in organized efforts to undermine the governance structure and treaty rights of

Native American tribes in the state of Washington, along with prominent "wise use" leader and key Arnold ally, Charles Cushman. Elements of this Anti-Indian movement, would later join forces with the property-rights wing of the "wise use" movement. For the relationship between the "wise use" and Anti-Indian movements, see Rudolph Ryser, *The Anti-Indian Movement on the Tribal Frontier*, Center for World Indigenous Studies, Kenmore, WA., 1992.

[10] In addition to utilizing Alan Gottlieb's political contacts and fundraising acumen, Arnold took advantage of communications media networks developed by the Center for the Defense of Free Enterprise. According to the dust jacket of Arnold's 1987 *Ecological Wars*, Arnold's weekly newspaper column was "distributed" by the American Press Syndicate (a CDFE project that claims the involvement of some 400 newspapers with an audience of eight million readers); and a weekly radio show hosted by Arnold, "Economics 101 on the Air," was carried by 400 radio stations across the country (presumably on CDFE's American Broadcasting network, which claims more than 600 participating radio stations and an audience of six million).

[11] Ron Arnold, *supra* endnote 4, p.4. Arnold's "wise use" strategy had to be carefully marketed to an industry audience that was, initially, more than a little reluctant.

[12] For an account of the Oregon Project, consult David Mazza, *God, Land and Politics: The Wise Use and Christian Right Connection in 1992 Oregon Politics*, Western States Center, Portland, 1993. See also the chapter on "wise use" in the Northwest timber industry by Tarso Luís Ramos in Echeverria and Eby, eds., *Let the People Judge: Wise Use and the Private Property Rights Movement*, Island Press, Washington, D.C., 1991.

[13] The Grannells' account is recorded in Alan R. Hayakawa, "More sophisticated Sagebrush Rebels broaden base," *The Oregonian*, 25 June 1989, p. Cl.

[14] Bohemia Inc. of Eugene, South Coast Lumber Co. of Brookings, and Hanel Lumber of Hood River each gave $15,000 to the Oregon Project; C&D Lumber Co., D.R. Johnson Lumber Co. and Herbert Lumber Co. gave a total of$15,000; Roseburg Forest Products and Sun Studs gave a joint gift of$15,000. Foster Church, "A time of passionate convictions," *The Oregonian*, 26 October 1990.

[15] The Oregon Project: A Special Project of the Western States Public Lands Coalition," undated.

[16] Alan R. Hayakawa, "More sophisticated Sagebrush Rebels broaden base," *The Oregonian*, 25 June 1989, p. Cl.

[17] See Gene Lawhorn, "Hopelessness and Hate In Our Timber Communities," *The Portland Alliance*, July 1991, p.2.

[18] The Oregon Wise Use Conference was no doubt one of the statewide wise use conferences planned at the 1988 Multiple Use Strategy Conference. Flyer titled "PARTIAL LIST OF SPEAKERS SCHEDULED TO APPEAR AT THE OREGON WISE USE CONFERENCE SATERDAY [*sic*], MAY 20, 1989," dated 24 April 1989. The event also was advertised in the May, 1989 issue of the *American Freedom Journal*, published by the American Freedom Coalition, the political arm of the Rev. Sun Mynag Moon's Unification Church. Both Ron Arnold and Alan Gottlieb have served as directors of the Washington State AFC chapter.

[19] In western Montana, for instance, large private timberland holders Plum Creek and Champion International abandoned sustained-yield forestry in the 1980s, choosing instead to liquidate their holdings. The massive over cutting that resulted would cost jobs as forests dwindled, and would create enormous pressure to step-up harvesting on public lands even though federal policy required that the Forest Service compensate for cuts on private lands by "locking up" adjoining federal forests. See, for instance, Richard Manning, *Last Stand: Logging Journalism, and the Case for Humility*, Peregrine Smith Books, Salt Lake City, 1991.

[20] "The Oregon Project: A Special Project of the Western States Public Lands Coalition," undated. (Circulated in April 1989.)

[21] "Efforts To Get in Summit Succeed," *The Oregon Project Weekly Report*, 15 June 1989, p.1.

[22] Kathie Durbin, "Timber towns in Oregon fight back," *The Oregonian*, 22 May 1989. These groups included Timber Resources Equal Economic Stability (TREES), Workers of Oregon Development (WOOD), Douglas Timber Operators, Southern Oregon Alliance for Resources (SOAR), Communities for a Great Oregon, Protect Industries Now Endangered (PINE), and the Yellow Ribbon Coalition.

[23] See, for example, Dana Tims, "Eugene reverberates with rumble of log trucks," *The Oregonian*, 4 June 1989.

[24] Jeff Mapes, "Protest draws thousands," *The Oregonian*, 14 April 1990, p. Al. Bill Grannell asserts that the Oregon Project "turned Bob Packwood around" on the wilderness issue. David Helvarg, *The War Against the Greens*, Sierra Club Books, San Francisco, 1994, p.162.

[25] *Id,* p. Al. See also Gene Lawhom, *supra* endnote 17.

[26] Mapes, *supra* endnote 24.

[27] Foster Church, "A time of passionate convictions," *The Oregonian,* 26 October 1990.

[28] *Id.* For an account of the origins of TREES, see Gene Lawhom, *supra* endnote 17, p. 2.

[29] According to Oregonian reporter Foster Church, "the statewide Oregon Project dissolved when organizers became dissatisfied with the degree of control the Grannells demanded." Foster Church, *supra* C2. For the Grannells' intentions to expand their efforts, see "What's Next for the Oregon Project?" *The Oregon Project Weekly Report,* 24 May, 1989, p.2.

[30] See David Mazza, *God, Land and Politics: The Wise Use and Christian Right Connection in 1992 Oregon Politics,* Western States Center, Portland, 1993. See also the chapter on "wise use" in the Northwest timber industry by Tarso Luís Ramos in Echeverria and Eby, *supra.*

[31] In 1992, during the Bush/Quayle reelection campaign, about 140 "Fly-In-For-Freedom" activists met at the White House with Bush's advisor on economic affairs, Michael Boskin, and other officials. Associated Press, "Help Promised for timber industry," *Wenatchee World* (Wenatchee, Wash.), 17 September 1992.

[32] Marc Cooper, "Fear brings perfectly decent Oregonians into the OCA fold," *Willamette Week,* 15-21 October 1992, p.1.

[33] Rachel Zimmerman,"Wise use adds environmentalists to the list of bogeymen,"*Willamette Week,* 15-21 October 1992, p. 1.

[34] The Grannells seemed somewhat more uncomfortable with right-wing activists of Arnold's ilk than does the OLC. (A partial explanation may be that they saw Arnold as competition and sought to defend their "turf.") Still, the OLC's relationship with Arnold is not always a comfortable one. As Arnold's ties to the Reverend Sun Myung Moon's Unification Church became more widely known following press revelations starting in 1989, the OLC publiclydistanced itself from CDFE. Nonetheless, as recently as October 1992 Arnold told a Portland newspaper, "I love the OLC. They've been very successfiil in promoting what I call our bumper-sticker philosophy: Man and nature can liveinproductive harmony." lle even claimed to plan sirategy with OLC leaders. In fact, the OLC's public distancing fromArnold may have been mostly for the sake of appearances. In response to press probes into the coalition's "Moonie connection," OLC public relations officer Jackie Lang arranged media training for members with none other thanWashington, D.C.-based Accuracy in Media, which is parent to the infamous right-wing campus "watchdog" group and *The Wise Use Agenda* signatory, Accuracy in Academia. Rachel Zimmerman, "Wise use adds environmentalists to the list of bogeymen," *Willamette Week,* 15-21 October 1992, p.1.

[35] A "thank you" letter to "individuals, Coalition groups and businesses that have provided support through financialor in-kind contributions" from the 10 July 1992 issue of the Oregon Lands Coalition newsletter, "Network News." The letter included a list of OLC supporters.

[36] An Alliance for America document dated December, 1992, states that the objectives of the group include: "1. To achieve representation in all 50 states. 2. To create a comprehensive communications network. 3. To hold members of Congress accountable for extreme environmental votes. 4. To establish an identity in the national press. 5. To develop and maintain communications with the U.S. Congress as well as state government legislatures." The brochure goes onto claim that the Alliance quickly made significant strides toward achieving these goals, such as establishing 50 state representation within the first seven months of operation.

[37] See Associated Press, "Logger and Fishermen picket CBS," *Statesman-Journal* (Salem, Ore.), 15 October 1992.

[38] David Howard, a property rights activist from Gloversville, N.Y., chairs the Alliance. He is also executive director of the Land Rights Foundation, publisher of *Land Rights Letter,* a widely read wise use/property rights newsletter. For a look at Alliance for America activities in New England, see William Kevin Burke, *The Scent of Opportunity: A Survey of the Wise Use/Propertv Rights Movement in New England,* Political Research Associates, Cambridge, Mass., 1992.

[39] CDFE 1993 conference flyer.

[40] At the 1993 Wise Use Leadership Conference, Perry Pendley picked up on the religious right rhetoric of Pat Buchanan's 1992 speech at the Republican National Convention and claimed that "There is a culture war in this country," in which a major battle should be that against "socialistic" environmentalists, or "watermelons~green on the outside and red on the inside." At this same event, Putting People First chair and conference M.C. Kathleen Marquardt derided the idea of statehood for Washington, D.C.

[41] Dan Budd, former Wyoming state senator and father to wise use leader Karen Budd-Falen.

[42] See David Postman, "Skousen under fire as he spreads ideology," Anchorage *Daily News*, 25 January 1987. See also, Robert Gottlieb and Peter Wiley, *America's Saints: The Rise of Mormon Power*, General Publishing Company, Toronto, 1984, p. 91.

[43] American Freedom Coalition corporate filings.

[44] For Cushman's anti-Indian activities, see Rudolph Ryser, *The Anti-Indian Movement on the Tribal Frontier*, Center for World Indigenous Studies, Kenmore, WA., 1992.

[45] See Alan Gottlieb, Ed., *The Wise Use Agenda: The Citizen's Policy Guide to Environmental Resource Issues*, Free Enterprise Press, Bellevue, WA., 1989.

[46] "EIR Special Report: Profiling the Environmentalist Conspiracy, 1965-1980," *Executive Intelligence Review*. Other attacks include "Man in the Rainbow," a LaRouchian-influenced tabloid television-style "expose" of alleged Greenpeace links with ecoterrorists, produced for Danish TV 2.

[47] *Ecoterrorism Watch* is published by Ecological Strategies, Inc., P.O. Box 214, Leesburg, VA 22075.

[48] This attack elicited the following response from a vice president of the national Wise Use group Alliance for America: "It seems quite evident that 'Wise Use' and 'LaRouche' are not in league... There is resonance to our arguments in places and the last thing I would propose is to silence the able cannons of [LaRouchian] Roger Maduro and crew." Brian Bishop, "WISEUSE and LaROUCHE... DO THEY RHYME?" *Land Rights Letter*, September 1995, p.9.

[49] Jon Margolis, "Odd trio could kill nature pact: Biodiversity treaty imperiled," *Chicago Tribune*, 30 September 1994, p. Al.

[50] Timothy Egan, "125 groups put their anti-environmental eggs in one basket to fight 'the perfect bogeyman,'" *New York Times* News Service, January 1992.

[51] Sponsors of NFLC conferences include: Boise Cascade, Darby Lumber, Idaho Farm Bureau Federation, Intermountain Forest Industry Association, Louisiana-Pacific, Meridian Gold, Montana Mining Association, Montana Woolgrowers Association, New Mexico Cattlegrowers' Association, New Mexico Wool Growers, People for the West!, Security Pacific Bank, Stolze Conner Lumber, Washington State Farm Bureau, West One Bank, W.I. Forest Products, Stolze Land and Lumber, Idaho Farm Bureau Federation, Washington Cattlemen's Association, and the Washington Wheat Growers Association. John Craig, "Coalition pushing 'wise use' plan," *The Spokesman-Review* (Spokane Co., Wash.), 20 November 1993. "County government is conference focus," *Western Livestock Journal*, 6 July 1992. J. Todd Foster, "Founder of Wise Use Movement to speak on private property rights," *Spokesman Review* (Spokane Co., Wash.), 10 June 1993. J. Todd Foster, "Speaker: National forests created for logging," *Spokesman Review* (Spokane Co., Wash.), 13 June 1993. Ben Long, "Group pushes local control of federal land," *The Daily Interlake* (Montana), 4 March 1993. Don Schwennesen, "Flathead's iffy about local governments on federal land," *Missoulian* (Montana), 10 March 1993, p. B4. Howard Hutchinson, "Last Chance for the West," *New Mexico Farm and Ranch*, June 1992. Caren Cowan Bremer, "Your County Government may be your last line of protection," *New Mexico Stockman*, April 1992, p. 63. Greg Lakes, "Counties wrangle for new code of the West," *Missoulian*, 16 February 1992. Erik Smith, "N.M. lawyer brings land-use rights ideas to Pasco," *Tri-Ci(y Herald* (Franklin, Co., Wash.), 12 September 1993. Katherine Hedland, "Group clamoring for more county involvement," *Daily News* (Latah, Whitman Co.s, Wash.), 19 May 1994.

[52] In January 1994 Idaho First District Judge James R. Michaud ruled that the Catron County-type ordinances, as adopted by Boundary County, Idaho, violated both state and federal constitutions. "Opinion and Order re: Plaintiffs' Motion for Summary Judgement and Defendants' Motion to Dismiss" in the case of *Boundarv Backpackers, et al, vs. Boundarv Countv*, case no. CV 93-9955, Judge James R. Michaud presiding, p.16.

[53] 42 U.S.C. BA4331(b)(4).

[54] Karen Budd, "County Governments and Federal Lands (Part II), *Federal Lands Update*, May 1991, p. 4.

[55] In a handout titled "Suggestions for Organizing Your County for Proper Self Government", the National Federal Lands Conference advises that, "As part of your comprehensive land use plan, you must research the full history of your county and the history of as many ranches, mines, lumber mills, etc. and record this information in your economic impact statement and comprehensive land use plan."

[56] "Protecting Community Stability — List of Citations," distributed by the National Federal Lands Conference. Regarding "coordination," Budd cites 36 CFR BA 221.3 (a) (1) and 36 CFR BA 219.7(a).

[57] Karen Budd Speech to Idaho Conservation District, December 1991, as cited in "Opinion and

Order re: Plaintiffs' Motion for Summary Judgement and Defendants' Motion to Dismiss" in the case of *Boundary Backpackers, et al, vs. Boundary County*, case no. CV 93-9955, Judge James R. Michaud presiding, p.18.

[58] Chris Junghans, "Land-use plan's authority outlined," *Lewistown News-Argus* (Fergus Co., Montana), 17 May 1992.

[59] Karen Budd, *supra* endnote 57.

[60] Staff, "Garfield residents meet to counter feds," *Great Falls Tribune*, 12 July 1993.

[61] See, for instance, *A Season of Discontent*, Montana Human Rights Network, Helena, Montana, May 1994).

[62] Jim Faulkner, "Why There Is A Need for the Militia in America," "*Federal Lands Update*, October, 1994.

[63] *Id.*

[64] *Id.*

[65] National Federal Lands Conference brochure. According to Arnold's Center for the Defense of Free Enterprise, Hage heads the group's Legal Defense Fund, and Pollot's Boise-based Stewards of the Range is a division of the CDFE.

[66] See, for instance, James Ridgeway and Jeffrey St. Clair, "Where the Buffalo Roam: The Wise Use Movement Plays on Every Western Fear," *Village Voice*, 11 July 1995. While Arnold repudiates the tactics of Nye County, Nevada commissioner Dick Carver, he disassociates himself from, but does not go as far as to condemn, the National Federal Lands Conference.

[67] *USA Today*, 24 April 1995.

[68] Vince Bielski, "Dispute Over 'Wise Use' Pits a Private Eye Against Lawyer," *Daily Journal*, 25 July 1995.

[69] Dan Yurman, "Visions of Blood and Fishes Swim in Political Circles," Econet Western Lands Gopher Service, 3 March 1995.

[70] Keith Schneider, "Bomb Echoes Extremists' Tactics," *New York Times*, 26 April 1995.

[71] See Kenneth S. Stern, *A Force Upon the Plain*, Simon & Shuster, New York, 1996, p.126.

[72] For instance, Carver spoke at the *Jubilee* "Jubilation" conference held at the Red Lion Inn in Bakersfield, California in August, 1994. Steve Sebelius, *Las Vegas Sun*, 26 August 1994; "Coalition Challenges Racism," *Turning the Tide*, September-October 1994.

[73] Steve Sebelius, *Las Vegas Sun*, 26 August 1994.

[74] Paul Hall, "Nevada Seeks to Boot Feds," in *Resource Update*, newsletter of the Stevens County Citizens Coalition, February 1995.

[75] Carver address to Snohomish County Property Rights Alliance and Everett Freedom Forum, 30 March 1995.

[76] "Super Spectacular Joint Meeting," flyer announcing Carver's appearance in Snohomish, sponsored by the Snohomish County Property Rights Alliance and The Freedom Forum.

[77] For an excellent analysis of takings case law, see Glenn P. Sugameli, "Takings Issues In Light of *Lucas V. South Carolina Coastal Council: A* Decision Full of Sound and Fury Signifying Nothing," *Virginia Environmental Law Journal*, vol.12, p. 439. A number of resource books on takings law and legislation have been produced by environmental organizations. One of the most useful of these is the "Takings Briefing Book" produced by the National Wildlife Federation, February 1994.

[78] The regulatory takings movement appears to have its origins in the libertanan school of legal thought associated with the University of Chicago and epitomized by professor Richard Epstein. Epstein's 1985 book, *Takings: Private Property and the Power of Eminent Domain*, provided the impetus for regulatory takings as a legal/legislative strategy. It is useful to examine Epstein's writings, for although proponents of regulatory takings legislation frequently argue that the scope of such laws would be finite, Epstein openly asserts that his position on regulatory takings "invalidates much of the twentieth century legislation," including: civil rights legislation; the National Labor Relations Act; Social Security; minimum wage laws; and virtually all government entitlement programs. In fact, Epstein proposes to challenge the entire New Deal as "inconsistent with the principles of limited government and with the constitutional provisions designed to secure that end." Richard Epstein, *Takings: Private Property and the Power of Eminent Domain*, Harvard University Press, Cambridge, Mass., 1985.

[79] James Wallace and Scott Maier, "Anti-environmentalists gain headway in Olympia," *Seattle Port-Intelligencer*, 18 March 1992.

[80] *Id.*

[81] See Tarso Luís Ramos, *The Wise Use Movement in Washington State*, Western States Center,

Portland, 1996.

[82] Rudolph Ryser, *supra.*

[83] For instance, the August 1995 edition of the *Oregon Observer* features articles promoting the national wise use group Alliance for America, attacking the Oregon Department of Fish and Wildlife as "The Agency from Hell," and declaring that "Truth Is An Endangered Species." The same issue carries advertisements for the "Oregon Militia" and the "Sovereign American's Handbook." Staff writer Ed Snook was chief petitioner for a wise use regulatory takings ("property rights") ballot measure in 1993.

[84] The meeting was held in Maltby, Washington on 11 February, 1995.

[85] Campaign contribution figures are from "Timber Industry Contributions in Western States," Western States Center, Helena, Montana, 1993.

[86] See "Plum Creek Admits Political Interference,"*Western Horizons* Newsletter of Western States Center and Montana State AFL-CIO's Wise Use Public Exposure Project, April 1993, p. 5.

[87] In the Oregon Lands Coalition newsletter, *Network News*, undated. Newsletter content suggests that it was published in the winter of 1993.

[88] Associated Press, "Help Promised for timber industry," *Wenatchee World* (Wenatchee, Wash.), 7 September 1992. According to the AP story, around 140 members of the Alliance for America met at the White House with Bush's advisor on economic affairs, Michael Boskin, and other officials.

[89] Presidential candidate Bill Clinton's announcement that he would hold a summit on forest issues in the early days of a Clinton administration may have kept Bush from making a moresignificant tour of the Northwest.

[90] Kristine Thomas, "Bush seeks jobs balance," *News-Review* (Roseburg, Ore.), 15 September 1992, p.1.

[91] Staff, "Bush OKs limited blowdown harvest," *Peninsula Daily News* (Port Angeles, Wash.), 10 September 1992.

[92] Marc Cooper, "Fear brings perfectly decent Oregonians into the OCA fold," *Willamette Week*, 15-21 October 1992, p.1.

[93] Rachel Zimmerman, "Wise use adds environmentalists to the list of bogeymen," *Willamette Week*, 15-21 October 1992, p. 1.

[94] Subcommittee on the Civil Service; U.S. House of Representatives; "Interference in Environmental Programs by Political Appointees: the Improper Treatment of a Senior Executive Service Official," 30 December 1992.

[95] *American Freedom Journal*, August 1989. Published by the American Freedom Coalition.

[96] David Hackett, "Commodity Groups Explain Meeting on Yellowstone Plan," *Star-Tribune*, 24 October 1992; Michael Milstein, "Victims of Political Report charges conspiracy in handling of ex-Park Service employee, 'Vision' plan," *Billings Gazette*, 7 January 1993, p. Al. For more on the Yellowstone Vision Document see Edwin Bender, *The Wise Use Movement in Wyoming*, Western States Center, Portland, 1996.

[97] Subcommittee on the Civil Service; U.S. House of Representatives; "Interference in Environmental Programs by Political Appointees: the Improper Treatment of a Senior Executive Service Official," 30 December 1992.

[98] Michael Milstein, *supra* endnote 96.

[99] Edwin Bender, "Whose Park Is It Anyway?," story of "Dateline NBC" broadcast 21 July 1992.

[100] Richard Stapleton, "On the Western Front: Dispatches from the war with the Wise Use Movement," *National Parks* magazine, January/February 1993, p. 34.

[101] Charles Cushman, National Inholders Association newsletter, 20 November, 1990, p. 1.

[102] Bob Ekey, "'Wise Use' or 'Me First!'," *Greater Yellowstone Report*, Vol.9, No.4, Fall 1992, p. 4-6.

[103] Richard Stapleton, "Greed vs. Green/How the Wise Use Movement employs corporate money and questionable tactics to stake its claim to public lands," *National Parks* magazine, November/December 1992, p. 35.

[104] Edwin Bender, "Whose Park Is It Anyway?,"*supra* endnote 99.

[105] Charles Cushman, Remarks to 1993 Wise Use Leadership Conference, John Ascuaga's Nugget, Sparks Nevada.

[106] Remarks at Forest Service Northwest Region Regional Leadership Conference session, "Working with Angry Communities," Portland, Ore., 7 February 1995.

[107] Bill Loftus, "An endangered dish," *Lewiston Tribune* (Lewiston, Idaho), 29 October 1993.

[108] For a good summary of the Chenoweth campaign, see Sidney Blumenthal, "Her Own Private Idaho," *The New Yorker*, 10 July 1995.

[109] Michael Wickline, "Chenoweth condemns bombing of Forest Service manager's can," *Lewiston Morning Tribune*, 8 August 1995. Sidney Blumenthal, *id*; "Sheriffs question bill to strip feds' authority," *Lewiston Morning Tribune*, 19 June 1995; Ken Miller, "Federal agents' growing 'arsenal' upsets Chenoweth," *Idaho Statesman*, 3 August 1995.

[110] Associated Press, "Orofino candidates propose requiring almost everyone in town to own a gun," 31 October 1995; Associated Press, "Chenoweth's 'no guns for feds' proposal puzzles Idaho police," *Idaho State Journal*, 2 July 1995.

[111] "Correction," *Lewiston Morning Tribune*, 10 August 1995; Sidney Blumenthal, *supra*. endnote 108.

[112] Associated Press, "Chenoweth: I want nothing from white supremacist groups," Kootenai County, Idaho Press, 16 June 1995.

[113] Sidney Blumenthal, *supra* endnote 108.

[114] Land Rights Network Press Release, "Cushman Denounces Bombing in Nevada, Urges Investigation of Earth First," 4 April 1995.

[115] Rob Taylor and Tina Kelley, "Forest dispute threatens to ignite," *Seattle Post-Intelligencer*, 3 April 1995, p. A3.

[116] Jonathan Brinckman, "Federal lands workers feel under siege," *Idaho Statesman*, 9 August 1995; Tim Anderson, Carson City Bombing: Ranchers defend county movement," *Reno Gazette Journal*,8 August 1995.

[117] *Id.*

[118] See, for instance, James Ridgeway and Jeffrey St. Clair, "This Land Is Our Land," *Village Voice*, 20 June 1995.

[119] Forest Service memo, dated 22 March 1995.

[120] "Sticking to Our Agenda," from "The Private Sector," newsletter of the Center for the Defense of Free Enterprise, Summer 1995. Ron Arnold, editor.

[121] Barry Clau sen and DanaRae Pomeroy, *Walking on the Edge* , Washington Contract Loggers Association, Olympia, WA., 1994.

[122] The book is distributed by Merril Press, owned by Alan Merril Gottlieb. Arnold gets credit in the book for the cover design. A former graphic designer for Boeing, Arnold operates a firm called Northwood Studios, which has designed the jackets for a number of wise use publications associated with the Center for the Defense of Free Enterprise.

[123] Flyer for Gold Hill Resources "Town Meeting" held 12 April 1994 in Potlatch, Idaho; J. Todd Foster, "Fighting Ecoterrorism: North Idaho loggers to get help from experts," *Spokesman Review*, 10 April 1994.

[124] Ellen Gary, testimony at congressional hearing, "America Under the Gun: The Militia Movement and Hate Groups in America," 11 July 1995.

[125] James Ridgeway and Jeffrey St. Clair, *supra* endnote 66.

[126] See Kathy Durbin, "The Battle for Okanogan County," *Seattle Weekly*, I I January 1995.

[127] Loretta Callahan, "The High Priest of Property Rights," *Columbian*, 17 May 1992, p. Al.

[128] Jill Hamburg, "The Lone Ranger," *Cahjornia Magazine*, November 1990.

[129] "Clean Air, Clean Water, Dirty Politics," from *60 Minutes* broadcast, June 1993.

Passacantando: How Industry Combats Efforts to Protect Our Climate

[1] Orie L. Loucks, "Comprehensive Estimate of Combined Costs of Fossil Fuel Emissions," Center for Sustainable Systems Studies, Miami University, OH. A Congressional Seminar presentation, 20 April 1998, Washington, DC.

[2] *The Washington Post*, 24 April, 1998.

[3] EPA web site: http://www.epa.gov/airprogm/oar/primer/health.htm.

[4] EPA web site. "EPA's Upate Clean Air Standards: A Common Sense Primer," September, 1997.

[5] Minnesota Public Utilities Commission on March 15, 1995, Transcript, p. 89.

[6] Western Fuels Association, Inc., *Western Fuels Annual Report, 1995*, p. 13.

[7] Ross Gelbspan, *The Heat is On*, Addison-Wesley Publishing, 1997, p. 41.

[8] Minnesota Public Utilities Commission, Transcript, p. 111.

[9] *Id.*, p. 112.

[10] Gelbspan, *supra* endnote 7, p. 45.

[11] Steve Wilson, "Can We Take the Chance Global Warming is a Sham?"*Arizona Republic*, November 24, 1995, p. A2.

[12] Information Council for the Environment. Advertising copy.
[13] *Id.*
[14] *The Columbus Dispatch,* September 17, 1996.
[15] Prepared statement of W. David Montgomery before the Senate Committee on Energy and Natural Resources, September 17, 1996.
[16] *Id.*
[17] *Id.*
[18] Charles River Associates Incorporated, "International Impact Assessment Model: Results for North America," October 24, 1996.
[19] *IPCC Climate Change 1995: Impacts, Adaptations, and Mitigation, Summary for Policymakers,* p.4
[20] *Ozone Protection in the United States,* World Resources Institute, November 1996, p.12
[21] *Id,* p. 15.
[22] "Industrial Group Plans to Battle Climate Treaty," *The New York Times,* April 26, 1998, p. A1.
[23] Chris Ball, Ozone Action, Director of Outreach. Personal notes from meeting, June 1, 1998.

Brown: Popular Epidemiology

[1] L. Kaplan, "'No More than an Ordinary X-Ray': A Study of the Hanford Nuclear Reservation and the Emergence of the Health Effects of Radiation as Public Problem," PHD dissertation, Heller School, Brandeis University, 1991.
[2] R. Bullard, "Anatomy of Environmental Racism and the Environmental Justice Movement," in R. Bullard, ed., *Confronting Environmental Racism: Voices from the Grassroots,* South End Press, Boston, 1992, pp. 15-39.
[3] P. Brown and S. Masterson-Allen, "Citizen Action on Toxic Waste Contamination: A New Type of Social Movement," *Society and Natural Resources,* No. 7, 1994, pp. 269-86.
[4] A. Szasz, *Ecopopulism: Toxic Waste and the Movement for Environmental Justice,* University of Minnesota Press, Minneapolis, 1994.
[5] P. Brown and F. Ferguson, "Making A Big Stink: Women's Work, Women's Relationships, and Toxic Waste Activism," *Gender & Society,* No. 9, 1995, pp. 14-72.
[6] A. Levine, *Love Canal: Science, Politics, and People,* Lexington Books, Lexington, MA., 1982.
[7] T.J. Blocker and D. Eckberg, "Environmental Issues as Women's Issues: General Concerns and Local Hazards," *Social Science Quarterly,* No. 70, 1989, pp. 58-93.
[8] M.Belenky, B. M. Clinchy, N. R. Goldberger and J. M. Tarule, *Women's Ways of Knowing: The Development of Self, Voice, and Mind* Basic Books, New York, 1986.
[9] Mohai, P. and B. Bryant, "Environmental Racism: Reviewing the Evidence," in B. Bryant and P. Mohai, eds., *Race and the Incidence of Environmental Hazards,* Westview, Boulder, CO., 1992, pp. 162-76.
[10] P. Brown, "Race, Class, and Environmental Health: A Review and Systematization of the Literature," *Environmental Research* No. 69, 1995, pp. 1-30.
[11] D. Pellow, "Environmental Justice and Popular Epidemiology: Symbolic Politics and Hidden Transcripts," paper presented at Annual Meeting of the American Sociological Association, Los Angeles, August, 1994.
[12] T. Kuhn, *The Structure of Scientific Revolutions,* University of Chicago Press., Chicago, 1963.
[13] Massachusetts Institute of Technology, 'Center For Environmental Health Sciences at MIT', Cambridge, MA: Massachusetts Institute of Technology, 1990.
[14] S. Wing, 'Limits of Epidemiology,' *Medicine and Global Survival,* No. 1, 1994, pp. 74-86.
[15] N. Krieger, 'The Making of Public Health Data: Paradigms, Politics, and Policy,' *Journal of Public Health Policy* No.13, 1992, pp. 412-27.
[16] See M.R. Edelstein, *Contaminated Communities: The Social and Psychological Impacts of Residential Toxic Exposure,* Westview, Boulder, 1988; S.R. Couch and J. S. Kroll-Smith, eds., *Communities at Risk: Collective Responses to Technological Hazards,* Peter Lang, New York, 1991; K. Erickson, *A New Species of Trouble: Explorations in Disaster Trauma, and Community,* Norton, New York, 1994.
[17] M. Bashem, *Cancer in the Community: Class and Medical Authority,* Smithsonian Institution Press.Washington, D.C., 1993.
[18] S. Kroll-Smith and H. H. Floyd, *Bodies in Protest: The Struggle Over Medical Knowledge,* New York University Press, New York, 1997.
[19] P. Brown, "Naming and Framing: The Social Construction of Diagnosis and Treatment," *Journal of Health and Social Behavior* extra issue, 1995a, pp. 3-52.
[20] P. Brown, "Toxic Waste Contamination and Popular Epidemiology: Lay and Professional Ways

of Knowing," *Journal of Health and Social Behavior* Vol. 33, 1992, pp. 267-81.
[21] U. Beck, *Risk Society*, Sage, Newbury Park, CA., 1992.
[22] K. Erikson, *A New Species of Trouble: Explorations in Disaster Trauma, and Community*, Norton, New York, 1994.

DeBonis: Environmental Organizing & Public Service Employees

[1] C. Brown. and C. Harris, Summary sheet: preliminary results of a study of values and change in the USDA Forest Service. Department of Resource Recreation and Tourism, College of Forestry, Wildlife, and Range Science, University of Idaho, Moscow, Idaho, 1990.
[2] Quote is from an article in *Forest Watch* in May 1991, titled "Judge Dwyer does it again. Opinion rebukes forest service for 'remarkable series of violations.'"

Robert V. Eye: A Healthy Environment

Epigraph. Clarence Darrow, "Who Knows Justice?" *Scribner's Magazine*. February, 1932, pp. 310-11
[1] *Griswold v. Connecticut*, 381 U.S. 478 (1965).
[2] Barry Commoner, *Making Peace with the Planet*, Pantheon Books, New York, 1990, pp. 19-40.
[3] *Kansas Water Quality Assessment Report*, Kansas Department of Health & Environment, December 1996.
[4] *EPA Journal*, March/April, 1992, pp.6-8.
[5] *EPA Environmental Justice Annual Report, 1994*, pp. 6-7.
[6] *KGE v. Eye et al.*, 246 Kan. 419, 789P.2d 1161 (1990).
[7] *First Annual Council on Environmental Quality Report*, U.S. Government Printing Office, 1970.

Orr & Ehrenfeld: None So Blind

[1] R. Bailey, *The True State of the Planet*, The Free Press, New York, 1995.
[2] F.G. Easterbrook, *A Moment on the Earth*, Viking, New York, 1995
[3] A. Wildavsky, *But is it True?* University Press, Cambridge, Massachusetts, 1995.

Punzo: Complicity with Evil

[1] Dr. Karl Menninger, *Whatever Became of Sin?*, Hawthorne Books, New York, 1973, p. 118.
[2] Environment News Service (electronic edition), "U.S. to Sign Kyoto Protocol on Climate Change," November 13, 1998.

Scherff: Resources for Acitivsts

[1] Interview. Janice C. Shields, Ph.D., Consumer Rights Advocate, Coordinator, Business Accountability Project, Institute for Business Research, Inc., P.O. Box 19793, Washington, DC 20036
[2] David C. Korten, "Limits to the social responsibility of business" in *Share International* magazine, http//.www.shareintl.org. 1998
[3] Curt Anderson,"Gore Slows EPA Pesticide Review," CNN, May 28, 1998.
[4] Bob Sheafor, CBS Sunday Morning, 26 April 1998
[5] Jonathan D. Salant, "Campaign Finance Bill Sparks Lobbyists," CNN, May 28, 1998.
[6] Marilyn Snell, "The World of Religion According to Huston Smith," *Mother Jones*, November-December 1997, p. 42.
[7] Lois Gibbs, Interview on CBS Sunday Morning, August 23, 1998.

INDEX

biomagnification, 89
Bison (buffalo) 19, 24
Bloedel, MacMillan, 148
Blue Ribbon Coalition, 162-3, 165
Blyshal, Jeff & Marie, 112, 120
Budd-Falen, Karen, 156-8
Bureau of Land Management (BLM) (Dept. Of Interior) 24, 35, 70, 72-3, 77, 80-4, 165, 206
Bureau of Reclamation, 56, 62 -3
        Newlands Reclamation Act of 1902, 6
Burnside, Orvin, 99-100
Bush Administration, 40, 43-4
Bush, President George, 152, 162

California Farm Bureau Federation, 155
campaign finance reform, 235
cancer
        atrazine, 106
        breast, 88, 93
        links to formaldehyde, 105
        lung, 92
                and air pollution, 92
        ovarian, 88
        pancreatic, 193
        prevention project, 193
        research establishment, 101
carcinogens (human), 88, 91-3, 94, 101, 107, 119
Carter, President Jimmy, 68
Carver, Dick, 159-60, 164-5
Cathedral of St. John the Divine, 6
CCHW Center for Health, Environment & Justice, 238
Center for Defense of Free Enterprise, 149, 152-3, 155, 158
Center for Indoor Air Research, 98
Center for Public Integrity, 237-8
Center for Responsive Politics, 238
Center for Science in the Public Interest, 118
Center for the Study of Commercialism, 121
Centers for Disease Control and Prevention (CDC), 170, 186, 191
Central Valley Project Reform Act, 1992, 64
Chemical Industry Institute of Toxicology, 101
       industry, 55, 97, 99
Chenoweth, Rep. Helen, 147, 162, 164
chlorofluorocarbons (CFCs), 154, 177
Christ, 11-2
Christians, 11-2, 230
Christianity, 11
Christian Coalition, 12
Christian Identity, 159
Christian Patriot, 159
Christian Right, 164
Christian Voice, 154
churches, 5
citizens groups, 65
Civil Rights Act of 1964, 211
Civil War, 17, 24
Clamshell Alliance, 155